Methods in Enzymology

Volume 306

EXPRESSION OF RECOMBINANT GENES IN EUKARYOTIC SYSTEMS

METHODS IN ENZYMOLOGY

EDITORS-IN-CHIEF

John N. Abelson   Melvin I. Simon

DIVISION OF BIOLOGY
CALIFORNIA INSTITUTE OF TECHNOLOGY
PASADENA, CALIFORNIA

FOUNDING EDITORS

Sidney P. Colowick and Nathan O. Kaplan

Methods in Enzymology

Volume 306

Expression of Recombinant Genes in Eukaryotic Systems

EDITED BY

Joseph C. Glorioso

DEPARTMENT OF MOLECULAR GENETICS AND BIOCHEMISTRY
UNIVERSITY OF PITTSBURGH SCHOOL OF MEDICINE
PITTSBURGH, PENNSYLVANIA

Martin C. Schmidt

DEPARTMENT OF MOLECULAR GENETICS AND BIOCHEMISTRY
UNIVERSITY OF PITTSBURGH SCHOOL OF MEDICINE
PITTSBURGH, PENNSYLVANIA

ACADEMIC PRESS
San Diego London Boston New York Sydney Tokyo Toronto

This book is printed on acid-free paper. 

Academic Press
A Harcourt Science and Technology Company
525 B Street, Suite 1900, San Diego, California 92101-4495, USA
http://www.academicpress.com

Academic Press Limited
24-28 Oval Road, London NW1 7DX, UK
http://www.hbuk.co.uk/ap/

International Standard Book Number: 0-12-182207-9

PRINTED IN THE UNITED STATES OF AMERICA
99 00 01 02 03 04 MM 9 8 7 6 5 4 3 2 1

Table of Contents

Section I. Analysis of Gene Expression

Section II. Gene Expression Systems for Functional Analysis

Section III. Yeast Expression Systems

Section IV. RNA-Based Control of Recombinant Gene Expression

Section V. Small Molecule Control of Recombinant Gene Expression

Section VI. Viral Systems for Recombinant Gene Expression

Contributors to Volume 306

Article numbers are in parentheses following the names of contributors.
Affiliations listed are current.

CARMELA BEGER (12), *Departments of Medicine and Biology, University of California, San Diego, La Jolla, California 92093-0665*

M. A. BENDER (3), *Fred Hutchinson Cancer Research Center, and Department of Pediatrics, University of Washington School of Medicine, Seattle, Washington 98104*

JEFFREY L. BRODSKY (10), *Department of Biological Sciences, University of Pittsburgh, Pittsburgh, Pennsylvania 15260*

GEORG CADERAS (21), *The Salk Institute for Biological Studies, La Jolla, California 92037*

PAUL CANTALUPO (17), *Department of Biological Sciences, University of Pittsburgh, Pittsburgh, Pennsylvania 15260*

LARRY A. COURY (10), *Laboratory of Epithelial Cell Biology, Renal Electrolyte Division, Department of Medicine, University of Pittsburgh Medical Center, Pittsburgh, Pennsylvania 15213-2500*

OLIVIER DANOS (13), *Genethon, CNRS URA 1922, 91002 Evry cedex 2, France*

PIER PAOLO D'AVINO (7), *Howard Hughes Medical Institute, Department of Human Genetics, University of Utah, Salt Lake City, Utah 84112-5331*

JENNIFER L. DAVIS (13), *Cell Genesys, Inc., Foster City, California 94404*

RONALD W. DAVIS (1), *Department of Biochemistry, Stanford University School of Medicine, Stanford, California 94305-5307*

ANTOINE DUMOULIN (6), *Department of Biology, Northeastern University, Boston, Massachusetts 02115*

STANLEY FIELDS (5), *Departments of Genetics and Medical Genetics, University of Washington, Seattle, Washington 98195-7360*

STEVEN FIERING (3), *Department of Microbiology, Dartmouth Medical School, Hanover, New Hampshire 03755*

MITCHELL E. GARBER (21), *The Salk Institute for Biological Studies, La Jolla, California 92037*

SABINE GEISSE (2), *Novartis Pharma Inc., CH-4002 Basel, Switzerland*

VIDYA GOPALAKRISHNAN (19), *Department of Molecular Biology and Genetics, The Johns Hopkins University School of Medicine, Baltimore, Maryland 21205*

PAUL R. GROSS (13), *Chiron Corporation, Emeryville, California 94608*

MARK GROUDINE (3), *Fred Hutchinson Cancer Research Center, and Department of Radiation Oncology, University of Washington School of Medicine, Seattle, Washington 98104*

BRIAN C. HORSBURGH (20), *NeuroVir Inc., Vancouver, British Columbia, Canada V6T 1Z3*

MARIA M. HUBINETTE (20), *NeuroVir Inc., Vancouver, British Columbia, Canada V6T 1Z3*

W. TERRY JENKINS (6), *Department of Chemistry, Indiana University, Bloomington, Indiana 47405*

KATHERINE A. JONES (21), *The Salk Institute for Biological Studies, La Jolla, California 92037*

SALEEM A. KHAN (19), *Department of Molecular Genetics and Biochemistry, University of Pittsburgh School of Medicine, Pittsburgh, Pennsylvania 15261*

HANS P. KOCHER (2), *Novartis Pharma Inc., CH-4002 Basel, Switzerland*

BRIAN KRAEMER (5), *Department of Biochemistry, University of Wisconsin, Madison, Wisconsin 53706-1544*

MARTIN KRÜGER (12), *Departments of Medicine and Biology, University of California, San Diego, La Jolla, California 92093-0665*

SIMON LABBÉ (8), *Department of Biological Chemistry, University of Michigan Medical School, Ann Arbor, Michigan 48109-0606*

DAVID MACKEY (18), *McArdle Laboratory for Cancer Research, University of Wisconsin, Madison, Wisconsin 53706*

JAMES M. MANNING (6), *Department of Biology, Northeastern University, Boston, Massachusetts 02115*

LOIS R. MANNING (6), *Department of Biology, Northeastern University, Boston, Massachusetts 02115*

BERT W. O'MALLEY (16), *Baylor College of Medicine, Houston, Texas 77030*

KENNETH R. PETERSON (11), *Department of Biochemistry and Molecular Biology, University of Kansas Medical Center, Kansas City, Kansas 66160*

JAMES M. PIPAS (17), *Department of Biological Sciences, University of Pittsburgh, Pittsburgh, Pennsylvania 15260*

ROY POLLOCK (15), *Ariad Pharmaceuticals, Cambridge, Massachusetts 02139-4234*

VICTOR M. RIVERA (15), *Ariad Pharmaceuticals, Cambridge, Massachusetts 02139-4234*

MICHAEL G. ROSENFELD (4), *Howard Hughes Medical Institute, University of California, San Diego, School of Medicine, La Jolla, California 92093-0648*

STUART A. ROSENFELD (9), *Applied Biotechnology, DuPont Pharmaceuticals Company, Wilmington, Delaware 19880*

M. TERESA SÁENZ-ROBLES (17), *Department of Biological Sciences, University of Pittsburgh, Pittsburgh, Pennsylvania 15260*

MARK SCHENA (1), *Department of Biochemistry, Stanford University School of Medicine, Stanford, California 94305-5307*

DHRUBA SENGUPTA (5), *Departments of Genetics and Medical Genetics, University of Washington, Seattle, Washington 98195-7360*

LAURA SHEAHAN (19), *Department of Molecular Genetics and Biochemistry, University of Pittsburgh School of Medicine, Pittsburgh, PA 15261*

SAIFUDDIN SHEIKH (19), *Department of Molecular Genetics and Biochemistry, University of Pittsburgh School of Medicine, Pittsburgh, PA 15261*

MARK W. SORNSON (4), *Howard Hughes Medical Institute, University of California, San Diego, School of Medicine, La Jolla, California 92093-0648*

BILL SUGDEN (18), *McArdle Laboratory for Cancer Research, University of Wisconsin, Madison, Wisconsin 53706*

FRANCIS SVERDRUP (19), *Sequitur, Inc., Waltham, Massachusetts 02453*

DENNIS J. THIELE (8), *Department of Biological Chemistry, University of Michigan Medical School, Ann Arbor, Michigan 48109-0606*

JAMES D. THOMPSON (14), *Ribozyme Pharmaceuticals Inc., Boulder, Colorado 80301*

CARL S. THUMMEL (7), *Howard Hughes Medical Institute, Department of Human Genetics, University of Utah, Salt Lake City, Utah 84112-5331*

SOPHIA Y. TSAI (16), *Baylor College of Medicine, Houston, Texas 77030*

FRANK TUFARO (20), *NeuroVir Inc., and University of British Columbia, Vancouver, British Columbia, Canada V6T 1Z3*

GERALD VAN HORN (19), *Department of Molecular Genetics and Biochemistry, University of Pittsburgh School of Medicine, Pittsburgh, PA 15261*

YAOLIN WANG (16), *Schering-Plough Research Institute, Kenilworth, New Jersey 07033*

PING WEI (21), *The Salk Institute for Biological Studies, La Jolla, California 92037, and Human Genome Sciences, Rockville, Maryland 20850*

MARVIN WICKENS (5), *Department of Biochemistry, University of Wisconsin, Madison, Wisconsin 53706-1544*

ROBERT M. WINSLOW (6), *Department of Medicine, Veterans Association Medical Center, University of California, San Diego, California 92110*

ELIZABETH A. WINZELER (1), *Department of Biochemistry, Stanford University School of Medicine, Stanford, California 94305-5307*

FLOSSIE WONG-STAAL (12), *Departments of Medicine and Biology, University of California, San Diego, La Jolla, California 92093-0665*

MARK L. ZEIDEL (10), *Laboratory of Epithelial Cell Biology, Renal Electrolyte Division, Department of Medicine, University of Pittsburgh Medical Center, Pittsburgh, Pennsylvania 15213-2500*

BEILIN ZHANG (5), *Department of Biochemistry, University of Wisconsin, Madison, Wisconsin 53706-1544*

Preface

The ability to express recombinant proteins in eukaryotic cells has greatly increased our ability to study protein function in a variety of settings. Initial work in this field focused on gaining an understanding of the cis-acting sequences necessary for mRNA expression, processing, and translation. As the field has matured, the more sophisticated task of finely regulating gene expression has taken center stage. Researchers want to be able to limit expression to particular tissues and to determine the timing and level of expression. In addition, exciting new technologies are being developed that use small RNA molecules to regulate expression of endogenous genes. In this volume of *Methods in Enzymology,* we have brought together a number of exciting new techniques for regulated gene expression in eukaryotic cells.

We hope that this volume will prove to be a valuable resource for those who are looking for new methods to express recombinant proteins in eukaryotic systems. We realize that no single volume can be complete, and certainly there are methods for recombinant protein expression not represented here. However, we have tried to bring together many novel and creative methods currently being used in many top laboratories.

We thank the authors for the time and effort they devoted to describing the current methods for recombinant protein expression in use in their laboratories.

Joseph C. Glorioso
Martin C. Schmidt

METHODS IN ENZYMOLOGY

VOLUME I. Preparation and Assay of Enzymes
Edited by SIDNEY P. COLOWICK AND NATHAN O. KAPLAN

VOLUME II. Preparation and Assay of Enzymes
Edited by SIDNEY P. COLOWICK AND NATHAN O. KAPLAN

VOLUME III. Preparation and Assay of Substrates
Edited by SIDNEY P. COLOWICK AND NATHAN O. KAPLAN

VOLUME IV. Special Techniques for the Enzymologist
Edited by SIDNEY P. COLOWICK AND NATHAN O. KAPLAN

VOLUME V. Preparation and Assay of Enzymes
Edited by SIDNEY P. COLOWICK AND NATHAN O. KAPLAN

VOLUME VI. Preparation and Assay of Enzymes (*Continued*)
Preparation and Assay of Substrates
Special Techniques
Edited by SIDNEY P. COLOWICK AND NATHAN O. KAPLAN

VOLUME VII. Cumulative Subject Index
Edited by SIDNEY P. COLOWICK AND NATHAN O. KAPLAN

VOLUME VIII. Complex Carbohydrates
Edited by ELIZABETH F. NEUFELD AND VICTOR GINSBURG

VOLUME IX. Carbohydrate Metabolism
Edited by WILLIS A. WOOD

VOLUME X. Oxidation and Phosphorylation
Edited by RONALD W. ESTABROOK AND MAYNARD E. PULLMAN

VOLUME XI. Enzyme Structure
Edited by C. H. W. HIRS

VOLUME XII. Nucleic Acids (Parts A and B)
Edited by LAWRENCE GROSSMAN AND KIVIE MOLDAVE

VOLUME XIII. Citric Acid Cycle
Edited by J. M. LOWENSTEIN

VOLUME XIV. Lipids
Edited by J. M. LOWENSTEIN

VOLUME XV. Steroids and Terpenoids
Edited by RAYMOND B. CLAYTON

VOLUME XVI. Fast Reactions
Edited by KENNETH KUSTIN

VOLUME XXXVI. Hormone Action (Part A: Steroid Hormones)
Edited by BERT W. O'MALLEY AND JOEL G. HARDMAN

VOLUME XXXVII. Hormone Action (Part B: Peptide Hormones)
Edited by BERT W. O'MALLEY AND JOEL G. HARDMAN

VOLUME XXXVIII. Hormone Action (Part C: Cyclic Nucleotides)
Edited by JOEL G. HARDMAN AND BERT W. O'MALLEY

VOLUME XXXIX. Hormone Action (Part D: Isolated Cells, Tissues, and Organ Systems)
Edited by JOEL G. HARDMAN AND BERT W. O'MALLEY

VOLUME XL. Hormone Action (Part E: Nuclear Structure and Function)
Edited by BERT W. O'MALLEY AND JOEL G. HARDMAN

VOLUME XLI. Carbohydrate Metabolism (Part B)
Edited by W. A. WOOD

VOLUME XLII. Carbohydrate Metabolism (Part C)
Edited by W. A. WOOD

VOLUME XLIII. Antibiotics
Edited by JOHN H. HASH

VOLUME XLIV. Immobilized Enzymes
Edited by KLAUS MOSBACH

VOLUME XLV. Proteolytic Enzymes (Part B)
Edited by LASZLO LORAND

VOLUME XLVI. Affinity Labeling
Edited by WILLIAM B. JAKOBY AND MEIR WILCHEK

VOLUME XLVII. Enzyme Structure (Part E)
Edited by C. H. W. HIRS AND SERGE N. TIMASHEFF

VOLUME XLVIII. Enzyme Structure (Part F)
Edited by C. H. W. HIRS AND SERGE N. TIMASHEFF

VOLUME XLIX. Enzyme Structure (Part G)
Edited by C. H. W. HIRS AND SERGE N. TIMASHEFF

VOLUME L. Complex Carbohydrates (Part C)
Edited by VICTOR GINSBURG

VOLUME LI. Purine and Pyrimidine Nucleotide Metabolism
Edited by PATRICIA A. HOFFEE AND MARY ELLEN JONES

VOLUME LII. Biomembranes (Part C: Biological Oxidations)
Edited by SIDNEY FLEISCHER AND LESTER PACKER

VOLUME LIII. Biomembranes (Part D: Biological Oxidations)
Edited by SIDNEY FLEISCHER AND LESTER PACKER

VOLUME LIV. Biomembranes (Part E: Biological Oxidations)
Edited by SIDNEY FLEISCHER AND LESTER PACKER

VOLUME LV. Biomembranes (Part F: Bioenergetics)
Edited by SIDNEY FLEISCHER AND LESTER PACKER

VOLUME LVI. Biomembranes (Part G: Bioenergetics)
Edited by SIDNEY FLEISCHER AND LESTER PACKER

VOLUME LVII. Bioluminescence and Chemiluminescence
Edited by MARLENE A. DELUCA

VOLUME LVIII. Cell Culture
Edited by WILLIAM B. JAKOBY AND IRA PASTAN

VOLUME LIX. Nucleic Acids and Protein Synthesis (Part G)
Edited by KIVIE MOLDAVE AND LAWRENCE GROSSMAN

VOLUME LX. Nucleic Acids and Protein Synthesis (Part H)
Edited by KIVIE MOLDAVE AND LAWRENCE GROSSMAN

VOLUME 61. Enzyme Structure (Part H)
Edited by C. H. W. HIRS AND SERGE N. TIMASHEFF

VOLUME 62. Vitamins and Coenzymes (Part D)
Edited by DONALD B. MCCORMICK AND LEMUEL D. WRIGHT

VOLUME 63. Enzyme Kinetics and Mechanism (Part A: Initial Rate and Inhibitor Methods)
Edited by DANIEL L. PURICH

VOLUME 64. Enzyme Kinetics and Mechanism (Part B: Isotopic Probes and Complex Enzyme Systems)
Edited by DANIEL L. PURICH

VOLUME 65. Nucleic Acids (Part I)
Edited by LAWRENCE GROSSMAN AND KIVIE MOLDAVE

VOLUME 66. Vitamins and Coenzymes (Part E)
Edited by DONALD B. MCCORMICK AND LEMUEL D. WRIGHT

VOLUME 67. Vitamins and Coenzymes (Part F)
Edited by DONALD B. MCCORMICK AND LEMUEL D. WRIGHT

VOLUME 68. Recombinant DNA
Edited by RAY WU

VOLUME 69. Photosynthesis and Nitrogen Fixation (Part C)
Edited by ANTHONY SAN PIETRO

VOLUME 70. Immunochemical Techniques (Part A)
Edited by HELEN VAN VUNAKIS AND JOHN J. LANGONE

VOLUME 71. Lipids (Part C)
Edited by JOHN M. LOWENSTEIN

VOLUME 72. Lipids (Part D)
Edited by JOHN M. LOWENSTEIN

VOLUME 108. Immunochemical Techniques (Part G: Separation and Characterization of Lymphoid Cells)
Edited by GIOVANNI DI SABATO, JOHN J. LANGONE, AND HELEN VAN VUNAKIS

VOLUME 109. Hormone Action (Part I: Peptide Hormones)
Edited by LUTZ BIRNBAUMER AND BERT W. O'MALLEY

VOLUME 110. Steroids and Isoprenoids (Part A)
Edited by JOHN H. LAW AND HANS C. RILLING

VOLUME 111. Steroids and Isoprenoids (Part B)
Edited by JOHN H. LAW AND HANS C. RILLING

VOLUME 112. Drug and Enzyme Targeting (Part A)
Edited by KENNETH J. WIDDER AND RALPH GREEN

VOLUME 113. Glutamate, Glutamine, Glutathione, and Related Compounds
Edited by ALTON MEISTER

VOLUME 114. Diffraction Methods for Biological Macromolecules (Part A)
Edited by HAROLD W. WYCKOFF, C. H. W. HIRS, AND SERGE N. TIMASHEFF

VOLUME 115. Diffraction Methods for Biological Macromolecules (Part B)
Edited by HAROLD W. WYCKOFF, C. H. W. HIRS, AND SERGE N. TIMASHEFF

VOLUME 116. Immunochemical Techniques (Part H: Effectors and Mediators of Lymphoid Cell Functions)
Edited by GIOVANNI DI SABATO, JOHN J. LANGONE, AND HELEN VAN VUNAKIS

VOLUME 117. Enzyme Structure (Part J)
Edited by C. H. W. HIRS AND SERGE N. TIMASHEFF

VOLUME 118. Plant Molecular Biology
Edited by ARTHUR WEISSBACH AND HERBERT WEISSBACH

VOLUME 119. Interferons (Part C)
Edited by SIDNEY PESTKA

VOLUME 120. Cumulative Subject Index Volumes 81–94, 96–101

VOLUME 121. Immunochemical Techniques (Part I: Hybridoma Technology and Monoclonal Antibodies)
Edited by JOHN J. LANGONE AND HELEN VAN VUNAKIS

VOLUME 122. Vitamins and Coenzymes (Part G)
Edited by FRANK CHYTIL AND DONALD B. MCCORMICK

VOLUME 123. Vitamins and Coenzymes (Part H)
Edited by FRANK CHYTIL AND DONALD B. MCCORMICK

VOLUME 124. Hormone Action (Part J: Neuroendocrine Peptides)
Edited by P. MICHAEL CONN

VOLUME 125. Biomembranes (Part M: Transport in Bacteria, Mitochondria, and Chloroplasts: General Approaches and Transport Systems)
Edited by SIDNEY FLEISCHER AND BECCA FLEISCHER

VOLUME 126. Biomembranes (Part N: Transport in Bacteria, Mitochondria, and Chloroplasts: Protonmotive Force)
Edited by SIDNEY FLEISCHER AND BECCA FLEISCHER

VOLUME 127. Biomembranes (Part O: Protons and Water: Structure and Translocation)
Edited by LESTER PACKER

VOLUME 128. Plasma Lipoproteins (Part A: Preparation, Structure, and Molecular Biology)
Edited by JERE P. SEGREST AND JOHN J. ALBERS

VOLUME 129. Plasma Lipoproteins (Part B: Characterization, Cell Biology, and Metabolism)
Edited by JOHN J. ALBERS AND JERE P. SEGREST

VOLUME 130. Enzyme Structure (Part K)
Edited by C. H. W. HIRS AND SERGE N. TIMASHEFF

VOLUME 131. Enzyme Structure (Part L)
Edited by C. H. W. HIRS AND SERGE N. TIMASHEFF

VOLUME 132. Immunochemical Techniques (Part J: Phagocytosis and Cell-Mediated Cytotoxicity)
Edited by GIOVANNI DI SABATO AND JOHANNES EVERSE

VOLUME 133. Bioluminescence and Chemiluminescence (Part B)
Edited by MARLENE DELUCA AND WILLIAM D. MCELROY

VOLUME 134. Structural and Contractile Proteins (Part C: The Contractile Apparatus and the Cytoskeleton)
Edited by RICHARD B. VALLEE

VOLUME 135. Immobilized Enzymes and Cells (Part B)
Edited by KLAUS MOSBACH

VOLUME 136. Immobilized Enzymes and Cells (Part C)
Edited by KLAUS MOSBACH

VOLUME 137. Immobilized Enzymes and Cells (Part D)
Edited by KLAUS MOSBACH

VOLUME 138. Complex Carbohydrates (Part E)
Edited by VICTOR GINSBURG

VOLUME 139. Cellular Regulators (Part A: Calcium- and Calmodulin-Binding Proteins)
Edited by ANTHONY R. MEANS AND P. MICHAEL CONN

VOLUME 140. Cumulative Subject Index Volumes 102–119, 121–134

VOLUME 159. Initiation and Termination of Cyclic Nucleotide Action
Edited by JACKIE D. CORBIN AND ROGER A. JOHNSON

VOLUME 160. Biomass (Part A: Cellulose and Hemicellulose)
Edited by WILLIS A. WOOD AND SCOTT T. KELLOGG

VOLUME 161. Biomass (Part B: Lignin, Pectin, and Chitin)
Edited by WILLIS A. WOOD AND SCOTT T. KELLOGG

VOLUME 162. Immunochemical Techniques (Part L: Chemotaxis and Inflammation)
Edited by GIOVANNI DI SABATO

VOLUME 163. Immunochemical Techniques (Part M: Chemotaxis and Inflammation)
Edited by GIOVANNI DI SABATO

VOLUME 164. Ribosomes
Edited by HARRY F. NOLLER, JR., AND KIVIE MOLDAVE

VOLUME 165. Microbial Toxins: Tools for Enzymology
Edited by SIDNEY HARSHMAN

VOLUME 166. Branched-Chain Amino Acids
Edited by ROBERT HARRIS AND JOHN R. SOKATCH

VOLUME 167. Cyanobacteria
Edited by LESTER PACKER AND ALEXANDER N. GLAZER

VOLUME 168. Hormone Action (Part K: Neuroendocrine Peptides)
Edited by P. MICHAEL CONN

VOLUME 169. Platelets: Receptors, Adhesion, Secretion (Part A)
Edited by JACEK HAWIGER

VOLUME 170. Nucleosomes
Edited by PAUL M. WASSARMAN AND ROGER D. KORNBERG

VOLUME 171. Biomembranes (Part R: Transport Theory: Cells and Model Membranes)
Edited by SIDNEY FLEISCHER AND BECCA FLEISCHER

VOLUME 172. Biomembranes (Part S: Transport: Membrane Isolation and Characterization)
Edited by SIDNEY FLEISCHER AND BECCA FLEISCHER

VOLUME 173. Biomembranes [Part T: Cellular and Subcellular Transport: Eukaryotic (Nonepithelial) Cells]
Edited by SIDNEY FLEISCHER AND BECCA FLEISCHER

VOLUME 174. Biomembranes [Part U: Cellular and Subcellular Transport: Eukaryotic (Nonepithelial) Cells]
Edited by SIDNEY FLEISCHER AND BECCA FLEISCHER

VOLUME 175. Cumulative Subject Index Volumes 135–139, 141–167

Volume 211. DNA Structures (Part A: Synthesis and Physical Analysis of DNA)
Edited by David M. J. Lilley and James E. Dahlberg

Volume 212. DNA Structures (Part B: Chemical and Electrophoretic Analysis of DNA)
Edited by David M. J. Lilley and James E. Dahlberg

Volume 213. Carotenoids (Part A: Chemistry, Separation, Quantitation, and Antioxidation)
Edited by Lester Packer

Volume 214. Carotenoids (Part B: Metabolism, Genetics, and Biosynthesis)
Edited by Lester Packer

Volume 215. Platelets: Receptors, Adhesion, Secretion (Part B)
Edited by Jacek J. Hawiger

Volume 216. Recombinant DNA (Part G)
Edited by Ray Wu

Volume 217. Recombinant DNA (Part H)
Edited by Ray Wu

Volume 218. Recombinant DNA (Part I)
Edited by Ray Wu

Volume 219. Reconstitution of Intracellular Transport
Edited by James E. Rothman

Volume 220. Membrane Fusion Techniques (Part A)
Edited by Nejat Düzgüneş

Volume 221. Membrane Fusion Techniques (Part B)
Edited by Nejat Düzgüneş

Volume 222. Proteolytic Enzymes in Coagulation, Fibrinolysis, and Complement Activation (Part A: Mammalian Blood Coagulation Factors and Inhibitors)
Edited by Laszlo Lorand and Kenneth G. Mann

Volume 223. Proteolytic Enzymes in Coagulation, Fibrinolysis, and Complement Activation (Part B: Complement Activation, Fibrinolysis, and Nonmammalian Blood Coagulation Factors)
Edited by Laszlo Lorand and Kenneth G. Mann

Volume 224. Molecular Evolution: Producing the Biochemical Data
Edited by Elizabeth Anne Zimmer, Thomas J. White, Rebecca L. Cann, and Allan C. Wilson

Volume 225. Guide to Techniques in Mouse Development
Edited by Paul M. Wassarman and Melvin L. DePamphilis

Volume 226. Metallobiochemistry (Part C: Spectroscopic and Physical Methods for Probing Metal Ion Environments in Metalloenzymes and Metalloproteins)
Edited by James F. Riordan and Bert L. Vallee

VOLUME 227. Metallobiochemistry (Part D: Physical and Spectroscopic Methods for Probing Metal Ion Environments in Metalloproteins)
Edited by JAMES F. RIORDAN AND BERT L. VALLEE

VOLUME 228. Aqueous Two-Phase Systems
Edited by HARRY WALTER AND GÖTE JOHANSSON

VOLUME 229. Cumulative Subject Index Volumes 195–198, 200–227

VOLUME 230. Guide to Techniques in Glycobiology
Edited by WILLIAM J. LENNARZ AND GERALD W. HART

VOLUME 231. Hemoglobins (Part B: Biochemical and Analytical Methods)
Edited by JOHANNES EVERSE, KIM D. VANDEGRIFF, AND ROBERT M. WINSLOW

VOLUME 232. Hemoglobins (Part C: Biophysical Methods)
Edited by JOHANNES EVERSE, KIM D. VANDEGRIFF, AND ROBERT M. WINSLOW

VOLUME 233. Oxygen Radicals in Biological Systems (Part C)
Edited by LESTER PACKER

VOLUME 234. Oxygen Radicals in Biological Systems (Part D)
Edited by LESTER PACKER

VOLUME 235. Bacterial Pathogenesis (Part A: Identification and Regulation of Virulence Factors)
Edited by VIRGINIA L. CLARK AND PATRIK M. BAVOIL

VOLUME 236. Bacterial Pathogenesis (Part B: Integration of Pathogenic Bacteria with Host Cells)
Edited by VIRGINIA L. CLARK AND PATRIK M. BAVOIL

VOLUME 237. Heterotrimeric G Proteins
Edited by RAVI IYENGAR

VOLUME 238. Heterotrimeric G-Protein Effectors
Edited by RAVI IYENGAR

VOLUME 239. Nuclear Magnetic Resonance (Part C)
Edited by THOMAS L. JAMES AND NORMAN J. OPPENHEIMER

VOLUME 240. Numerical Computer Methods (Part B)
Edited by MICHAEL L. JOHNSON AND LUDWIG BRAND

VOLUME 241. Retroviral Proteases
Edited by LAWRENCE C. KUO AND JULES A. SHAFER

VOLUME 242. Neoglycoconjugates (Part A)
Edited by Y. C. LEE AND REIKO T. LEE

VOLUME 243. Inorganic Microbial Sulfur Metabolism
Edited by HARRY D. PECK, JR., AND JEAN LEGALL

VOLUME 244. Proteolytic Enzymes: Serine and Cysteine Peptidases
Edited by ALAN J. BARRETT

VOLUME 245. Extracellular Matrix Components
Edited by E. RUOSLAHTI AND E. ENGVALL

VOLUME 246. Biochemical Spectroscopy
Edited by KENNETH SAUER

VOLUME 247. Neoglycoconjugates (Part B: Biomedical Applications)
Edited by Y. C. LEE AND REIKO T. LEE

VOLUME 248. Proteolytic Enzymes: Aspartic and Metallo Peptidases
Edited by ALAN J. BARRETT

VOLUME 249. Enzyme Kinetics and Mechanism (Part D: Developments in Enzyme Dynamics)
Edited by DANIEL L. PURICH

VOLUME 250. Lipid Modifications of Proteins
Edited by PATRICK J. CASEY AND JANICE E. BUSS

VOLUME 251. Biothiols (Part A: Monothiols and Dithiols, Protein Thiols, and Thiyl Radicals)
Edited by LESTER PACKER

VOLUME 252. Biothiols (Part B: Glutathione and Thioredoxin; Thiols in Signal Transduction and Gene Regulation)
Edited by LESTER PACKER

VOLUME 253. Adhesion of Microbial Pathogens
Edited by RON J. DOYLE AND ITZHAK OFEK

VOLUME 254. Oncogene Techniques
Edited by PETER K. VOGT AND INDER M. VERMA

VOLUME 255. Small GTPases and Their Regulators (Part A: Ras Family)
Edited by W. E. BALCH, CHANNING J. DER, AND ALAN HALL

VOLUME 256. Small GTPases and Their Regulators (Part B: Rho Family)
Edited by W. E. BALCH, CHANNING J. DER, AND ALAN HALL

VOLUME 257. Small GTPases and Their Regulators (Part C: Proteins Involved in Transport)
Edited by W. E. BALCH, CHANNING J. DER, AND ALAN HALL

VOLUME 258. Redox-Active Amino Acids in Biology
Edited by JUDITH P. KLINMAN

VOLUME 259. Energetics of Biological Macromolecules
Edited by MICHAEL L. JOHNSON AND GARY K. ACKERS

VOLUME 260. Mitochondrial Biogenesis and Genetics (Part A)
Edited by GIUSEPPE M. ATTARDI AND ANNE CHOMYN

VOLUME 261. Nuclear Magnetic Resonance and Nucleic Acids
Edited by THOMAS L. JAMES

VOLUME 262. DNA Replication
Edited by JUDITH L. CAMPBELL

Section I

Analysis of Gene Expression

[1] Fluorescence-Based Expression Monitoring Using Microarrays

By ELIZABETH A. WINZELER, MARK SCHENA, and RONALD W. DAVIS

Introduction

The amount of DNA sequence available to researchers, along with the need to access this great wealth of information experimentally, has been increasing exponentially. Miniature nucleic acid arrays, often called "DNA chips" or "microarrays," offer opportunities to collect much of the same data that can be obtained with standard molecular biology hybridization methods but in a highly parallel fashion. These microarrays contain dense collections of nucleic acids [either polymerase chain reaction (PCR) products or oligonucleotides] that are either synthesized or deposited at fixed spatial locations on specially prepared glass slides. When labeled DNA or RNA samples are hybridized to the microarrays, the abundance of specific target sequences in the sample can be estimated based on the observed signal intensity at the physical position of the complementary probe or probes. Pico- to femtomoles of thousands of different nucleic acid probes can be arranged at densities of 400 to 250,000 elements/cm^2. With this miniaturization and through the use of fluorescence the simultaneous analysis of entire genomes becomes possible.

Although microarrays show great promise as tools for genotyping, mapping, and resequencing,[1,2] an equally important application for the microarray is measuring transcript abundance.[3,4] Microarrays have been used to simultaneously measure the mRNA expression levels for every gene in *Saccharomyces cerevisiae* under several different growth conditions,[5–7] to characterize the differences between normal and metastatic tissues,[8] and

[1] M. Chee, R. Yang, E. Hubbell, A. Berno, X. C. Huang, D. Stern, J. Winkler, D. J. Lockhart, M. S. Morris, and S. P. Fodor, *Science* **274,** 610 (1996).

[2] J. G. Hacia, L. C. Brody, M. S. Chee, S. P. Fodor, and F. S. Collins, *Nature Genet.* **14,** 441 (1996).

[3] M. Schena, D. Shalon, R. W. Davis, and P. O. Brown, *Science* **270,** 467 (1995).

[4] D. J. Lockhart, H. Dong, M. C. Byrne, K. T. Follettie, M. V. Gallo, M. S. Chee, M. Mittmann, C. Wang, M. Kobayashi, H. Horton, and E. L. Brown, *Nature Biotechnol.* **14,** 1675 (1996).

[5] L. Wodicka, H. Dong, M. Mittmann, M.-H. Ho, and D. J. Lockhart, *Nature Biotechnol.* **15,** 1359 (1997).

[6] J. L. DeRisi, V. R. Iyer, and P. O. Brown, *Science* **278,** 680 (1997).

[7] R. J. Cho, M. J. Campbell, E. A. Winzeler, L. Steinmetz, A. Conway, L. Wodicka, T. G. Wolfsberg, A. E. Gabrielian, D. Landsman, D. J. Lockhart, and R. W. Davis, *Mol. Cell* (1998).

[8] J. DeRisi, L. Penland, P. O. Brown, M. L. Bittner, P. S. Meltzer, M. Ray, Y. Chen, Y. A. Su, and J. M. Trent, *Nature Genet.* **14,** 457 (1996).

0076-6879/99 $30.00

as a vehicle for gene discovery.[9,10] They can also be used to probe genome content when DNA instead of RNA or cDNA is hybridized.[5,11] Global gene expression profiles will be used in diagnostics and to look for drug targets.[12] Data generated by array-based expression experiments will be essential for understanding genetic regulatory networks and integrated cellular responses.

As the availability of fluorescence-based microarray technology grows, more applications for these tools will be discovered and new methods for generating microarrays will be developed. The different types, their different manufacturing processes, and different applications have been reviewed extensively elsewhere.[13–15] This article describes the two basic types of microarrays, protocols for fluorescently labeling messenger RNA from eukaryotic cells for hybridization to microarrays, and considerations involved in experimental design and data analysis.

Nomenclature

The term DNA microarray (sometimes called cDNA microarray) will be used to describe ordered collections of plasmid clones or DNA fragments that have been attached to glass at densities of greater than 100 probe elements/cm^2. Oligonucleotide microarrays will be distinguished from DNA microarrays, not because their manufacture is necessarily different, but because hybridization conditions are generally different. Oligonucleotide or DNA arrays may also be fabricated on porous materials, such as nitrocellulose or nylon membranes, but the spot density is generally lower. Although these filter arrays can also be used to monitor expression using a large number of probes, the target is usually radiolabeled and discussion of such is outside the scope of this article. In addition, the word "target" will describe the molecule whose characteristics are unknown, in this case the messenger RNA or cDNA in solution, and the word "probe" will describe the molecules that are affixed to the microarray. Unlike conventional Southern or Northern methods, the target, not the probe, bears the label.

[9] M. Schena, D. Shalon, R. Heller, A. Chai, P. O. Brown, and R. W. Davis, *Proc. Natl. Acad. Sci. U.S.A.* **93,** 10614 (1996).

[10] R. A. Heller, M. Schena, A. Chai, D. Shalon, T. Bedilion, J. Gilmore, D. E. Woolley, and R. W. Davis, *Proc. Natl. Acad. Sci. U.S.A.* **94,** 2150 (1997).

[11] D. Lashkari, J. DeRisi, J. McCusker, A. Namath, C. Gentile, S. Hwang, P. Brown, and R. Davis, *Proc. Natl. Acad. Sci. U.S.A.* **94,** 13057 (1997).

[12] L. H. Hartwell, P. Szankasi, C. J. Roberts, A. W. Murray, and S. H. Friend, *Science* **278,** 1064 (1997).

[13] A. Marshall and J. Hodgson, *Nature Biotechnol.* **16,** 27 (1998).

[14] M. Johnston, *Curr. Biol.* **8,** R171 (1998).

[15] G. Ramsay, *Nature Biotechnol.* **16,** 40 (1998).

To construct DNA microarrays, nucleic acid probes (usually 500 to 2000 bases in size) are generated by PCR-amplifying plasmid library inserts (using primers complementary to the vector portion of the library) or portions of genomic DNA (using primers designed specifically for the open reading frames or genes of interest). The PCR products are then deposited, usually using a robotic microspotting device, at defined locations on a glass slide.[16] The robotic spotting device is equipped with a set of tips that pick up small amounts of PCR products (~0.2 μl) from a microtiter plate by capillary action and then sequentially dispense the product at specific locations on multiple slides by touching the tips to the slide surface in a way that is equivalent to the mechanism by which ink is released from a quill pen when the tip touches paper. When fine spotting pins are used, spots containing as little as 1 ng of nucleic acid can be arrayed at densities of up to 5000 elements/cm^2. Because each PCR reaction may produce a few micrograms of probe, thousands of microarrays may be produced from a single set of PCR reactions. Other robotic microfabrication technologies, such as ink jetting, in which the probe solution is placed in a piezoelectric controlled print head that can release microscopic droplets, are under development.[17]

After the PCR products are spotted on the microarray, they are fixed covalently to the surface and denatured. Fluorescently labeled mRNA, cDNA, or cRNA is hybridized to the microarray. The fluorescence intensity at each location on the slide is determined with a confocal microscope that has been modified to scan the slide surface or with a charge-coupled device (CCD) camera.[18] Detection by mass spectrometry or radioisotope is also possible, but will not be considered here. The scanned image is then analyzed. An example of a scanned image of a DNA microarray is shown in Fig. 1 (see color insert). Messenger RNA was prepared from human Jurkat cells and hybridized to an array containing 248 human and *Arabidopsis* cDNAs using the procedures described here. As the amount of probe on the array exceeds the amount of target, the observed signal at any given position is a good estimate of the abundance of cognate target in the sample.

Oligonucleotide Microarrays

Oligonucleotide microarrays consist of shorter (less than 50 bases) nucleic acid fragments of known sequence, covalently attached to a derivatized glass slide at defined locations. The oligonucleotides can be either prefabricated and then attached using microdeposition technology (robotic printing

[16] D. Shalon, S. J. Smith, and P. O. Brown, *Genome Res.* **6,** 639 (1996).

[17] A. M. Castellino, *Genome Res.* **7,** 943 (1997).

[18] M. Schena and R. Davis, *in* "PCR Methods Manual" (M. Innis, D. Gelfand, and J. Sninsky, eds.), p. 445. Academic Press, San Diego, 1999.

or ink jetting) or synthesized *in situ* using techniques such as photolithography or ink jetting of the individual dA, dC, dG, or dT phosphoramidite monomers.[17] Affymetrix (Santa Clara, CA) manufactures high-density oligonucleotide microarrays using a light-directed combinatorial method.[19,20] With this method, photoreactive (versus chemically reactive) phosphoramidite monomers are used in oligonucleotide synthesis. A series of physical photomasks that protect portions of the glass surface is designed based on the sequences of the oligonucleotides that will be synthesized. By shining a mercury light through the different photomasks [no more than 100 (4×25) masks are need to make 25-mers], the synthesis can be controlled in a spatially addressable fashion, allowing the production of microarrays whose sequence diversity is limited only by the resolution of the mask (up to 400,000 different 25-mer probes in a 1.6-cm^2 area). Each element on the array contains $\sim 10^7$ molecules of a particular probe sequence in a 50-μm^2 area. One of the chief advantages of *in situ* methods over amplification/microdeposition methods is that the sequences are selected from databases using software, and no physical handling of the probes is required, reducing the probability of probe cross-contamination or mix-up significantly.

As with DNA microarrays, the mRNA (or cDNA) is labeled and hybridized to the microarray. Hybridization occurs at lower temperatures, and protocols usually call for the fragmentation of the labeled material before hybridization to reduce the amount of secondary structure that forms at the lower temperatures. The microarrays are washed and scanned, and hybridization is detected by fluorescence.

Analysis of Gene Expression Using Microarrays

Preparation of Microarrays

Microarrays designed to use fluorescence detection and containing probe sequences from different organisms can be purchased from suppliers such as Affymetrix (Santa Clara, CA) or Synteni (Fremont, CA) and many new producers are entering the market. The manufacture of certain types of microarrays may be technically inaccessible to most laboratories. However, if a robotic spotting device is available, DNA microarrays can be fabricated in the laboratory. These instruments can be built using commercially available parts (see http://cmgm.stanford.edu/pbrown/) or can be purchased from companies such as Molecular Dynamics (Sunnyvale, CA) or

[19] S. P. A. Fodor, J. L. Read, M. C. Pirrung, L. Stryer, A. T. Lu, and D. Solas, *Science* **251,** 767 (1991).

[20] A. C. Pease, D. Solas, E. J. Sullivan, M. T. Cronin, C. P. Holmes, and S. P. Fodor, *Proc. Natl. Acad. Sci. U.S.A.* **91,** 5022 (1994).

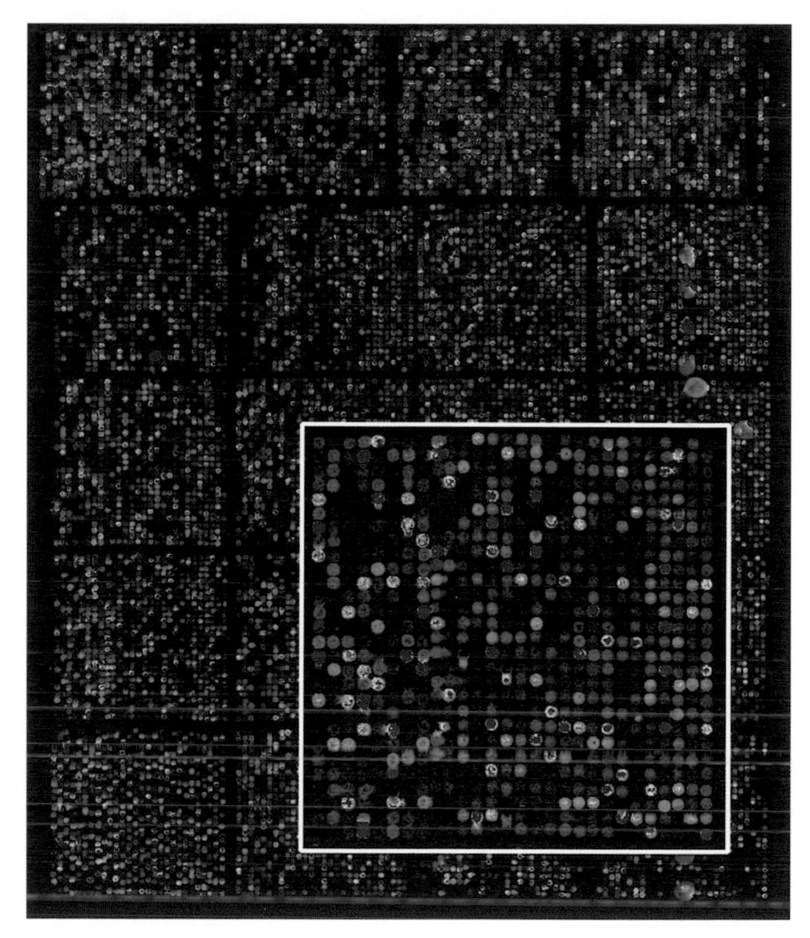

FIG. 1. Scanned image of a gene expression microarray. A fluorescent scan of a microarray prepared by mechanical microspotting is shown. A total of 240 human blood cDNAs and 8 *Arabidopsis* controls were amplified by PCR, purified, and arrayed on silylated microscope slides in duplicate at 200-μm spacing. The 496 element microarray was then hybridized with a fluorescent probe prepared by enzymatic incorporation of Cy3-dCTP into cDNA by single-round reverse transcription of poly(A)$^+$ mRNA from cultured Jurkat cells, essentially as described in Table III. The microarray was scanned at 10-μm resolution with a confocal laser detection system. Fluorescent data are depicted in a rainbow pseudocolor palette for ease of visualization. Elements complementary to abundant messages are red, whereas black/dark blue elements correspond to genes having low or undetectable transcript levels.

Telechem International (Sunnyvale, CA). Generating arrays with a spotting device, although repetitive, is not prohibitively difficult (Tables I and II). An example of a scanned image of an array generated with a robotic spotting device is shown in Fig. 1. Hundreds or thousands of PCR reactions need to be performed using either a large number of different templates or a large number of different primers. Clones from any plasmid library (ordered or unordered cDNA, genomic DNA, etc.) can be used as templates. In principle unamplified plasmid DNA could also be spotted, but the presence of vector sequence present in all elements could increase background.

If the genomic DNA is to be used as template, different pairs of oligonucleotide primers need to be designed and synthesized for each element

TABLE I
PCR AMPLIFICATION OF CLONES FROM GENOMIC DNA OR cDNA LIBRARIES

1. Assemble PCR reactions in 96-well plate

Reagent	Volume (μl)
10× PCR buffer (15 mM Mg^{2+})	10.0
dNTP cocktail (2 mM each)	10.0
Primer 1 (100 pmol/μl)[a]	1.0
Primer 2 (100 pmol/μl)	1.0
Genomic or plasmid DNA (10 ng/μl)[b]	1.0
H_2O	76.0
Taq DNA polymerase (5 U/μl)	1.0
	100 μl

2. Amplify targets in 96- or 384-well format using 30 rounds of PCR (94°, 30 sec; 55°, 30 sec; 72°, 60 sec)
3. Purify using PCR product purification kit and elute products with 100 μl of 0.1× TE (pH 8.0)
4. Dry products to completion in Speed-Vac
5. Resuspend each PCR product in 7.5 μl 5× SSC (0.3–1.0 mg/ml DNA)
6. Transfer to flat bottom 384-well plate (Nunc) for arraying

Suggested materials
- PCR primers modified with a 5′-amino modifier C6 (Glen Research)
- 96-well thermal cycler (PCR system 9600-Perkin Elmer, Norwalk, CT)
- 96-well PCR plates (MicroAmp 96-well Perkin Elmer)
- *Taq* DNA polymerase (Stratagene, La Jolla, CA)
- PCR product purification kit (Telechem International)
- Flat-bottom 384-well plates (Nunc, Naperville, IL)

[a] Use of generic primer pairs (~21-mers) to vector sequences allows high-throughput processing.

[b] Plasmid DNA can be prepared by alkaline lysis and purified. The 96-well REAL prep (Qiagen) facilitates rapid preparation.

TABLE II
MICROARRAYING AND SLIDE PROCESSING

1. Obtain silylated (free aldehyde) microscope slides (CEL Associates, Houston, TX)
2. Print cDNAs using microspotting device according to manufacturer's instructions
3. Allow printed arrays to dry overnight in slide box[a]
4. Soak slides twice in 0.2% (w/v) sodium dodecyl sulfate (SDS) for 2 min at room temperature with vigorous agitation[b]
5. Soak slides twice in doubly distilled H_2O for 2 min at room temperature with vigorous agitation
6. Transfer slides into doubly distilled H_2O at 95–100° for 2 min to allow DNA denaturation
7. Allow slides to dry thoroughly at room temperature (~5 min)
8. Transfer slides into a sodium borohydride solution[c] for 5 min at room temperature to reduce free aldehydes
9. Rinse slides three times in 0.2% SDS for 1 min each at room temperature
10. Rinse slides once in doubly distilled H_2O for 1 min at room temperature
11. Submerge slides in doubly distilled H_2O at 95–100° for 2 sec[d]
12. Allow the slides to air dry and store in the dark at 25° (stable for >1 year)

Suggested materials
  Microspotting device (Telechem International, Molecular Dynamics, Cartesian)
  Microscope slides (CEL Associates)

[a] Drying increases cross-linking efficiency. Several days or more is acceptable.
[b] This step removes salt and unbound DNA.
[c] Dissolve 1.0 g $NaBH_4$ in 300 ml phosphate-buffered saline. Add 100 ml 100% ethanol to reduce bubbling. Prepare *just prior* to use!
[d] Heating the slides aids greatly in the drying process.

on the microarray. The Whitehead Institute Primer program (http://www.genome.wi.mit.edu/genome_software/), which uses a nearest neighbor algorithm to calculate melting temperatures, has been used successfully to pick primers for the amplification of open reading frames from the yeast genome.[11,21] Using this program, greater than 94% of PCR reactions generated useable products,[11] as detected by analyzing a small amount of product on a gel. Ideally, primer pairs should be chosen so that they have similar melting temperatures and so that the resulting PCR products are of similar sizes. When possible, primers should be synthesized in the same format as used for the PCR reactions (microtiter plates, either 96 or 384 well). Computer scripts can be written that automate the process of primer selection for groups of sequences or for an entire genome.

The success of PCR reactions can be checked by analyzing the products on an agarose gel or by examining the microarray elements once they have been spotted: fluorescein-labeled dNTPs can be included in the PCR

[21] W. Rychlik, W. J. Spencer, and R. E. Rhoads, *Nucleic Acids Res.* **18,** 6409 (1990).

reactions and the array can be scanned for fluorescein to determine the reaction success. PCR reactions that fail are noted and repeated, although failures are generally arrayed along with successes in order to ease handling and documentation (if space is not an issue). It is generally not necessary to quantitate the amount of product if two-color hybridization (described later) will be performed. In addition, it is assumed that even with inefficient reactions, the amount of product will exceed the amount of target. PCR reactions are spotted onto glass that is derivatized chemically by treatment with reactive aldehydes (Table II) or polycations, such as polylysine.[6]

Preparation of Polyadenylated mRNA

Small differences in the environmental conditions to which a cell is exposed can have a profound impact on the global pattern of transcription.[5] If two growth conditions are to be compared, extreme caution should be taken to ensure that the cells from which the mRNA is isolated are treated equivalently. For example, cells should ideally be grown in the same batch of media, in tissue culture-treated plasticware, and should be harvested at similar densities if possible. Various methods can be used to rapidly isolate total RNA from cells, depending on the organism. Following isolation, the polyadenylated RNA is usually purified from total RNA on oligo(dT) resin. The amount of polyadenylated RNA needed will depend on the particular microarray, the labeling method, and the size of the hybridization chamber, but is generally between 0.5 and 10 μg. With 10 μg of labeled material, transcripts that are present at one copy per mammalian cell or one copy per every 20 yeast cells can be detected. If mRNA quantities are limited (0.5 μg or less), the mRNA sample can be amplified by an *in vitro* transcription step as described later.

Two-Color Analysis

Because microarrays have typically been manufactured using relatively simple instruments, ensuring that identical amounts of DNA are deposited at different locations has been difficult. This theoretical limitation can be overcome by using a two-color labeling strategy.[16] Here target derived from mRNA from one condition is labeled with one fluor, whereas target from a different condition is labeled with a second fluor. Similar amounts of labeled material (usually cDNA) from the two samples are cohybridized to the microarray and the fluorescence intensity at the two appropriate emission wavelengths is determined. A good estimate of the relative differences in abundance of a target in the two samples can be obtained by comparing the ratio of the fluorescence intensities at the two wavelengths. By always using the same reference sample, microarrays produced using different sets of PCR products and by different individuals can be compared. Two-color strategies have been employed for mutation screening with oligonucleotide microarrays[2] and could theoretically be used in expression analysis, reducing the number of hybridizations that would need to be performed.

Array Design

The design of arrays for gene expression experiments should include appropriate controls for signal linearity and specificity. Probes to targets whose abundance is well characterized and probes to genes from different organisms should be included on the microarray, for example, human probes on an *Arabidopsis* microarray or bacterial genes on a yeast microarray. Target for these control probes can be generated by cloning the probe into a vector containing a phage promoter, and a poly(A) tract, and then generating polyadenylated mRNA by runoff *in vitro* transcription using the phage polymerase. This target can be added at various concentrations or at various times during the mRNA isolation and hybridization. Probes to ribosomal genes and to nontranscribed regions of the genome may also be included. In addition, controls that allow different scans to be normalized (as described later) should be arrayed. These may include multiple probes to genes whose expression is not expected to vary in the experiment (in yeast, probes complementary to the genes encoding actin and the TATA-binding protein have been used) or spots of total genomic DNA (for microarray experiments).[11]

Direct Labeling of Messenger RNA for Hybridization to Oligonucleotide Microarrays

One of the simplest methods for generating target is to label mRNA directly. At least two different methods have been reported: the mRNA is first fragmented to 30–50 base sizes by precipitating the RNA, resuspending the sample in a buffer containing magnesium ions [40 m*M* Tris–acetate (pH 8.1), 100 m*M* potassium acetate, 30 m*M* magnesium acetate] and then heating the sample to 94° for 35 min. The RNA fragments are then kinased by diluting the fragmentation reaction 2-fold and adding ATP to 8 μM, dithiothreitol (DTT, to 3 m*M*), bovine serum albumin (BSA, 10 μg/ml), and polynucleotide kinase. The reaction is incubated for 2 hr at 37°. Then a biotinylated oligoribonucleotide (5′-biotin-AAAAAA 3′) is ligated directly to fragmented, kinased mRNA using T4 RNA ligase in a buffer [50 m*M* Tris–HCl (pH 7.6), 10 m*M* DTT, and 1 m*M* ATP] containing a 10-fold molar excess of 5′-biotin-AAAAAA 3′.[4]

Heat-denatured mRNA (or total RNA) can also be incubated with a biotinylated psoralen derivative (35 μM, from Schleicher and Schuell,

Keene, NH) in diethyl pyrocarbonate (DEPC)-treated H_2O.[22] Psoralen will intercalate RNA and can be cross-linked to the mRNA by exposing the RNA-containing solution to ultraviolet light (350 nm) for 3 hr. After cross-linking, the excess psoralen is removed by three extractions with water-saturated *n*-butanol. The RNA is ethanol precipitated and fragmented in a magnesium buffer as described earlier. This method has been used to label prokaryotic transcripts that are difficult to purify away from ribosomal RNA.[22]

The biotin-labeled fragments are hybridized to the microarray. After washing, the label can be visualized by staining for 15 min with a solution containing 6× SSPE-T [0.9 *M* NaCl, 60 m*M* NaH_2PO_4 (pH 7.6), 6 m*M* EDTA, 0.005% Triton], 1 mg/ml acetylated BSA, and 2 μg/ml streptavidin *R*-phycoerythrin conjugate (Molecular Probes, Eugene, OR). The major disadvantages of these approaches are that larger quantities of mRNA are needed, that ribosomal RNA is labeled along with the messenger RNA, possibly creating higher backgrounds, and that solutions need to be kept RNase free at all steps.

Direct Labeling of cDNA for Hybridization to Oligonucleotide Microarrays

Another simple method that can be used when significant amounts of mRNA are available is direct labeling of cDNA.[7] Approximately 10–20 μg polyadenylated RNA is first converted to single-stranded cDNA using reverse transcriptase and an oligo(dT)21-primer essentially as described in Table III, but scaling everything proportionally and using unlabeled dCTP nucleotides. The cDNA is then phenol : chloroform extracted and precipitated by the addition of 0.5 volumes of 7.5 *M* ammonium acetate and 2 volumes of ethanol. The pellet is washed with 70% (v/v) ethanol and resuspended in 35 μl of a buffer (10 m*M* Tris–acetate (pH 7.5), 10 m*M* magnesium acetate, and 50 m*M* potassium acetate) containing 1.5 m*M* $CoCl_2$. The single-stranded cDNA is fragmented by adding ~0.75 units DNase I to the sample and incubating for 5 min at 37°. Following digestion, the DNase I is inactivated by transferring the microfuge tube to a boiling water bath for 15 min. It is important that the fragment sizes be checked by analyzing 1 μl of the reaction on a 2% agarose gel* as different batches of DNase I may vary in potency, and underdigestion or overdigestion may produce a poor signal. Optimally, the digested fragments should be about 50 bases in size. DNase I is also very sensitive to contaminants; for a clean

[22] A. de Saizieu, U. Certa, J. Warrington, C. Gray, W. Keck, and J. Mous, *Nature Biotechnol.* **16,** 45 (1998).

* For maximum sensitivity, SYBR-II green (Molecular Probes), a dye that binds single-stranded nucleic acids, is recommended instead of ethidium bromide.

digestion, the starting material should be relatively free of salts, particularly EDTA. The fragmented cDNA is then end labeled with biotin by the addition of 25 units (U) terminal transferase and 1 μl (1 mM) biotin-N6-ddATP (NEN, Boston, MA). The entire sample is hybridized to the array (usually overnight at 42°). The hybridization kinetics are slightly different than when RNA is hybridized and less stringent washes (15 min at 42° versus 50° in 0.5× SSPE-T) may be necessary. After hybridization, the samples are stained with a phycoerythrin–streptavidin conjugate as described earlier. This same method can be used to label genomic DNA whose hybridization pattern can serve as an excellent control for the cDNA hybridization pattern. Difficulties with this method include obtaining a reproducible DNase I cutting pattern, verifying that all mRNA has been converted to cDNA, and that large amounts of cDNA are required.

TABLE III
TARGET PREPARATION

1. Assemble following mix in microcentrifuge tube:

Reagent	Volume (μl)
Total mRNA (1.0 μg/μl)[a]	5.0
Control mRNA cocktail (0.5 ng/μl)[b]	1.0
Oligo(dT) 21-mer (1.0 μg/μl)	4.0
H_2O (DEPC treated)	17.0
	27.0 μl

2. Denature mRNA for 3 min at 65°. Anneal oligo(dT) to mRNA for 10 min at 25°. Add

5× First strand buffer[c]	10.0
10× Dithiothreitol (DTT, 0.1 M)	5.0
RNase block (20 U/μl)	1.5
dATP, dGTP, dTTP cocktail (25 mM each)	1.0
dCTP (1 mM)	2.0
Cy3-dCTP (1 mM)[d]	2.0
SuperScript II reverse transcriptase (200 U/μl)	1.5
	50.0 μl total

3. Reverse transcribe polyadenylated RNA for 2 hr at 37°
4. Add 5.0 μl of 2.5 M sodium acetate and 110 μl 100% ethanol at 25°[e]
5. Centrifuge for 15 min at 25° in microfuge to pellet cDNA/mRNA hybrids[f]
6. Remove and discard supernatant and carefully wash pellet with 0.5 ml 80% (v/v) ethanol[g]
7. Dry pellet in Speed-Vac and resuspend in 10.0 μl 1× TE (pH 8.0)[h]
8. Boil sample 3 min to denature cDNA/mRNA hybrids. Chill on ice immediately
9. Add 2.5 μl 1 N NaOH and incubate for 10 min at 37° to degrade mRNA
10. Neutralize cDNA mixture by adding 2.5 μl 1 M Tris–HCl (pH 6.8) and 2.0 μl 1 M HCl

TABLE III (*Continued*)

11. Add 1.7 μl 2.5 *M* sodium acetate and 37 μl 100% ethanol
12. Centrifuge for 15 min at full speed in microfuge to pellet cDNA
13. Remove and discard supernatant and wash pellet with 0.5 ml 80% ethanol[i]
14. Dry pellet in Speed-Vac and resuspend in 6.5 μl H_2O
15. Add 2.5 μl 20× SSC[j] and 1.0 μl 2% SDS
16. Heat at 65° for 0.5 min to dissolve target mixture
17. Centrifuge for 2 min in microfuge at high speed to pellet trace debris[k]
18. Transfer supernatant to new tube
19. Final target concentration should be ~0.5 μg/μl per fluor in 5× SSC with 0.2% SDS

Suggested materials
- StrataScript RT-PCR kit (Stratagene)
- Oligo(dT) 21-mer (treated with 0.1% DEPC to inactivate ribonucleases)
- 100 m*M* dATP, dCTP, dGTP, dTTP (Pharmacia, Piscataway, NJ)
- 1 m*M* Cy3-dCTP (Amersham, Piscataway, NJ)
- 1 m*M* Cy5-dCTP (Amersham)
- 1 m*M* fluorescein-12-dCTP (DuPont, Boston, MA)
- SuperScript II RNase H-reverse transcriptase (GIBCO-BRL, Gaithersburg, MD)

[a] Total mRNA purified from total RNA using Oligotex-dT (Qiagen)
[b] Control mRNAs from *in vitro* transcription are chosen, based on the experiment.
[c] 5× first strand buffer: 250 m*M* Tris–HCl (pH 8.3), 375 m*M* KCl, 15 m*M* KCl.
[d] To label mRNA with other fluors, substitute Fl12- or Cy5-dCTP in the reaction.
[e] Chilling or use of >2 volumes of ethanol results in the unwanted precipitation of free label.
[f] Pellet product on one side of the tube, then remove supernatant from the other side.
[g] To prevent loss of pellet, centrifuge 1 min before removing 80% ethanol.
[h] To prevent loss of pellet, centrifuge 1 min before removing 80% ethanol. The product often smears up the side of the tube.
[i] To prevent loss of pellet, centrifuge 1 min before removing 80% ethanol.
[j] 20× SSC buffer: 3 *M* NaCl, 0.3 *M* sodium citrate (pH 7.0).
[k] Tiny particles interfere with hybridization, which is carried out under a coverslip.

Generating Labeled Target by in Vitro Transcription

When only limited amounts of cDNA are available, amplification steps may be necessary.[4] First strand cDNA synthesis is performed essentially as described in Table III, except that the polyadenylated messenger RNA is primed for reverse transcription with an oligo(dT) primer that contains a phage T7 promoter sequence at its 5′ end.[23,24] The second strand of the cDNA is then synthesized with *Escherichia coli* DNA polymerase I.[4,25] Approximately 0.5 μg of the double-stranded cDNA is used as template

[23] J. Eberwine, H. Yeh, K. Miyashiro, Y. Cao, S. Nair, R. Finnell, M. Zettel, and P. Coleman, *Proc. Natl. Acad. Sci. U.S.A.* **89,** 3010 (1992).
[24] R. N. Van Gelder, M. E. von Zastrow, A. Yool, W. C. Dement, J. D. Barchas, and J. H. Eberwine, *Proc. Natl. Acad. Sci. U.S.A.* **87,** 1663 (1990).
[25] U. Gubler and B. J. Hoffman, *Gene* **25,** 263 (1983).

for an *in vitro* transcription reaction using T7 RNA polymerase in a buffer containing 7.5 m*M* ATP and GTP, 5.6 m*M* UTP and CTP, and 1.875 m*M* biotin-6-UTP and biotin-11-CTP. The *in vitro*-transcribed RNA is then ethanol precipitated and fragmented using magnesium ions as described earlier before hybridization to the microarray. It is important that the T7 promoter-oligo(dT) primer be high-performance liquid chromatography (HPLC) purified to ensure that it is full length, as oligonucleotide synthesis is performed 3′ to 5′ and contamination with truncated oligonucleotides will result in poor *in vitro* transcription yields. This method results in a large amplification of the amount of starting material (up to 50-fold). Its disadvantages are that it is more complex and time-consuming, that it employs extra enzymatic steps, that regions toward the 3′ ends of genes may be overrepresented in the final product, and that all solutions must be kept RNase free.

Microarray Target Preparation by Single-Round Reverse Transcription

All methods described previously have been used to prepare target for hybridization to oligonucleotide microarrays, but should work just as well for DNA microarrays. DNA microarray hybridization targets are commonly generated by a single round of reverse transcription in the presence of any of several fluorescently labeled nucleotides (Cy3, Cy5, fluorescein) using an oligo(dT) primer (Table III). Less starting material is required for hybridization because hybridization is generally performed in a very small volume (under a coverslip). Random nonamers or hexamers can also be used instead of oligo(dT) to prime cDNA synthesis,[11] permitting the detection of RNA species that are not polyadenylated.

Hybridization and Washing

The labeled target is hybridized to a microarray for times ranging from 2 to 20 hr. Arrays purchased from commercial suppliers may come enclosed in a hybridization cassette. Custom water-tight hybridization chambers, just large enough to enclose a microscope slide, can also be fabricated in a machine shop or purchased (Telechem International). Oligonucleotide hybridizations are typically performed between 42° and 50°, whereas DNA microarray hybridizations are performed between 55° and 70°. The linear range of the microarray can sometimes be extended by performing a second hybridization for a longer or shorter time. Different buffers have been used, including 6× SSPE-T [0.9 *M* NaCl, 60 m*M* NaH_2PO_4 (pH 7.6), 6 m*M* EDTA, 0.005% Triton] and 5× SSC [150 m*M* NaCl, 15 m*M* sodium citrate (pH 7.0)]. Dedicated wash stations can be purchased from some microarray suppliers. These instruments flush the arrays with buffers that have been heated to the correct temperature, thus standardizing and easing the washing process. The DNA microarrays can also be washed by immersion in a beaker containing a stir bar and a buffer at the appropriate concentration and temperature (Table IV). Wash times, temperatures, and stringencies will depend on the experiment (whether RNA or DNA is being hybridized) and may need to be determined empirically. Some protocols call for a short, low-stringency wash (15 min to 1 hr in 0.5× SSPE + 0.005% Triton X-100). An example is given in Table IV. Blocking agents such as unlabeled fragmented genomic (10 μg/ml) DNA may also be included.

Scanning Microarrays

Scanning systems can be purchased from manufacturers such as Hewlett Packard, General Scanning, or Molecular Dynamics. Scanners generally contain a motor for moving either the array or a mirror, a laser, confocal

TABLE IV
HYBRIDIZATION AND WASHING

1. Place microarray in hybridization cassette[a]
2. Add 5.0 μl of 5× SSC + 0.2% SDS to bottom of cassette for humidification[b]
3. Aliquot 5.0 μl of fluorescent target (see Table III) onto edge of microarray
4. Cover target droplet with 22-mm^2 glass coverslip using forceps[c]
5. Seal cassette containing microarray
6. Submerge hybridization cassette in water bath set at 62°
7. Hybridize for 6 hr at 62°
8. Following hybridization, remove microarray from hybridization cassette and place microarray *immediately* into beaker containing 400 ml 1× SSC and 0.1% SDS[d]
9. Wash microarray by gentle buffer agitation for 5 min at room temperature[e]
10. Transfer microarray to second beaker containing 400 ml 0.1× SSC and 0.1% SDS
11. Wash microarray by gentle buffer agitation for 5 min at room temperature
12. Rinse microarray briefly in third beaker containing 0.1× SSC to remove SDS
13. Allow microarrays to air dry[f]
14. Scan microarrays for fluorescence emission

[a] Cassettes can be purchased from Telechem International.
[b] Prevents dessication of the fluorescent sample.
[c] Coverslips must be dust and particle free to allow even seating on the array. Air bubbles trapped under the coverslip exit after several minutes at 62°.
[d] Use a 600-ml Pyrex beaker containing a magnetic stir bar. Alternately, wash stations can be purchased from Telechem International.
[e] Buffer agitation is accomplished by placing the beaker on a stir plate.
[f] Cy3 and Cy5 are scanned dry.

lenses, filter sets, and a photomultiplier tube (PMT) for collecting the fluorescent emission. Fluorescent microscopes can also be modified to scan the surface of a slide.[16] The number of photons received by the photomultiplier at every position on the array is stored on a personal computer in files that can be read into the various data analysis programs (Affymetrix GENECHIP, Imagene microarray software, Biodiscovery, AIS/BMS). These programs either use element coordinate information supplied by the user or allow the user to place a grid over the scanned image. The programs return an intensity value for every pixel in every array element in the scanned image.

Data Analysis

The analysis of data from the scanned image is a challenging problem associated with array-based experiments and one that is often given little consideration. The more mundane questions are what constitutes signal above background and what is the linear range for a particular hybridization. Different fluors may give different levels of backgrounds. Variations in scanner laser alignment may give different overall intensities. If background values are not set correctly, errors in estimates of both relative and absolute abundance may arise.

Several approaches have been used to calculate background levels. Some users place a border around a particular element on the array in the area where there is no bound nucleic acid and then average the intensity values from the border region.[11] This should work well theoretically, except that in some cases the immobilized nucleic acids work as a blocking agent, such that regions on the array where there is no nucleic acid may have a higher level of fluorescence than regions containing nonhybridizing nucleic acids. Alternately, the pixel intensity values for the entire element can be ranked from lowest to highest. The background level can be set to the intensity of the pixel for which 95% of the values are higher. In both cases, statistical methods are used to exclude outlying data points from the analysis.

At Affymetrix, scientists synthesize a second oligonucleotide probe that contains a single base mismatch at the central location on the probe relative to the perfect match for every element on the microarray. This mismatch probe is placed adjacent to the perfect match probe.[4] Oligonucleotide probes are very sensitive to the presence of mismatches, and hybridization at the mismatch position is usually reduced greatly. The mismatch signal can then be subtracted as background from the perfect match signal.

Different methods also exist for calculating the signal for each element on the array. After excluding outliers, one can compute the mean of all pixels for an element that are above background or one can rank the intensities of all pixels in the array element and then use the intensity for which 75% of the values are smaller.

Scan-to-Scan Normalization

Different scans may have higher or lower intensities, depending on the scanner, the fluor being used, or any of a number of other different factors. These differences can be normalized using at least two different methods.

First, the hybridization intensity from appropriate controls whose signal is not expected to vary can be used to normalize the scan intensity. This method works best when a large number of these controls is used. Second, if the microarrays contain a very large number of elements and wholesale differences in mRNA levels are not expected, the average overall hybridization intensity for all the elements on the microarray can be calculated and used to normalize signal strength.

Obtaining absolute estimates of target abundance by hybridization has always been a difficult problem. Different DNA sequences may have different melting temperatures and thus different hybridization properties, and it has been difficult to ensure that identical amounts of probe are deposited at all locations on the microarray. By using a two-color strategy for microarray hybridizations, estimates of *relative* abundance can be obtained when the hybridization signal is in the linear range. This is expected for most probes, as in theory much more probe is on the microarray than target. The experimental design should include a number of control probes whose target can be added at different amounts in order to determine the linear range for a series of hybridizations. More quantitative estimates could theoretically be obtained by carefully selecting probes to have similar hybridization properties and by improving the manufacturing process (e.g., by ensuring that equimolar amounts of probe are applied at each position on the microarray).

Oligonucleotides are even more sensitive to sequence-specific differences in annealing behavior than longer PCR products, creating an even larger barrier to measuring target abundance. Lockhart *et al.*[4] have shown that quantitative estimates of transcript abundance can be obtained by using a number of different oligonucleotide probes chosen using a set of heuristic rules. When fluorescently labeled target is hybridized, the fluorescent signal is quantitative (within a factor of two) when the background-corrected signal from 20 oligonucleotide probes, of different sequence but all complementary to an mRNA, is integrated. However, the number of probes needed for a quantitative estimate of target abundance can be expected to drop as algorithms for predicting melting temperatures and hybridization behavior for particular probes are refined. Despite these somewhat daunting considerations associated with data analysis, microarrays will probably have their widest application as a screening tool: most researchers will be completely satisfied to learn which genes are induced or repressed under their favorite growth condition or genetic background and may not be too concerned about whether induction is 10- or 15-fold.

Discussion

The labeling protocols presented here have all been tested, but not necessarily optimized. Many variations on the protocols given here are feasible: poly(A) purification steps might be eliminated, different fluors may work as well or better than those described here, and PCR amplification of cDNA may be possible. If microarrays are purchased from a commercial supplier, descriptions of methods will most likely be provided along with the product.

A potential problem for all microarray experiments is cross-hybridization. Because microarrays generally have longer probes than do oligonucleotide arrays, some have argued that their specificity is greater and that the potential for cross-hybridization is lower. However, the probability that a 25-mer will be duplicated randomly in the human genome is very low ($p = 2.0 \times 10^{-6}$). However, genome sequence is not a random collection of A, C, G, and T. Coding regions that have very similar or identical nucleotide sequence are distributed throughout the yeast genome, creating some potential for cross-hybridization. Such duplications can be expected or have been observed in the genomes of other organisms. The power of microarrays is therefore enhanced when the complete genome sequence is known and the specific sequences that may cross-hybridize can be known *a priori.*

Sophisticated approaches to dealing with the large amount of data that is produced in microarray-based experiments and novel ways to query the data, particularly if microarrays are to be used for purposes other than for screening, also need to be developed. An experiment using an array in which probes to all of the genes in the yeast genome are present generates a minimum of 6000 pieces of data.

Finally, it should be kept in mind that transcript abundance should not be equated with protein abundance or even promoter strength. Different mRNAs will have different half-lives and different decay directions (5′ to 3′, or vice versa). Even if the measured signal levels reflect the levels of a translationally competent mRNA, the encoded proteins may be regulated posttranslationally. Microarrays will have their widest application in conjunction with other tools, such as mass spectrometry.

[2] Protein Expression in Mammalian and Insect Cell Systems

By SABINE GEISSE and HANS P. KOCHER

Introduction

Almost three decades ago the first experiments of transfer of naked DNA into mammalian cells were performed. Since then these techniques have been explored thoroughly and refined into what is now called "eukaryotic expression systems."

These transfer techniques actually cover a large diversity of systems, which can be arbitrarily grouped into the following categories: (1) gene transfer via transfection of plasmids, leading either to stably integrated copies of the transgene into the host genome (stable expression) or to episomally replicating plasmids, which are gradually lost, unless selective pressure via antibiotics is applied (transient expression) and (2) gene transfer via infection: Several systems based on recombinant DNA or RNA viruses (including retroviruses) are available, which allow expression of transgenes again on a stable or transient basis.

The transfer of genetic material into suitable recipient cells is used for many purposes, ranging from expression cloning and mutational analysis to gene or cell therapy. In this article the focus of discussion will be devoted entirely to gene transfer to achieve recombinant protein production. Table I summarizes the most frequently employed expression systems for this purpose.

Prominent examples of the different types of systems will be described in theoretical and practical detail later. A few general remarks should precede the overview, however.

One of the key factors for success in recombinant protein expression is optimal cell culture maintenance. The recipient cell lines should exhibit logarithmic growth at the time point of transfection/infection, and the absence of mycoplasma and other contaminating agents needs to be controlled regularly. The use of antibiotics and antimycotics in the cell culture medium is frequently recommended; however, we prefer to cultivate our cell lines without antibiotics except for the selection of recombinants. Care should also be taken with respect to the age of the culture; switching routinely to younger passages from frozen stocks is certainly advisable. Except for the insect cell lines kept at 28°, all other mammalian cells are

0076-6879/99 $30.00

TABLE I
EXPRESSION SYSTEMS FOR RECOMBINANT PROTEIN PRODUCTION

Transfection systems based on stable integration of transgene	Transfection systems based on transient expression of transgene	Expression systems based on viral infection
CHO, BHK, NIH 3T3, HEK 293 cells Myeloma cells (Sp2/0, J558L, NS/O) MEL cells *Drosophila* Schneider S2 cells	COS cells HEK.EBNA cells	Baculovirus Alphavirus: Sindbis virus, Semliki Forest virus Vaccinia virus

cultivated at 37°/5% (v/v) CO_2 in a humidified atmosphere and are subcultured twice per week.

Finally, to enhance the frequency of stable integration events in Chinese hamster ovary (CHO) and myeloma cells, the recombinant expression plasmid should be linearized by restriction enzyme digestion at an appropriate site in the vector to facilitate integration.

Expression in Chinese Hamster Ovary Cells

Among all cell lines suitable for recombinant protein production, the CHO cell line established in 1957 by Puck is by far the most popular. Many features contribute to its popularity: good growth characteristics in serum-free and serum-containing media, ease of transfectability, genetic stability, and extensive characterization in terms of endogenous viral load and permissiveness for adventitious agents/viruses.[1]

In addition, CHO cells were shown to be highly suitable for the induction of gene amplification mechanisms with the aim of increasing productivity. One such mechanism relies on the inhibition of dominant amplification markers such as glutamine synthetase or adenosine deaminase by the addition of specific potent enzyme inhibitors such as methionine sulfoximine or 2′-deoxyformycin, respectively.[2,3]

Alternatively, recessive amplification markers can be employed; wild-type dihydrofolate reductase (dhfr) is the most commonly used.[4] A deriva-

[1] F. M. Wurm, *in* "Mammalian Cell Biotechnology in Protein Production" (H. Hauser and R. Wagner, eds.), p. 87. W. de Gruyter Verlag, Berlin, New York, 1997.

[2] M. I. Cockett, C. R. Bebbington, and G. T. Yarranton, *BioTechnology* **8,** 662 (1990).

[3] R. J. Kaufman, P. Murtha, D. E. Ingolia, C.-Y. Yeung, and R. E. Kellems, *Proc. Natl. Acad. Sci. U.S.A.* **83,** 3136 (1986).

[4] R. E. Kellems, *Curr. Opin. Biotechnol.* **2,** 723 (1991).

tive of the original CHO K1 cell line established in 1980 by Urlaub and Chasin[5] by chemical mutagenesis harbors a nonfunctional *dhfr* gene (CHO DUK X B11 cells). Due to mutation, these cells are unable to convert folate to tetrahydrofolate, a biocatalyst required for the *de novo* synthesis of purines and pyrimidines. Survival of cells is sustained if a supply of precursors (adenosine, deoxyadenosine, and thymidine) is provided by media components for utilization in the salvage pathway of purine and pyrimidine synthesis. Hence, cell culture medium devoid of these precursors is used for the primary selection of cells after cotransfection of a functional *dhfr* gene in conjunction with the gene of interest.

Gene amplification is induced by adding amethopterin (methotrexate, MTX) in stepwise increments to the culture medium. This folic acid analog, or antifolate, binds and inhibits dihydrofolate reductase stoichiometrically, forcing the cells to undergo genomic rearrangements and subsequent gene amplification for survival. Starting with low concentrations of MTX (in the nanomolar range), the concentration can be increased sequentially to several micromoles per liter of medium, giving rise to amplified stretches of DNA containing several hundred copies of the originally transfected plasmid(s) per cell.[6]

The underlying mechanistic details of gene amplification have been investigated and discussed for many years without a final, conclusive result being reached. An in-depth overview on current knowledge can be found in an article published by Wurm.[1]

This article discusses only a few points of essential importance for the success of experiments aiming at dhfr-linked, MTX-mediated gene amplification. First of all, a carefully designed and constructed expression plasmid is essential. It has been shown that the gene of interest and the selectable marker can be cotransfected on two individual plasmids.[7] As the amplified genomic locus is much larger in size than the original plasmids, coamplification of the two plasmids is likely to occur.[8,9] A more sophisticated approach involves coexpression of the two genes from the same plasmid via a bicistronic messenger RNA. If a promotorless selectable marker gene, e.g., the *dhfr* gene is placed 3′ of the gene of interest, transcription is driven solely by the 5′ upstream promotor. Initiation of translation of the 3′ gene is, however, very inefficient in comparison to the 5′ gene of

[5] G. Urlaub and L. A. Chasin, *Proc. Natl. Acad. Sci. U.S.A.* **77,** 4216 (1980).

[6] R. J. Kaufman, *Methods Enzymol.* **185,** 537 (1990).

[7] R. J. Kaufman, L. C. Wasley, A. J. Spiliotes, S. D. Gossels, S. A. Latt, G. R. Larsen, and R. M. Kay, *Mol. Cell. Biol.* **5,** 1750 (1985).

[8] M. Carroll, M. L. DeRose, P. Gaudray, C. M. Moore, D. R. Needham-Vandevanter, D. D. von Hoff, and G. M. Wahl, *Mol. Cell. Biol.* **8,** 1525 (1988).

[9] E. Heard, S. V. Williams, D. Sheer, and M. Fried, *Proc. Natl. Acad. Sci. U.S.A.* **88,** 8242 (1991).

interest, resulting in a 40- to 300- fold difference in protein quantity between the two gene products. As sufficient quantities of dhfr are the prerequisite for the survival of cells in methotrexate-containing medium, drug resistance is linked directly to enhanced protein production from the 5′ transgene.[10,11] The strong selection forces applied favor and induce genetic instability, however, frequently resulting in genomic rearrangements with concurrent loss of the 5′ transgene. For this reason, a new generation of bicistronic vectors was created featuring an internal ribosomal entry site element (IRES) upstream of the selectable marker to facilitate its translation.[12,13]

Second, the stepwise increase of MTX in the medium can be performed either on cell pools or on individually selected clones. The latter procedure is disadvantageous, however, because the vigorous gene amplification treatment induces the heterogeneity of cells associated with loss of the clonal phenotype during each round of amplification.[14]

Once the highest degree of gene amplification, reflected in maximal titers of recombinant protein, has been reached, it is essential to perform a cloning experiment in order to retrieve a stable, homogeneous cell population for further protein production.

Whether protein production, especially on large scale, can or should be done under continued selective pressure or in the absence of methotrexate is controversial.[15,16] The physical location of the amplified sequences and the degree of amplification achieved have profound effects on the stability of the production clone. Maintaining constant selective pressure requires an empirical decision on a case-by-case basis.

Example 1: Expression Experiments Performed in CHO Cells

1. Maintenance of Cell Culture. CHO DUK X B11 cells should be kept in MEM Alpha medium containing ribonucleosides and deoxyribonucleosides [MEM Alpha (+), Life Technologies, Rockville, MD] supplemented with 10% (v/v) fetal calf serum (FCS). If CHO K1 cells are used, they can be grown in either (1:1) Dulbecco's modified Eagle's medium (DMEM)/Ham's F12 medium or Ham's F12 medium alone, both enriched by 10% FCS. All these media contain proline, which eliminates the need for supplementing the media for growth of the proline-auxotroph CHO cells.

[10] E. Boel, K. L. Berkner, B. A. Nexo, and T. W. Schwartz, *FEBS Lett.* **219,** 181 (1987).
[11] R. J. Kaufman, P. Murtha, and M. V. Davies, *EMBO J.* **6,** 187 (1987).
[12] M. V. Davies and R. J. Kaufman, *Curr. Opin. Biotechnol.* **3,** 512 (1992).
[13] P. S. Mountford and A. G. Smith, *Trends Genet.* **11,** 179 (1995).
[14] R. E. Kellems, "Gene Amplification in Mammalian Cells." Dekker, New York, 1993.
[15] U. H. Weidle, P. Buckel, and J. Wienberg, *Gene* **66,** 193 (1988).
[16] M. G. Pallavicini, P. S. DeTeresa, C. Rosette, J. W. Gray, and F. M. Wurm, *Mol. Cell. Biol.* **10,** 401 (1990).

2. *Transfection.* Several techniques are suitable for the transfection of CHO cells, e.g., calcium phosphate-mediated gene transfer, lipofection, and electroporation. Whatever method is chosen, optimal transfection efficiencies, as judged by protein levels expressed and cell viability, need to be determined empirically with respect to the quantity of plasmid used, the quantity of reagent used for lipofection, and electroporation conditions. Once optimal conditions have been defined, it is advisable to establish several transfection pools in parallel to increase the probability of establishment of a high-producer cell line.

3. *Primary Selection.* Selection is usually started 48 hr after transfection. IF CHO K1 cells are used, a selectable marker conferring antibiotic resistance should be cotransfected to allow selection of positive transfectants (e.g., neomycin, hygromycin, or puromycin resistance genes).

In the case of CHO DUK X cells intended for subsequent gene amplification experiments, primary selection can be performed by selecting for growth in MEM Alpha medium depleted of nucleosides [MEM Alpha (−)], i.e., for *dhfr* expression. Alternatively, an antibiotic resistance gene can also be cotransfected for initial selection, followed by a switch to *dhfr* selection. It is also possible to pursue a simultaneous double-selection strategy with different marker combinations, with the aim to enrich the cell pools for high-level expressing cells.[17,18]

4. *Gene Amplification.* Once the cell pools subjected to primary selection have resumed logarithmic growth, methotrexate is added to the medium, usually at a starting concentration of 5–20 n*M*. This induces a severe crisis associated with cell death, which can last 3–4 weeks, until by outgrowth of cell islets resistant to this concentration the tissue culture flask is repopulated. The MTX concentration should not be augmented before the cell pools have fully recovered and are growing normally again, i.e., after two to three passages. Repeating the same procedure, the MTX concentration is increased in a stepwise fashion, e.g., in fivefold increments. During this procedure it is important to monitor production titers at each level of MTX, as these will vary because not all cell pools are equally amplifiable. It is also advisable to freeze some vials of cells at each MTX concentration as a backup.

5. *Cloning.* During the process of MTX-induced gene amplification, usually one cell pool will emerge producing the highest levels of recombinant protein at a maximal MTX concentration. From this population homogeneous clones are derived, either by limiting dilution cloning or (faster, but less well defined) by isolation of cell islets via cloning cylinders. Again

[17] M. Wirth, J. Bode, G. Zettlmeissl, and H. Hauser, *Gene* **73,** 419 (1988).
[18] M. J. Page and M. A. Sydenham, *BioTechnology* **9,** 64 (1991).

all clones require careful analyses of production titers, cell growth rates, and also stability of production of the recombinant protein over several weeks.

Expression Using Myeloma Cells

The most frequently used cell lines for recombinant expression, i.e., CHO cells, BHK cells, and HEK 293 cells, grow adherently and require tedious adaptation to suspension culture if large-scale production, for instance in roller bottles, spinner culture, or bioreactors, is to be used. An alternative as host cells are myeloma cell lines: they grow naturally in single cell suspension and can be adapted easily to serum-free growth. Originally used as the immortal fusion partner in the establishment of hybridomas, these often nonsecreting B cells retain their potential as "professional" secretory cells. The most commonly used myeloma cell lines for recombinant protein production are J558L [secreting immunoglobulin (Ig) λ1 light chains[19]], nonsecreting Sp2/0 Ag14[20,21] and NSO cells.[22–24]

For expression of engineered chimeric or humanized antibodies, these cell lines proved to be extremely useful. Other types of proteins were also expressed successfully in myeloma cells.[19,25]

Following cotransfection of dominant antibiotic selection markers such as neomycin or hygromycin, positive transfectants can be selected in media containing the corresponding antibiotic. Induction of gene amplification using glutamine synthetase[22] or dihydrofolate reductase[21] genes as markers is also possible.

Dhfr-mediated gene amplification is, however, complicated by the fact that these cells harbor a functional *dhfr* gene. To overcome this problem, a variant mouse *dhfr* gene carrying a point mutation with concomitant amino acid exchange (leucine to arginine at position 22 of the enzyme) is used.[26] If the mutant *dhfr* gene is introduced into myeloma cells, preferential amplification over the endogenous *dhfr* gene on MTX treatment is induced,

[19] A. Traunecker, F. Oliveri, and K. Karjalainen, *TIBTECH* **9,** 109 (1991).

[20] H. Dorai and G. P. Moore, *J. Immunol.* **139,** 4232 (1987).

[21] S. D. Gillies, H. Dorai, J. Wesolowski, G. Majeau, D. Young, J. Boyd, J. Gardner, and K. James, *BioTechnology* **7,** 799 (1989).

[22] C. R. Bebbington, G. Renner, S. Thomson, D. King, D. Abrams, and G. T. Yarranton, *BioTechnology* **10,** 169 (1992).

[23] C. Rossmann, N. Sharp, G. Allen, and D. Gewert, *Protein Express. Purif.* **7,** 335 (1996).

[24] W. Zhou, C.-C. Chen, B. Buckland, and J. Aunins, *Biotechnol. Bioeng.* **55,** 783 (1997).

[25] P. Lane, T. Brocker, S. Hubele, E. Padovan, A. Lanzavecchia, and F. McConnell, *J. Exp. Med.* **177,** 1209 (1993).

[26] C. C. Simonsen and A. D. Levinson, *Proc. Natl. Acad. Sci. U.S.A.* **80,** 2495 (1983).

as the mutant enzyme displays a 270-fold lower affinity for methotrexate than the wild-type *dhfr*. It should be noted that the starting concentration of methotrexate required is higher than for *dhfr*-minus CHO cells in order to silence the endogenous copies of the *dhfr* gene. Also, the maximal degree of amplification achievable with respect to copy number will be lower due to the insensitivity of the mutant *dhfr* toward MTX.[6]

Results expressed as product titers obtained using myeloma cells as recipients are impressive.[19,22–24,27] Special emphasis should be placed again on careful expression vector design to fully exploit the potential of these cells. Apart from strong viral promotors, expression plasmids featuring an immunoglobulin promotor/enhancer combination were shown to be extremely powerful in driving transcription in myeloma cell systems.[19]

Example 2: DHFR-Linked Expression in Sp2/0 Cells

1. Cell Culture Maintenance. Sp2/0 cells are easy to cultivate in RPMI 1640-based media supplemented with 10% FCS and 5×10^{-5} *M* 2-mercaptoethanol. They exhibit some sensitivity toward overgrowth and to too high dilutions, thereby requiring routine subculturing on the basis of cell count and viability determinations.

2. Transfection. The most suitable transfection method for suspension cells is certainly electroporation. Electroporation conditions yielding optimal transfection efficiencies need to be determined empirically. Again it is recommendable to transfect several cell pools in parallel.

3. Primary Selection. If a chimeric antibody molecule is to be produced, an antibiotic resistance gene can be linked to, e.g., the light chain sequences on one plasmid and the *dhfr* mutant gene to the heavy chain sequences on a second plasmid. Alternatively, if only a single chain molecule is desired, a selection plasmid such as pSV_2neo can be cotransfected. Depending on the cotransfected antibiotic resistance marker, primary selection can be performed using neomycin, hygromycin, or puromycin.

4. Gene Amplification Using Methotrexate. In essence, the gene amplification protocol used for myeloma cells follows the outline of the procedure for CHO cells, with two exceptions. The starting concentration of methotrexate is higher (200 n*M*) and usually maximal production titers are obtained in three rounds of amplification.

5. Cloning. A limiting dilution cloning experiment is performed subsequently using the highest producing cell pool to derive stable production clones.

[27] D. K. Robinson, J. Widmer, and K. Memmert, *J. Biotechnol.* **22,** 41 (1992).

Expression in Murine Erythroleukemic (MEL) Cells

One of the main contributors to the successful establishment of high-titer production cell lines is the so-called "position effect," i.e., integration of the expression plasmid into a transcriptionally highly active site of the genome. Unless gene targeting via homologous recombination is performed, this event occurs randomly, necessitating labor-intensive search for stable, possibly amplifiable, and highly active producer clones.

To circumvement this tedious procedure, an expression system has been described based on the use of "locus control regions" (LCRs or DCRs[28]). In nature, high-level expression from single copy genes (as, for instance, the immunoglobulin genes or the β-globin gene cluster) is mediated by potent genomic regulators. In the case of β-globin genes this "dominant control region" (DCR) is located 5′ upstream of the promotor/enhancer region and was identified originally by the presence of several DNase hypersensitive sites.[29] It is assumed currently that these control regions function in two ways: to create an "open" and thus DNase-sensitive chromatin configuration and to serve as a powerful enhancer of tissue-specific, position-independent gene transcription.[30]

Combining DNA sequences surrounding four of these hypersensitive sites, a relatively small, artificial DCR can be created to serve as part of the expression vector. On transfection into erythroid cells, the DCR is expected to govern and control high-level transcription irrespective of the integration site into the genome.[31] Expression plasmids for this type of system have been described previously.[28,32]

Mouse erythroleukemic cells[28,33] serve as suitable recipients for this type of expression system. The cell line we have used in the experiments described, MEL-E9, is a clonal derivative of MEL-NP5, a cell line derived from the spleen of mice infected with a replication-deficient spleen focus forming virus (SSFV). In contrast to the frequently employed Friend MEL cells, these cells do not shed viral particles, but are tumorigenic on injection into syngeneic mice. Also, they do not differentiate spontaneously, but can be induced to differentiate into hemoglobin-containing cells by treatment

[28] M. Needham, C. Gooding, K. Hudson, M. Antoniou, F. Grosveld, and M. Hollis, *Nucleic Acids Res.* **20,** 997 (1992).

[29] N. Dillon and F. Grosveld, *Trends Genet.* **9,** 134 (1993).

[30] R. Festenstein, M. Tolaini, P. Corbella, C. Mamalaki, J. Parrington, M. Fox, A. Miliou, M. Jones, and D. Kioussis, *Science* **271,** 1123 (1996).

[31] G. Blom van Assendelft, O. Hanscombe, F. Grosveld, and D. R. Greaves, *Cell* **56,** 969 (1989).

[32] P. Collis, M. Antoniou, and F. Grosveld, *EMBO J.* **9,** 233 (1990).

[33] P. A. Shelton, N. W. Davies, M. Antoniou, F. Grosveld, M. Needham, M. Hollis, W. J. Brammar, and E. C. Conley, *Receptors Channels* **1,** 25 (1993).

with dimethyl sulfoxide (DMSO), hemin, or hexamethylenebisacetamide (HMBA).[34,35]

One attractive feature of this system is the primary selection for positive transfectants in combination with a cloning procedure in semisolid medium, allowing the direct recovery of clonal populations immediately after selection.

Example 3: Production of Heterologous Proteins in MEL-E9 Cells

1. *Cell Culture Maintenance.* MEL-E9 cells are cultivated in Iscove's modified Dulbecco's medium (IMDM) supplemented with 10% fetal calf serum and 2 m*M* glutamine.

2. *Transfection.* Introduction of plasmids is performed by electroporation, although other transfection procedures are also suitable.[28] The cells are cotransfected with a mutant *dhfr* gene to allow for dominant selection against MTX (see later).

3. *Primary Selection and Cloning.* The transfected cell pools are cultivated for 2 days posttransfection to allow for recovery, followed by a switch of medium to IMDM plus 10% dialyzed fetal calf serum, including 100 n*M* methotrexate. After a cultivation period of another 4 days, the cell pools are transferred at appropriate dilutions into 10-cm petri dishes containing the same selective medium dissolved in 1.2% methocel. Incubation is continued for 2 weeks, after which time MTX-resistant colonies become visible by eye. These cell colonies can be isolated by picking and are transferred directly into 24-well plates for further expansion and analysis of product titers.

In this context it is worthwhile mentioning that we could not reproduce the claim of "high-level, position-independent expression" using this system. On screening of a relatively large number of clones, high producers could be identified, but the system required a considerable analytical effort. The variability in expression levels observed can be explained either by a copy number effect[28,36] or by variable transcription activation induced by the selection of clones with low levels of methotrexate.

Transient Expression: COS Cell System

When talking about transient expression of genes in mammalian cells, the COS cell system almost certainly comes to mind. COS cell lines, estab-

[34] L. Wolff and S. Ruscetti, *Science* **228,** 1549 (1985).

[35] L. Wolff, P. Tambourin, and S. Ruscetti, *Virology* **152,** 272 (1986).

[36] M. Needham, M. Egerton, A. Millest, S. Evans, M. Popplewell, G. Cerillo, J. McPheat, A. Monk, A. Jack, D. Johnstone, and M. Hollis, *Protein Express Purif.* **6,** 124 (1995).

lished in 1981 by Y. Gluzman, have found widespread application for the rapid analysis of functionality of plasmids and for expression cloning and production of recombinant proteins on a small scale.[37]

The most frequently used COS-1 and COS-7 cell lines are derivatives of the African green monkey kidney cell line CV-1, harboring a replication-defective simian virus 40 (SV40) genome. However, they do produce large amounts of SV40 large T antigen, which allows extrachromosomal replication of plasmids carrying an SV40 origin of replication.[38] On transfection of an expression plasmid featuring the gene of interest under the control of a strong, mostly viral, promotor and an SV40 ori, this construct replicates rapidly to high copy numbers, allowing the recovery of protein from cell supernatants or recombinant cells within 48–72 hr.[39] However, as the protein-processing machinery of the COS cells is overloaded by the large amount of DNA (hundreds to thousand of copies per cell have been reported), the cells suffer from cytopathic effect and die within days after transfection.[40]

This effect precludes the use of the COS cell system for prolonged periods of time or on a large scale, i.e., in fermentation. However, it has been shown that COS expression can be performed over extended time periods by frequent medium changes or in batch mode at an intermediate scale. Blasey and Bernard[41] reported transient COS expression by the cultivation of cells on microcarrier beads in spinner culture with multiple harvests of spent culture medium, resulting in the cumulative production of several milligrams of recombinant protein. We chose to simultaneously transfect larger quantities of cells in batches, grown adherently either on roller surface or on 500-cm^2 tissue culture plates, also giving rise to the recovery of milligram quantities of recombinant protein.[42]

Two parameters are of essential importance for the success of the COS cell system used on an extended scale: the optimal maintenance of the cell culture allowing the growth of large quantities of cells for transfection and a well-optimized transfection procedure facilitating the efficient uptake of plasmid DNA. A suitable transfection protocol for COS cells is summarized next.

[37] C. P. Edwards and A. Aruffo, *Curr. Opin. Biotechnol.* **4,** 558 (1993).

[38] Y. Gluzman, *Cell* **23,** 175 (1981).

[39] P. Mellon, V. Parker, Y. Gluzman, and T. Maniatis, *Cell* **27,** 279 (1981).

[40] R. D. Gerard and Y. Gluzman, *Mol. Cell. Biol.* **5,** 3231 (1985).

[41] H. D. Blasey and A. R. Bernard, *in* "Animal Cell Technology: Products of Today, Prospects for Tomorrow" (R. E. Spier, J. B. Griffiths, and J. B. Berthold, eds.), p. 331. Butterworth and Heinemann, Oxford, 1994.

[42] R. Ridder, S. Geisse, B. Kleuser, P. Kawalleck, and H. Gram, *Gene* **166,** 273 (1995).

Example 4: Transient Expression in COS Cells

1. Cell Culture Maintenance. COS-1 (or COS-7) cells are grown in Dulbecco's modified Eagle's medium (DMEM) enriched by 10% fetal calf serum.

2. Transfection. Several procedures for the introduction of plasmids into COS cells are suitable; on comparison, protoplast fusion and DEAE-dextran-mediated gene transfer seem to be the most efficient.[43] We have obtained the best results with DEAE-dextran-mediated gene transfer in combination with chloroquine treatment for 4 hr followed by DMSO shock for 2 min. It should be kept in mind that there are variations among different batches of DEAE-dextran; pretesting is therefore recommended.

Following transfection, incubation of cells is continued for 72 hr before harvest, as plasmid replication peaks after 48 hr, but protein production reaches its maximum after 72 hr posttransfection.[37,44] In these cases where the presence of 10% fetal calf serum is undesirable for protein purification, a medium change to DMEM without further additives can be performed 24 hr after transfection without a significant loss in productivity.

For transfection of COS cells on an enlarged scale, sufficient quantities of cells prior to transfection need to be available, which are then seeded onto 500-cm^2 tissue culture plates or roller surfaces. Once the cells have become attached (which may require several days of roller cultivation), the cell population is transfected in analogy to the protocol used for small-scale transfections. Large quantities of plasmid are required for these experiments; transfection of cells on roller surfaces needs 75–100 μg of plasmid per roller, whereas a 500-cm^2 tissue culture plate requires 70 μg of plasmid for optimal results.

Epstein–Barr Virus-Driven Expression in HEK 293 Cells

Another example of an expression system based on the extrachromosomal replication of DNA, which has been employed less frequently, is based on structural elements of Epstein–Barr virus (EBV). It has long been known that EBV can efficiently immortalize primary human B lymphocytes by infection. Consequently, many lymphocytic cell lines established *in vitro,* as well as many human tumor cell lines derived from tissue specimens, contain EBV in its latent form. In most cases the virus persists

[43] F.-W. Kluxen and H. Luebbert, *Anal. Biochem.* **208,** 352 (1993).
[44] J. J. Trill, A. R. Shatzman, and S. Ganguly, *Curr. Opin. Biotechnol.* **6,** 553 (1995).

episomally in the range of 5 to 50 copies per cell and replicates coordinately with cellular DNA synthesis, i.e., once per S phase.[45,46]

EBV nuclear antigen 1 (EBNA-1) has been identified as one of the major immortalizing genes of EBV. The EBNA-1 protein product binds to and activates the latent viral origin of replication, oriP, initiating viral replication.[47,48]

Transfection experiments revealed that expression plasmids featuring a minimal 2.6-kb EBNA-1 fragment plus a 1.7-kb oriP fragment can transform a variety of cell lines successfully; the rest of the 172-kb viral genome is dispensable for the extrachromosomal replication of plasmids.[49] In the absence of selective pressure, the episomal entity is gradually lost from the cells at a rate of 2–5% per generation. If, however, selective pressure is applied via an antibiotic resistance marker included in the expression vector, the episomes can be maintained stably over a period of months.[50,51] The EBV system therefore represents a borderline case, being transient with respect to integration events and stable in terms of production rates over long periods of time.

A suitable candidate as a recipient for EBV vectors as described earlier is the human embryonic kidney cell line HEK 293. Primary kidney cells were originally immortalized by exposure to sheared fragments of Adenovirus 5 DNA. Molecular analysis revealed that the resultant immortal 293 cell line contains 17% of the extreme left-hand portion of the adenovirus genome bearing the *E1a/E1b* genes.[52] The E1a protein has been shown to transactivate a large variety of other genes and viral promotors. For example, the cytomegalovirus (CMV) promotor can be activated through transcription factors such as ATF and CREB by E1a.[53,54]

An optimal expression plasmid for this system should therefore contain the CMV promotor driving the transgene of choice, the *EBNA-1* gene and the oriP of EBV for episomal replication. Alternatively, a HEK 293 cell line constitutively expressing *EBNA-1* can be used. An antibiotic resistance

[45] B. Sugden, K. Marsh, and J. Yates, *Mol. Cell. Biol.* **5,** 410 (1985).
[46] J. L. Yates, N. Warren, and B. Sugden, *Nature* **313,** 812 (1985).
[47] S. Lupton and A. J. Levine, *Mol. Cell. Biol.* **5,** 2533 (1985).
[48] D. Reisman, J. Yates, and B. Sugden, *Mol. Cell. Biol.* **5,** 1822 (1985).
[49] T. Chittenden, S. Lupton, and A. J. Levine, *J. Virol.* **63,** 3016 (1989).
[50] J. L. Yates and N. Guan, *J. Virol.* **65,** 483 (1991).
[51] G. Cachianes, C. Ho, R. F. Weber, S. R. Williams, D. V. Goeddel, and D. W. Leung, *BioTechniques* **15,** 255 (1993).
[52] F. L. Graham, J. Smiley, W. C. Russell, and R. Nairn, *J. Gen. Virol.* **36,** 59 (1977).
[53] C. M. Gorman, D. Gies, G. McCray, and M. Huang, *Virology* **171,** 377 (1989).
[54] D. M. Olive, W. Al-Mulla, M. Simsek, S. Zarban, and W. Al-Nakib, *Arch. Virol.* **112,** 67 (1990).

marker included on the plasmid allows selection for a stable maintenance of episomes.

Example 5: OriP-Driven Expression in HEK.EBNA Cells

1. Cell Culture Maintenance. The HEK.EBNA cell line, which stably carries a functional copy of the *EBNA-1* gene, is available commercially from Invitrogen (Carlsbad, CA). These cells grow well in DMEM supplemented with 10% fetal calf serum. Alternatively, wild-type HEK 293 cells can be used, which should be kept in minimum essential medium (MEM) with 10% fetal calf serum.

2. Transfection. Again, several transfection procedures are suitable, of which lipofection or calcium phosphate-mediated transfection works best in our hands. If lipofection is used, a comparison of different liposome preparations and careful optimization of the protocol (as suggested by the vendors) can enhance yield markedly. If transient production is attempted, cell culture supernatants are harvested 72 hr after transfection. The cells can then be passaged or fed with fresh culture medium to allow multiple harvests, but the episomal entities are lost fairly rapidly from the cells, as revealed by a drop in production titer.

3. Selection for Stable Episomes. To obtain a stable production cell line, antibiotic selection is started 48 hr after transfection. In the case of the HEK.EBNA cell line, selection by neomycin is not possible, as the *EBNA-1* gene was introduced into the cell line via selection with geneticin (G418). However, successful selection with hygromycin has been reported.[51,55] We have used the bleomycin derivative Zeocin, which works equally well in selecting stable cell populations.[56,57] Because HEK.EBNA cells adhere only very loosely to plastic surfaces, with a strong tendency to float, care should be taken to perform only a minimum of operations (such as passaging of the cells at a low dilution rate) during the ongoing selection process, until healthy-looking attached cells emerge. These cell pools can then be passaged and expanded regularly for production on an enlarged scale.

Protein Production via Infection: Baculovirus System

One of the drawbacks of gene transfer by conventional transfection technologies is certainly the limited uptake of foreign DNA into cells. Viruses have found mechanisms to circumvent this rate limitation and

[55] R. A. Horlick, K. Sperle, L. A. Breth, C. C. Reid, E. S. Shen, A. K. Robbins, G. M. Cooke, and B. L. Largent, *Protein Express. Purif.* **9,** 301 (1997).

[56] O. Genilloud, M. C. Garrido, and F. Moreno, *Gene* **32,** 225 (1984).

[57] P. Mulsant, A. Gatignol, M. Dalens, and G. Tiraby, *Somat. Cell. Mol. Genet.* **14,** 243 (1988).

deliver their genetic material into cells at a much higher percentage, which renders them particularly attractive as vehicles for gene transfer.

Several systems for protein production based on recombinant viruses are available, of which the baculovirus expression vector system (BEVS) is probably the most well known.

Baculoviruses comprise a large family of DNA viruses, with a very narrow host range confined mainly to arthropods. For recombinant protein production, two species of baculoviruses are mainly employed: *Autographa californica* nuclear polyhedrosis virus (AcNPV) and, less frequently, *Bombyx mori* nuclear polyhedrosis virus (BmPV). Initial experiments performed in the 1970s to construct recombinant viruses for gene transfer were based on observations that two of the most abundantly expressed proteins of the virus during the late phase of infection, polyhedrin and p10 protein, participate in the formation of so-called "occlusion bodies." These particles (polyhedra) accumulate inside the cell nucleus as a reservoir and function as a survival mechanism for the virus. For viral replication the polyhedrin and p10 genes are dispensable and can thus be replaced by transgenes of choice.[58] Accordingly, early expression protocols using the baculovirus system were based on the identification of virus populations exhibiting an occlusion body-negative phenotype.[59,60] As insertion of the gene of interest under control of the polyhedrin or p10 promotor was achieved by homologous recombination, occurring with a frequency of 1–3%, the search for recombinant viruses was a tedious and painstaking procedure.

No other expression system has undergone such a rapid and effective development and refinement of methodology than the baculovirus expression system during the last few years. Nowadays, linearized viral preparations are available commercially that contain a deletion in an essential gene adjacent to polyhedrin, in addition to a marker gene [usually *lacZ,* recently also green fluorescent protein (GFP)], which allows the easy identification of successful recombination events at a frequency of 80–100%.[61] Instead of cotransfection of viral DNA and recombinant transfer vector into insect cells, recombination can also be performed in bacteria with subsequent isolation and transfection of the recombinant bacmid.[62] Verification of recombination events is done by polymerase chain reaction (PCR)[63] as well as on the protein level, and virus stocks can be titrated by classical plaque

[58] G. W. Blissard and G. F. Rohrmann, *Annu. Rev. Entomol.* **53,** 127 (1990).

[59] G. E. Smith, M. D. Summers, and M. J. Fraser, *Mol. Cell. Biol.* **3,** 2156 (1983).

[60] G. D. Pennock, C. Shoemaker, and L. K. Miller, *Mol. Cell. Biol.* **4,** 399 (1984).

[61] P. A. Kitts and R. D. Possee, *BioTechniques* **14,** 810 (1993).

[62] V. A. Luckow, S. C. Lee, G. F. Barry, and P. O. Olins, *J. Virol.* **67,** 4566 (1993).

[63] W. P. Sisk, J. D. Bradley, L. L. Seivert, R. A. Vargas, and R. A. Horlick, *BioTechniques* **13,** 186 (1992).

assay or limiting dilution. An ELISA-based titration system has also been commercialized by Clontech (Palo Alto, CA).

In addition, two other facts contributed significantly to the success of the system. The availability of the entire sequence of the AcNPV virus[64] has allowed identification and characterization of many hitherto unknown viral genes. This has led not only to many clues on functionality and life cycle management of the virus, but also to the discovery of other viral promotors suitable for driving transcription of a transgene, e.g., the AcNPV basic protein promotor shown to be active at an earlier time point after infection.[65] The use of early promotors has an impact on product titers but, more importantly, on secondary modifications, such as glycosylation, in the insect cell system. It is believed that at the very late stages of infection, the processing machinery of the insect cell may be compromised or impaired to such an extent that incomplete processing—unrelated to an insufficient or modified insect cell repertoire of processing enzymes—occurs. In cases where highly glycosylated or otherwise modified proteins are to be produced, expression driven by a weaker earlier promotor may therefore be advantageous.[66–68]

Much developmental progress has also taken place in the field of insect cell culture. Over the years, many insect cell lines have been established and evaluated for suitability as hosts in protein production, but all major expression work is still done in IPLB-Sf21-AE/Sf9 or BTI-TN-5B1-4 (TN5, High Five) cells. However, insect cell media and cell culture maintenance have undergone significant improvements, also allowing large-scale production in serum-free media.[69,70]

In summary, the baculovirus expression system has become one of the major players in the field of protein expression, reflected in the vast number of published examples of the successful production of secreted, membrane-bound, and intracellularly expressed proteins. The lytic nature of the system renders it particularly attractive for the production of otherwise cell-toxic proteins; cell lysis, however, restricts the system to batch mode or, at best,

[64] M. D. Ayres, S. C. Howard, J. Kuzio, M. Lopez-Ferber, and R. D. Possee, *Virology* **202,** 586 (1994).

[65] B. C. Bonning, P. W. Roelvink, J. M. Vlak, R. D. Possee, and B. D. Hammock, *J. Gen. Virol.* **75,** 1551 (1994).

[66] P. Sridhar, A. K. Panda, R. Pal, G. P. Talwar, and S. E. Hasnain, *FEBS Lett.* **315,** 282 (1993).

[67] G. Chazenbalk and B. Rapoport, *J. Biol. Chem.* **270,** 1543 (1995).

[68] D. L. Jarvis, C. Weinkauf, and L. A. Guarino, *Protein Express. Purif.* **8,** 191 (1996).

[69] R. L. Tom, A. W. Caron, B. Massie, and A. A. Kamen, *in* "Methods in Molecular Biology" (C. D. Richardson, ed.), Vol. 39, p. 203. Humana Press, Totowa, NJ, 1995.

[70] J.-D. Yang, P. Gecik, A. Collins, S. Czarnecki, H.-H. Hsu, A. Lasdun, R. Sundaram, G. Muthukumar, and M. Silberklang, *Biotechnol. Bioeng.* **52,** 696 (1996).

semicontinuous production runs. This has prompted attempts to develop protocols for the stable transfection of insect cells with expression plasmids carrying the early active viral ie-1 promotor,[71,72] but a striking success with this approach has yet to be demonstrated.

Example 6: Expression of Proteins Using Baculovirus-Infected Insect Cells

1. Cell Culture Maintenance. For initial transfection and selection of recombinant virus and plaque assays, we have obtained the best results using IPLB-Sf21-AE cells cultivated on tissue culture flasks in Excell 401 medium (JRH Bioscience, Lenexa, KS) plus 10% FCS. Sf21 cells attach more firmly to plastic surfaces than Sf9 cells, which increases the detectability and quality of plaque formation.

For the production of recombinant proteins, Sf9 cells cultivated in either T.C.100 medium plus 10% FCS or in serum-free SF900 II medium on roller bottles (both from Life Technologies, Rockville, MD) are used conveniently. Alternatively, TN5 cells grown in suspension in spinner culture or Erlenmeyer shaker flasks using SF900 II medium or other commercially available serum-free media formulations can be employed as well.

Great care should be taken to avoid cross-contaminations of infected cultures with noninfected backup cells. Ideally, the infected cells should be separated physically from the backup cultures, either in a separate laboratory, or at least in separate incubators. As a minimal compromise, infected and noninfected cells should never be handled simultaneously under the same laminar flow bench.

2. Transfection. Several baculovirus systems providing linearized, deleted AcNPV virus DNA and various transfer vectors are available. Our system of choice is BacPAK 6 (Clontech) in combination with the corresponding transfer vectors BacPAK 8 and 9. The recombinant transfer plasmid is cotransfected together with viral DNA via lipofection into Sf21 cells according to the protocol supplied by the vendor. The supernatant is harvested after 5 days of incubation and is subjected to plaque purification to obtain homogeneous, clonal virus populations for further studies. An excellent detailed protocol for plaque assays and plaque purification is given in Gorman *et al.*[73]

3. Analysis of Recombinant Virus. Plaque picks obtained from the first plaque assay are amplified on T25 tissue culture flasks, and the supernatant

[71] K. A. Joyce, A. E. Atkinson, I. Bermudez, D. J. Beadle, and L. A. King, *FEBS Lett.* **335,** 61 (1993).

[72] J. R. McLachlin and L. K. Miller, *In Vitro Cell. Dev. Biol.* **33,** 575 (1997).

[73] L. A. King and R. D. Possee, "The Baculovirus Expression System, A Laboratory Guide." Chapman and Hall, 1992.

is subsequently analyzed for the presence of recombinant virus. A rapid method for the analysis of recombination events is performing PCR reactions on crude viral lysates, provided that the primers used in the reaction are very sensitive and gene specific.[63] Alternatively, viral DNA can be isolated first before PCR analysis.[74] The expression of protein should be controlled either by ELISA techniques or, for correct size, by Western blotting. In case no suitable antibody preparation is available, the desired protein can be expressed in a "tagged" version, using transfer vectors carrying a tag such as glutathione *S*-transferase (GST) or polyhistidine, which will aid in later purification of the protein as well.[75]

4. Establishment of High-Titer Virus Stocks and Kinetic Experiments. Once a recombinant virus preparation has been identified by PCR and/or protein analysis, a high-titer virus stock is prepared by infecting cells at a low multiplicity of infection (MOI) (usually 0.1–0.2). The well-infected stock is then recovered and titrated carefully as a basis for determining optimal production conditions. For this purpose, a kinetic experiment is set up. Cells or supernatants are harvested at different time points following infection with varying MOIs. Analysis of recombinant protein production in these samples will indicate how much virus is needed to obtain maximal yield after a given time of infection. Only this type of experiment, in combination with the careful titration of virus stocks, will guarantee the reproducibility of experiments during repeated production runs.

5. Production in Insect Cells. Finally, a working virus stock of sufficient volume needs to be prepared and titrated. Based on the results of the kinetic experiment, batch productions are performed; the scale is dependent on the availability of cell culture equipment. If virus or cell culture devices are limited, "continued batch production" may help in some instances: here, 10% of the infected cell culture is retained in the vessel during harvest and subsequently inoculated with fresh cells. We have had good results with a "triple batch procedure" in the case of secreted proteins, albeit not in every instance.

Comparative Protein Production in Different Systems: Expression of hu-LIF

In order to exemplify the overview on expression strategies discussed earlier, the following section is used as an example of the practical applica-

[74] A. Day, T. Wright, A. Sewall, M. Price-Laface, N. Srivastava, and M. Finlayson, *in* "Methods in Molecular Biology" (C. D. Richardson, ed.), Vol. 39, p. 143. Humana Press, Totowa, NJ, 1995.

[75] J. Nilsson, S. Stahl, J. Lundeberg, M. Uhlen, and P.-A. Nygren, *Protein Express. Purif.* **11,** 1 (1997).

tion of these systems. As a candidate for comparative expression we chose the gene coding for human leukemia inhibitory factor (hu-LIF). Leukemia inhibitory factor is a cytokine with pleiotropic functions[76]; from a structural point of view, it is a medium-sized protein, which is heavily glycosylated on seven potential N-linked glycosylation sites (20–67 kDa molecular mass, depending on source). Six cysteine residues in the amino acid sequence participate in the formation of three disulfide bridges.[77] These structural and biochemical characteristics render it a well-suited candidate for eukaryotic expression. The cDNA of the mature hu-LIF, derived from exons 2 and 3 of the single copy LIF gene, comprises 550 bp, which translate into a 180 amino acid protein.[78] This cDNA fragment was cloned into appropriate expression vectors for each system, and transfections, selections, and cloning were performed following the procedures outlined previously. For precise details on transfection conditions, plasmid quantities, and so on, see Geisse *et al.*,[79] and for details on the HEK.EBNA system, see Horlick *et al.*[55] A summary of methods applied is given in Table II. Titers of hu-LIF in cell culture supernatants were determined by sandwich ELISA using a biotinylated polyclonal rabbit antiserum for detection. The sensitivity of the ELISA ranged between 5 and 10 ng of hu-LIF/ml supernatant.

On expression in each system on a fairly large scale (1–2 liters in roller bottles), the hu-LIF protein was purified via immunoaffinity chromatography using immobilized rabbit polyclonal antiserum, and subsequently subjected to SDS–PAGE analysis. Results of the protein analysis are shown in Fig 1.

All purified protein preparations were also analyzed by N-terminal amino acid sequence determination and were found to be homogeneous, except for the CHO-derived material, where the first N-terminal amino acid was missing in 25% of the analyzed protein sample.

A summary of data on yields and time frames for each expression system is compiled in Table III.

Interpretation of Results and Practical Implications

For the production of a recombinant protein using a eukaryotic expression system, the choice of a system is largely dependent on three criteria: the quantity of recombinant protein to be produced and its future use (as

[76] D. J. Hilton and N. M. Gough, *J. Cell. Biochem.* **46,** 21 (1991).

[77] H. Gascan, A. Godard, C. Ferenz, J. Naulet, V. Praloran, M.-A. Peyrat, R. Hewick, Y. Jacques, J.-F. Moreau, and J.-P. Soulillou, *J. Biol. Chem.* **264,** 21509 (1989).

[78] D. J. Hilton, *TIBS* **17,** 72 (1992).

[79] S. Geisse, H. Gram, B. Kleuser, and H. P. Kocher, *Protein Express. Purif.* **8,** 271 (1996).

TABLE II
COMPARATIVE EXPRESSION: METHODOLOGY

Cells	Maintenance medium	Transfection method	Primary selection medium	Amplification medium	Cloning procedure	Production conditions
CHO DUK X B11	MEM Alpha(+)/ 10% FCS	Electroporation	MEM Alpha(+)/ 10% FCS, 0.35 mg/ml G418	MEM Alpha(−)/ 10% FCS dialyzed MTX: 20 > 100 > 500 > 1000 > 2500 n*M*	Limiting dilution	Adaptation to serum-free medium possible
Sp2/0 Ag14.10	RPMI 1640/10% FCS/2-ME	Electroporation	RPMI 1640/10% FCS/2-ME, 0.7 mg/ml G418	RPMI 1640/10% FCS dialyzed/ 2-ME MTX: 200 > 1000 > 2500 n*M*	Limiting dilution	Adaptation to serum-free medium possible
MEL-E9	IMDM/10% FCS	Electroporation	IMDM/10% FCS dialyzed, 100 n*M* MTX	—	Semisolid medium 1.2% methocel	Induction with 10^{-4} *M* hemin 2–3 days prior to harvest
COS-1	DMEM/10% FCS	DEAE-dextran/ chloroquine + DMSO shock	—	—	—	Change of medium to serum-free DMEM
HEK.EBNA	DMEM/10% FCS	Calcium phosphate mediated	DMEM/10% FCS, 0.1 mg/ ml zeocin	—	Limiting dilution (optional)	Adaptation to serum-free medium possible
Sf21	Excell 401/10% FCS	Lipofection	—	—	Plaque purification	Infection kinetic: 3 MOI/3 days optimal Production using Sf9 cells in T.C. 100/10% FCS or SF900 II

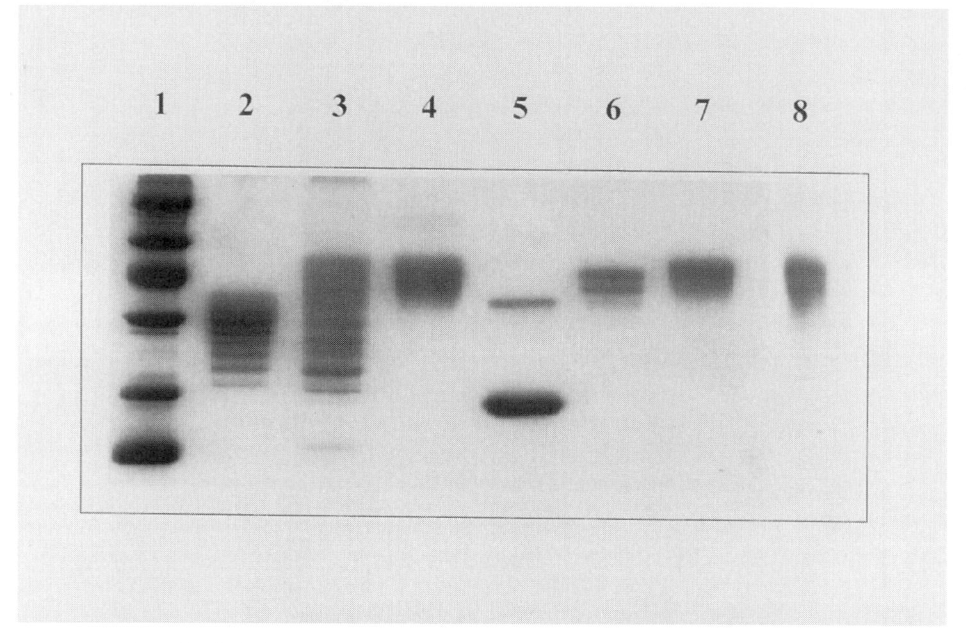

FIG. 1. 12.5% SDS–PAGE analysis. Lane 1: Molecular weight marker (Pharmacia, Piscataway, NJ); from top to bottom: 94K, 67K, 43K, 30K, 20.1K, 14.4K. Lane 2: 10 μg Sf9-LIF. Lane 3: 10 μg COS-LIF. Lane 4: 10 μg Sp2/0-LIF. Lane 5: 10 μg *Escherichia coli*-LIF (as reference). Lane 6: 10 μg HEK.EBNA-LIF. Lane 7: 10 μg MEL-LIF. Lane 8: 7.5 μg CHO-LIF. The weak band seen at approximately 40K in lane 5 represents most likely aggregated *E. coli*-derived LIF protein.

a research tool or as a biopharmaceutical), the time frame necessary for production, and the authenticity of the recombinant product.

As is clearly evident from the results shown in Table III, the six eukaryotic expression systems can be subdivided into three categories: systems giving rise to stable high-titer production clones on the basis of gene amplification using a well-characterized host cell line, the establishment of which requires considerable time (e.g., CHO cells and myeloma cells). In contrast, rapid expression systems (using COS and MEL cells) give rise to comparatively low yields. Expression using HEK.EBNA cells or infection of insect cells with baculoviruses falls into a third category, as remarkable quantities of recombinant protein can be produced in a relatively short time.

The quality of the recombinant protein produced is the second important aspect to consider. Natural human LIF (purified from a T lymphoma cell line or the human bladder carcinoma cell line 5637) is known to be glycosylated

TABLE III
COMPARATIVE EXPRESSION: YIELDS AND TIME FRAMES

Production system	CHO DUK X B11	Sp2/0 Ag14.10	MEL E9	COS-1	HEK.EBNA	Sf9
In T flask	1.5–2.0 mg/liter	7.5–13 mg/liter	1–3 mg/liter	—	2–6 mg/liter	—
In roller bottles	11–17 mg/liter	19–25 mg/liter	1–3 mg/liter	4–5 mg/liter	>10 mg/liter	12 mg/liter
Specific productivity (on T flask)	3–5 μg/10^6 cells/day	5–9 μg/10^6 cells/day	1.8–2.7 μg/10^6 cells/day	n.d.	n.d.	n.d.
Suitability for fermenter scale	Yes	Yes	Yes	No	Yes	Yes
Time frame from transfection to production of milligram quantities of protein	≈9 months	≈6–9 months	≤3 months	≈1–2 months	≈2 months	≈3 months

heavily and displays a broad band migrating at 43 kDa in SDS–PAGE.[77,80] As can be seen in Fig. 1, recombinant hu-LIF produced from stable cell lines or episomes is also highly glycosylated and fairly homogeneous. In contrast, hu-LIF expressed in COS cells shows incomplete protein processing as reflected by a smear of bands on the gel as a consequence of the overwhelming amount of plasmid DNA and subsequent protein present in the cells. This result is in accordance with data published previously on the expression of LIF in COS cells.[81]

Insect cells are capable of adding N- and O-linked carbohydrate moieties to protein backbones, but they lack the capacity to build highly branched, complex type sugars due to a shortage in processing enzymes.[82,83] Our results for hu-LIF expression in insect cells support these findings. A ladder of discrete bands of lower molecular weight in comparison to, e.g., CHO or Sp2/0 derived hu-LIF could be detected (Fig. 1), reflecting incomplete glycosylation as a consequence of incomplete processing, possibly in conjunction with the compromised state of the insect cell at the time of harvest.

In general, glycosylation of proteins is cell type specific and, equally important, dependent on the cell culture conditions.[84–87] The impact of variable or aberrant glycosylation of a recombinant protein on biological activity is critical and often not assessed readily. Differences in glycosylation pattern may not be reflected in the outcome of *in vitro* assays designed to determine biological activity. However, *in vivo,* where antigenicity, biological half-life, and clearance rates contribute to biological activity and efficacy, the degree and homogeneity of carbohydrate residues present may have a significant influence, as has been demonstrated elegantly for recombinant erythropoietin.[88,89]

[80] H. Gascan, I. Anegon, V. Praloran, J. Naulet, A. Godard, J.-P. Soulillou, and Y. Jacques, *J. Immunol.* **144,** 2592 (1990).

[81] J.-F. Moreau, D. D. Donaldson, F. Bennett, J. Witek-Giannotti, S.-C. Clark, and G. G. Wong, *Nature* **336,** 690 (1988).

[82] T. R. Davis and H. A. Wood, *In Vitro Cell. Dev. Biol.* **31,** 659 (1995).

[83] D. L. Jarvis and E. E. Finn, *Nature Biotechnol.* **14,** 1288 (1996).

[84] R. A. Dwek, *Biochem. Soc. Trans.* **23,** 1 (1995).

[85] D. C. James, R. B. Freedman, M. Hoare, O. Ogonah, B. C. Rooney, O. A. Larionov, V. N. Dobrovolsky, O. V. Lagutin, and N. Jenkins, *BioTechnology* **13,** 592 (1995).

[86] E. M. A. Curling, P. M. Hayter, A. J. Baines, A. T. Bull, K. Gull, P. G. Strange, and N. Jenkins, *Biochem. J.* **272,** 333 (1990).

[87] N. Jenkins and E. M. A. Curling, *Enzyme Microb. Technol.* **16,** 354 (1994).

[88] N. Imai, M. Higuchi, A. Kawamura, K. Tomonoh, M. Oh-Eda, M. Fujiwara, Y. Shimonaka, and N. Ochi, *Eur. J. Biochem.* **194,** 457 (1990).

[89] K. Yamaguchi, K. Akai, G. Kawanishi, M. Ueda, S. Masuda, and R. Sasaki, *J. Biol. Chem.* **266,** 20434 (1991).

This has prompted active research in search of "better" host cell lines, either by establishment of new cell lines (as, e.g., *Estigmena acrea* cells infected by recombinant baculovirus[90]) or by genetic engineering of existing cell lines (e.g., CHO cells,[91] BHK 21 cells,[92] or Sf9 cells[83]). The ultimate goal is to achieve production of a homogeneous recombinant protein that resembles its natural counterpart to the highest degree possible.

Conclusion

Prior to the start of expression experiments, the following four parameters should be evaluated carefully.

1. The nature of the protein to be expressed, i.e., size and structural features, such as disulfide bridges, glycosylation, or other secondary modifications.
2. The amount of protein needed in combination with the intended use (*in vitro, in vivo*, structural biology).
3. The design of a suitable expression vector for an expression system of choice, also including a potential "tagging" of the protein to facilitate on-line monitoring of production titers and purification.
4. The availability or establishment of appropriate assay systems (ELISA, Western blot, bioassay) and a tentative purification protocol for the recombinant protein.

If all these criteria are addressed and considered in advance, the choice of an appropriate expression system is straightforward. The enormous flexibility of nature in producing proteins in variable quantitites and quality, also in recombinant systems, will sometimes, however, necessitate expression experiments performed in more than one system to obtain the desired recombinant protein.

Concerning the expression protocols described in this article, it should be stressed that these represent basic outlines of procedures to which numerous modifications and additions can be found in the literature. Due to space limitation, not all of these variations could be addressed in this article.

[90] O. Ogonah, R. B. Freedman, N. Jenkins, K. Patel, and B. C. Rooney, *BioTechnology* **14,** 197 (1996).

[91] S. L. Minch, P. T. Kallio, and J. E. Bailey, *Biotechnol. Progr.* **11,** 348 (1995).

[92] E. Grabenhorst, A. Hoffmann, M. Nimtz, G. Zettlmeissl, and H. S. Conradt, *Eur. J. Biochem.* **232,** 718 (1995).

Acknowledgments

We gratefully acknowledge the contributions of Hermann Gram, Beate Kleuser, Mauro Zurini, and co-workers for the establishment of the described expression systems and, in particular, the excellent technical support of Agnes Patoux and Fabian Rohr. Many thanks also to Paul Ramage and Bertram Opalka for critically reviewing the manuscript.

[3] Analysis of Mammalian Cis-regulatory DNA Elements by Homologous Recombination

By STEVEN FIERING, M. A. BENDER, and MARK GROUDINE

Uses of Homologous Recombination

Sequence-specific DNA-binding proteins control the major DNA-dependent activities of replication and transcription. Such factors are also important for the equally vital DNA-dependent activities of imprinting, recombination, DNA repair, and chromosomal maintenance and segregation. To understand how these proteins function, their cognate recognition sites in the genome and their developmental/tissue-specific expression pattern must be determined. The modification of such regulatory regions and the determination of associated functional changes throughout normal mammalian development are essential for understanding the contribution of specific proteins and their binding sites to tissue-specific and developmentally regulated processes. Cis-acting regulatory elements have been analyzed *in vitro,* in transient transfection or stable integration assays in tissue culture cells, and in transgenic mice. More recently, because of the acknowledgment of the importance of chromosomal context when analyzing gene regulation, there have been substantial efforts to analyze cis-regulatory DNA elements by mutating these elements in endogenous loci using homologous recombination. Homologous recombination is less facile than other systems, but has the major advantage of providing definitive information about the functional role of an element in its normal chromosomal context. To fully appreciate the advantages of using homologous recombination (HR), it is necessary to understand the limitations of the other available approaches.

The ability to analyze specific interactions between purified proteins and nucleic acids makes biochemical analysis powerful, but these interactions often do not faithfully mimic the complexities seen in intact cells. For example, enhancer elements that are important for transcription in intact cells do not appear to be necessary to achieve high levels of expression in

0076-6879/99 $30.00

extracts, presumably in part because the influence of chromatin is difficult to reproduce.[1] Similarly, transcription from tissue-specific and developmentally regulated genes can be activated by extracts from unrelated cell types, demonstrating the difference in requirements for transcription in reconstituted systems. In general the connection between *in vitro* analysis and *in vivo* regulation is indirect. Clearly, however, biochemical analysis can define interactions that may be important, and these studies are a vital complement to *in vivo* work.

Transient expression assays are the most common approach used for the analysis of cis-acting regulatory transcriptional elements in tissue culture cells. These assays are rapid, inexpensive, and, through the use of immortalized cells of different tissue origin, can begin to provide insights into regulation in different cellular milieus. Unfortunately, some regulatory elements do not function in unintegrated plasmid DNA, suggesting that aspects of chromatin structure or other epigenetic modification must be different under the two conditions. For example, hypersensitive sites 3 and 4 of the β-globin locus control region and elements from the immunoglobulin (Ig) locus or the human CD2 gene show enhancer activity only after genomic integration into erythroid or lymphoid cells, respectively.[2–8] Similarly, a subset of matrix attachment regions have been shown to influence transcription only after integration into the genome[9,10] or passage through the germ line.[11] Thus, being rapid and easy, transient expression assays are well suited for screening potentially interesting regulatory regions but often do not accurately reflect the complexities associated with tissue- and stage-specific chromatin organization.

Analysis of stably integrated constructs in cell lines or transgenic mice offers the advantage of determining the function of regulatory elements after integration into the genome and assimilation into chromatin.

[1] P. J. Laybourn and J. T. Kadonaga, *Science* **257,** 1682 (1992).

[2] D. H. Tuan, W. B. Solomon, I. M. London, and D. P. Lee, *Proc. Natl. Acad. Sci. U.S.A.* **86,** 2554 (1989).

[3] W. C. Forrester, U. Novak, R. Gelinas, and M. Groudine, *Proc. Natl. Acad. Sci. U.S.A.* **86,** 5439 (1989).

[4] P. Fraser, J. Hurst, P. Collis, and F. Grosveld, *Nucleic Acids Res.* **18,** 3503 (1990).

[5] A. M. Moon and T. J. Ley, *Blood* **77,** 2272 (1991).

[6] B. A. Hug, A. M. Moon, and T. J. Ley, *Nucleic Acids Res.* **20,** 5771 (1992).

[7] L. Madisen and M. Groudine, *Genes Dev.* **8,** 2212 (1994).

[8] R. Festenstein, M. Tolaini, P. Corbella, C. Mamalaki, J. Parrington, M. Fox, A. Miliou, M. Jones, and D. Kioussis, *Science* **271,** 1123 (1996).

[9] A. Stief, D. Winter, W. Stratling, and A. Sippel, *Nature* **341,** 343 (1989).

[10] D. Schubeler, C. Mielke, K. Maass, and J. Bode, *Biochemistry* **35,** 11160 (1996).

[11] W. Forrester, C. van Genderen, T. Jenuwein, and R. Grosschedl, *Science* **265,** 1221 (1994).

Transgenic mice provide the further advantage of permitting analysis of expression of sequences in various tissues throughout development as well as after passage through the germ line when regulatory epigenetic modifications may occur. However, transgenes integrate randomly into the genome, and the site of integration itself can have powerful effects on expression, making the interpretation of functional analysis difficult. This phenomenon of position effects can influence the level of expression (stable position effect) as well as the probability that a given cell within an apparently uniform population will express the construct (variegating position effects). Several loci have regulatory elements known as locus control regions (LCRs) that can substantially overcome both stable and variegating position effects when they are intact, but mutational analysis of these elements in randomly integrated transgenes leads to position effects and associated difficulties in the interpretation of results.[8,12] In addition, transgenes in transgenic mice usually integrate in multicopy arrays, and this duplication of cis-regulatory elements has, not surprisingly, been shown to influence the functional activity of the transgenes.[13,14]

One approach to improving the use of transgenes as a model system for gene regulation studies is to use larger transgenes. Larger transgenes have been made from yeast and bacterial artificial chromosomes and ligated cosmids.[15–17] The initial assumption was that a transgene flanked by a large enough region of DNA from the endogenous locus would assure formation of a chromatin domain that behaves identically to the endogenous locus, regardless of the integration site. When integrated as single copies, these large transgenes are less susceptible to position effects than smaller transgenes, but are not completely immune.[12,18] However, these large transgene constructs undergo frequent rearrangements, making them difficult to work with and necessitating fastidious documentation of structure. Although a useful way to analyze human regulatory sequences, there are potential cross-species differences that can influence the results when analyzing human cis-regulatory elements in a mouse.

[12] E. Milot, J. Strouboulis, T. Trimborn, M. Wijgerde, E. de Boer, A. Langeveld, K. Tan-un, W. Vergeer, N. Yannoutsos, F. Grosveld, and P. Fraser, *Cell* **87,** 105 (1996).

[13] J. Ellis, K. Tan-un, A. Harper, D. Michalovich, N. Yannoutsos, S. Philipsen, and F. Grosveld, *EMBO J.* **15,** 562 (1996).

[14] D. Garrick, S. Fiering, D. Martin, and E. Whitelaw, *Nature Genet.* **18,** 56 (1998).

[15] K. M. Gaensler, N. Kitamura, and K. Y. W., *Proc. Natl. Acad. Sci. U.S.A.* **90,** 11381 (1993).

[16] J. Strouboulis, N. Dillon, and F. Grosveld, *Genes Dev.* **6,** 1857 (1992).

[17] K. R. Peterson, C. H. Clegg, C. Huxley, B. M. Josephson, H. S. Haugen, T. Furukawa, and G. Stamatoyannopoulos, *Proc. Natl. Acad. Sci. U.S.A.* **90,** 7593 (1993).

[18] S. Porcu, M. Kitamura, E. Witkowska, Z. Zhang, A. Mutero, C. Lin, J. Chang, and K. Gaensler, *Blood* **90,** 4602 (1997).

Because of the limitations of these approaches, the most definitive way to assess the function of cis-regulatory elements may be to mutate them at their endogenous loci by homologous recombination in embryonic stem (ES) cells and generate mutant mice. This assures that all relevant regulatory sequences are present in their normal configuration, avoids position and species-related concerns, and allows a detailed developmental analysis of the mutant phenotype in all tissues. However, this approach cannot address the function of human cis-regulatory elements directly. Therefore, for the analysis of human regulatory elements, HR modification of the endogenous locus in tissue culture cells may be the most rigorous approach.

Systems for Mutational Analysis of Endogenous Cis-regulatory Elements

All systems for mutating endogenous cis-regulatory elements in mammalian cells start with the mutation of the locus via HR in tissue culture cells. The choice of cell line is crucial because (1) the rate of HR varies dramatically between cell types and (2) the cells must be of, or able to give rise to, cells of a lineage and developmental stage appropriate for assaying the mutated regulatory region.

Primary cells can be modified by HR in culture with frequencies comparable to other cells.[19,20] Primary cells can be derived from a variety of species and tissues, giving some flexibility to the cellular environment in which the regulatory regions are assayed. Unfortunately, primary cells cannot be derived from all tissues, have a limited expansion capability, and have a limited potential to differentiate in culture. Although some primary cells can be immortalized, the alterations in growth and other cellular processes associated with immortalization may reduce or eliminate the advantages of using primary cells in the first place. In addition, because primary cells generally have two copies of the gene of interest, it may be necessary to alter two alleles to analyze the phenotype accurately.

Transformed cell lines can be used for homologous recombination. Although there are reports of cell lines that have high rates of homologous recombination,[21,22] similar to ES cells, most cell lines have not been tested thoroughly for their efficiency of HR. Primary and transformed lines usually have at least two copies of a gene, so multiple alleles need to be targeted

[19] J. Brown, W. Wei, and J. Sedivy, *Science* **277,** 831 (1997).

[20] S. Williams, F. Ousley, L. Vitez, and R. DuBridge, *Proc. Natl. Acad. Sci. U.S.A.* **91,** 11943 (1994).

[21] J. M. Buerstedde and S. Takeda, *Cell* **67,** 179 (1991).

[22] H. Zheng and J. Wilson, *Nature* **344,** 170 (1990).

to observe recessive phenotypes. We have found a technique known as "stepping up" to be useful in avoiding multiple targetings. Depending on the locus, cell type, and selectable marker used, cultures heterozygous for an HR event can be screened for subclones that have become homozygous for the targeted allele through mitotic recombination or chromosomal loss and duplication. This can be done efficiently by looking for differential growth in increasing concentrations of selective media.[23] However, this has been done primarily in ES cells and there has been little experience in other cell types. Unfortunately, if a gene is required for survival of the cell line, a homozygous null mutation will not be recoverable and will essentially appear to be a failed experiment. Despite these caveats, there are examples of productive HR-based mutational analysis of cis-regulatory elements performed in transformed cells, particularly those that are functionally hemizygous, such as pre-B cells for analysis of an immunoglobulin locus[24,25] and somatic cell hybrids that contain a single copy of a chromosomal locus of interest.[26–28]

Many of the limitations of using transformed or primary cells can be overcome by using ES cells where the rate of HR is often high and mutants can be bred to homozygosity after the generation of heterozygous mutant mice. Once in a mouse line, the phenotype can be assessed in all tissues throughout development and after passage through the germ line. In the case of homozygous lethality, phenotypic analyses of the mutants through development can give information regarding the mechanism of the HR-generated mutant phenotype. In addition, the large number of polymorphic alleles in the mouse makes it possible to study the transcription of many homozygous lethal mutations in a heterozygous background. The HR-derived mutant mouse strain can be crossed with a strain carrying a polymorphism in the gene of interest, allowing a quantitation of expression from the mutant allele using the wild-type polymorphic allele as an internal control.[29,30]

[23] R. Mortensen, D. Conner, S. Chao, A. Geisterfer-Lowrance, and J. G. Seidman, *Mol. Cell. Biol.* **12,** 2391 (1992).

[24] A. Oancea and M. Shulman, *Int. Immunol.* **6,** 1161 (1994).

[25] A. Oancea, M. Berru, and M. Shulman, *Mol. Cell. Biol.* **17,** 2658 (1997).

[26] C. G. Kim, E. M. Epner, W. C. Forrester, and M. Groudine, *Genes Dev.* **6,** 928 (1992).

[27] S. Fiering, C. G. Kim, E. E. Epner, and M. G. Groudine, *Proc. Natl. Acad. Sci. U.S.A.* **90,** 8469 (1993).

[28] E. Dicken, E. Epner, S. Fiering, R. E. K. Fournier, and M. Groudine, *Nature Genet.* **12,** 174 (1996).

[29] S. Fiering, E. Epner, K. Robinson, Y. Zhuang, A. Telling, M. Hu, D. I. K. Martin, T. Enver, T. Ley, and M. Groudine, *Genes Dev.* **9,** 2203 (1995).

[30] B. Hug, R. L. Wesselschmidt, S. Fiering, M. A. Bender, M. Groudine, and T. J. Ley, *Mol. Cell. Biol.* **16,** 2906 (1996).

One approach to analyzing human regulatory sequences is to use somatic cell hybrids. Avian leukosis virus-transformed chicken pre-B-cell lines have been reported to perform HR in a very high proportion of selectable marker-expressing cells.[21] Mammalian chromosomes have been shown to undergo gene targeting at high frequencies (generally above 10% of cells resistant to the positive selection) after micro-cell-mediated fusion into one such avian line (DT-40).[28] Therefore, desired modifications of the mammalian chromosome can be made in DT-40, and the altered chromosome can be transferred to a mammalian cell line in which the phenotypic consequences of the mutation can be determined.[28] Although improving the HR frequency at some loci, this system requires multiple micro-cell fusion-mediated chromosomal transfers, and the analysis is limited by the characteristics of the cell line used to assay the phenotype.[31] It has been demonstrated that ES cells can acquire a human chromosome or chromosome fragment by somatic cell fusion and still contribute to all tissues of a transgenic mouse, including the germ line.[32] This observation opens the possibility that human chromosomes transferred into ES cells may be modified efficiently in the ES cell. Following mutation of the locus of interest, the phenotypic consequences of the targeted mutation could be assayed by *in vitro* differentiation of the ES cells or, if possible, by making a mouse line from the cells.

Effect of Selectable Markers on Gene Regulation after Homologous Recombination

The basic procedures for making an HR construct and selecting cells that have undergone HR have been reviewed extensively. This article focuses on an important aspect of HR that must be modified for the analysis of cis-regulatory elements: the insertion of selectable markers. For studies of cis element function, the presence of a selectable marker is problematic, as markers have been shown to influence the transcription or proper function of genes in several loci. The most convincing direct demonstrations of this phenomena are provided by experiments in which the selectable marker introduced by HR is removed subsequently by a site-specific recombinase and the phenotype is analyzed with and without the selectable

[31] A. Reik, A. Telling, G. Zitnik, D. Cimbora, E. Epner, and M. Groudine, *Mol. Cell. Biol.* **18,** 5992 (1998).

[32] K. Tomizuka, H. Yoshida, H. Uejima, H. Kugoh, K. Sato, A. Ohguma, M. Hayasaka, K. Hanaoka, M. Oshimura, and I. Ishida, *Nature Genet.* **16,** 133 (1997).

marker.[27,29,30,33–37] In each of these experiments, different phenotypes were observed in the presence and absence of the selectable marker.

Indirect evidence for selectable marker interference is provided by HR experiments in which unexpected phenotypes are observed when a selectable marker is simply inserted into a locus in the absence of other mutations.[26,38–40] Further indirect evidence for selectable marker-based interference is provided by the disruption in expression of neighboring genes by an HR-mediated mutation.[41,42]

Selectable marker interference has been clearly observed in the β-globin locus,[26,27,29,30] α-globin locus,[40,42] the Ig heavy chain locus,[38] the κ light chain locus,[35,36,39] the T-cell receptor (TCR) β locus,[34] the granzyme locus,[33] and the GATA-1 locus.[37] These data are consistent with marker-mediated interference from the Hox locus and the MRF4 locus.[41]

Many of the loci in which marker interference has been observed have multiple genes and/or demonstrate complex regulation. It is likely that the phenomena of selectable marker interference will extend to other less complex loci. In fact, analysis of the apparently less complex GATA-1 locus also shows disruption of regulation by the expression of a selectable marker gene.[37] The region surrounding this locus, like many others targeted by HR, is not well characterized, thus the regulation of nearby genes may be influenced unknowingly by the presence of a selectable marker.

The effect that selectable markers have on a locus is not predictable. Gene expression in the T-cell receptor β locus[43] and the Ig heavy chain locus[24] is influenced by the insertion of a selectable marker into one but not another site in the same locus. In addition, in loci in which multiple HR-mediated modifications were performed, the same selectable marker in different sites and associated with different mutations had quantitatively

[33] C. Pham, D. MacIvor, B. Hug, J. Heusel, and T. Ley, *Proc. Natl. Acad. Sci. U.S.A.* **93,** 13090 (1996).

[34] J. Bories, J. Demengeot, L. Davidson, and F. Alt, *Proc. Natl. Acad. Sci. U.S.A.* **93,** 7871 (1996).

[35] J. Gorman, N. van der Stoep, R. Monroe, M. Cogne, L. Davidson, and F. Alt, *Immunity* **4,** 241 (1996).

[36] Y. Xu, L. Davidson, F. Alt, and D. Baltimore, *Immunity* **4,** 377 (1996).

[37] R. Shivdasani, Y. Fujiwara, M. McDevitt, and S. Orkin, *EMBO J.* **16,** 3965 (1997).

[38] J. Chen Young, F. A. Bottaro, V. Stewart, R. Smith, and F. Alt, *EMBO J.* **12,** 4635 (1993).

[39] S. Takeda, Y. Zou, H. Bluethmann, D. Kitamura, U. Muller, and K. Rajewsky, *EMBO J.* **12,** 2329 (1993).

[40] A. Bernet, personal communication, 1997.

[41] E. Olson, H. Arnold, P. Rigby, and B. Wold, *Cell* **85,** 1 (1996).

[42] A. Leder, C. Daugherty, B. Whitney, and P. Leder, *Blood* **90,** 1275 (1997).

[43] G. Bouvier, F. Watrin, M. Naspetti, C. Verthuy, P. Naquet, and P. Ferrier, *Proc. Natl. Acad. Sci. U.S.A.* **93,** 7877 (1996).

different effects on the locus.[27,29,30,33,44] Generally, the insertion leads to a reduction in expression; however, insertions in Ig and TCR loci have led to either an inhibition[34,35,39] or an enhancement of rearrangement.[45]

Over time, data derived concerning the influence of selectable marker insertion on endogenous loci may form a cohesive body that will contribute to our understanding of genomic regulation and allow the prediction of the effect of a selectable marker anywhere in the genome. However, at this time it is not possible to predict the influence of the selectable marker for any given HR event. Thus, to define the role of particular cis-regulatory elements in the expression of specific genes accurately, it is mandatory that the selectable marker be removed from the locus. In addition, it would seem prudent to incorporate systems for the removal of selectable markers from HR events that are designed to "knock out" a gene, as the selectable marker may affect the expression of nearby genes and thus alter the phenotype.[41] Although the feasibility of doing HRs in ES cells without selectable markers has been suggested,[46] this approach has not yet been widely applied, and currently the utilization of site-specific recombinases such as Cre or Flp is the most frequently exploited approach for removal of the selectable marker from an HR.

Removing Selectable Marker with Flp or Cre Site-Specific Recombinases

Removal of the selectable marker by a site-specific recombinase after an HR reaction is conceptually straightforward; however, the strategy must be included in the initial design of the experiment. Site-specific recombinases are enzymes that catalyze a DNA recombination reaction between two sequence-specific target sites.[47] The Flp recombinase from *Saccharomyces cerevisiae* and the Cre recombinase from *Escherichia coli* phage P1 have both been shown to function efficiently in mammalian cells and are each encoded by a single polypeptide. These enzymes are members of the λ family of integrases and are very similar in the reactions they carry out, the mechanism of recombination, and the structure of the target sites that they recognize. Their target sites are small (34 bp for Cre and 48 bp for Flp) and are made up of inverted repeats separated by a spacer sequence. The spacer gives the target site its orientation and this orientation determines the product of the reaction. When two target sites have the same

[44] M. Bender, S. Fiering, and M. Groudine, manuscript in preparation, 1999.

[45] P. Sieh and J. Chen, personal communication, 1997.

[46] N. Templeton, D. Roberts, and M. Safer, *Gene Ther.* **4,** 700 (1997).

[47] N. J. Kilby, M. R. Snaith, and J. A. H. Murray, *Trends Genet.* **9,** 413 (1993).

orientation and are linked in cis, recombination will produce a site-specific excision of all DNA between the target sites and a closed circle with the deleted DNA (Fig. 1A). A single target site will be left at the deletion site and another target site will be on the closed circle. In contrast, if the two target sites are linked in cis and have the opposite orientation, the recombinase reaction will invert the sequence between the target sites (Fig. 1B).

In order to use these recombinases to delete the selectable marker after an HR reaction, the HR construct must be made with the appropriate recombinase target sites (FRT sites for Flp and loxP sites for Cre) flanking the selectable marker and in the same orientation. After isolating targeted clones, expression of the recombinase deletes the marker. This simple approach has been used in many of the examples cited earlier in which the selectable marker was shown to influence expression of the gene or locus undergoing modification.

Recombinase-mediated excisions can be performed after transient or stable expression of Flp or Cre in cell lines or trangenic mice. Although both recombinases work well, we currently recommend the Cre/loxP system because of a higher efficiency of recombination and a greater availability of reagents and experience. The difference in efficiency at least partially reflects the difference in temperature optimum between the enzymes.[48] Cre, originating from *E. coli* phage P1, works optimally at 37°, the temperature of mammalian cell culture. Flp, which is derived from the 2-μm circle of *S. cerevisiae,* has a lower optimal temperature. The commonly utilized Flp expression plasmid pOG44[49] has been shown to contain a mutation that exacerbates the temperature sensitivity of Flp.[48] Correction of this mutation increased Flp protein activity at 37°.[48] In transgenic mice, the Flp recombinase has not been tested extensively,[50] whereas the Cre recombinase has proven efficacious in many situations.[35,44,50,51,52,53] For transfection into cultured cells, Cre is more dependable, but there are situations where Flp may be preferred (see later). In addition, use of both Flp and Cre target sites in one locus has generated multiple mutations from a single initial HR event.[31] Regardless of the recombinase used or the means of expressing the recombinase, the standard recombinase-mediated deletion reaction will

[48] F. Bucholz, L. Ringrose, P. Angrand, F. Rossi, and A. Stewart, *Nucleic Acids Res.* **24,** 4256 (1996).

[49] S. O'Gorman, D. T. Fox, and G. M. Wahl, *Science* **251,** 1351 (1991).

[50] S. Dymecki, *Proc. Natl. Acad. Science U.S.A.* **93,** 6191 (1996).

[51] M. Lakso, J. Pichel, J. Gorman, B. Sauer, Y. Okamoto, E. Lee, F. Alt, and H. Westphal, *Proc. Natl. Acad. Sci. U.S.A.* **93,** 5860 (1996).

[52] P. Orban, D. Chui, and J. Marth, *Proc. Natl. Acad. Sci. U.S.A.* **89,** 6861 (1992).

[53] F. Schwenk, U. Baron, and K. Rajewsky, *Nucleic Acids Res.* **23,** 5080 (1995).

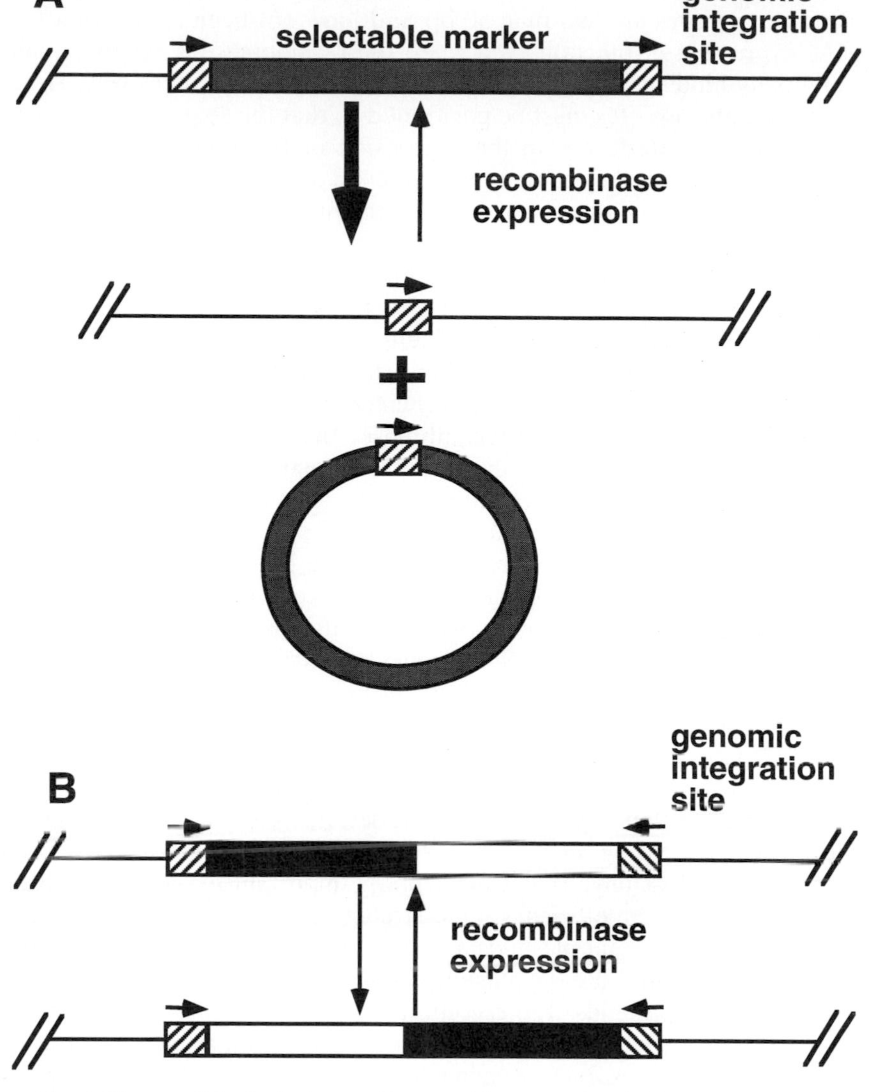

FIG. 1. Diagram of the basic site-specific recombinase reaction. (A) The deletion/insertion reaction. The hatched box with an arrow over it represents the recombinase target site and its orientation. Vertical arrows indicate that this reversible reaction favors the deletion. (B) The inversion reaction is mediated by the recombination of two oppositely oriented cis-linked recombinase target sites.

leave a single recombinase target site in the genome at the deletion site. These target sites are less than 50 bp and have not been shown to affect gene expression or function even when the recombinase is expressed and potentially binding, although this has not been studied extensively. However, recombinase sites must be positioned so that the sequence of the site itself does not interfere with the expression or function of genes in the locus. For example, they cannot be in an exon or they will disrupt the gene coding sequence, and they cannot disrupt the promoter architecture or they will modify expression.

Recombinase-mediated deletions of selectable markers have been performed in tissue culture cells after the transient expression of Cre or Flp.[27,29,36,43,54–56] Both the number of cells expressing the recombinase and the activity of the recombinase are important for obtaining subclones with the selectable marker excised. The efficiency of the transfection, thus the percentage of cells expressing recombinase, must be optimized as there is generally no way to select for excision of the marker directly.

A generous amount of recombinase expression plasmid (over 20 μg) improves the probability of obtaining subclones with the marker excised. Our experience is that the Cre-mediated deletion occurs in 25–90% of the plasmid-expressing cells [the number of cells expressing plasmid is determined by a test transfection with a β-gatactoside (β-Gal) expression plasmid and an X-Gal stain]. Similar analyses have revealed that the Flp-mediated deletion occurs in roughly 1–10% of the cells that express the plasmid. It is important to note that these figures do not reflect the potential increase in Flp-mediated deletion efficiency that may accompany the modifications to the Flp gene that make it more effective at 37°.[48]

Isolating subclones that have had the selectable marker deleted can be facilitated by focusing analysis on progeny of cells known to have been transfected successfully. This is particularly important for cell lines with a very low transfection efficiency. One strategy depends on the observation that when two different plasmids are cotransfected, cells that take up and express one plasmid are significantly more likely to have taken up and expressed both plasmids. The recombinase-expressing plasmid is cotransfected with a marker plasmid that can be used to isolate transfected cells on the basis of fluorescence, utilizing flow cytometry (FACS), or by expression of a surface antigen and adherence to a recognizing antibody. Appropriate markers include green fluorescent protein (GFP), β-Gal, or a surface

[54] H. Gu, Y. R. Zou, and K. Rajewsky, *Cell* **73,** 1155 (1993).
[55] S. Jung, K. Rajewsky, and A. Radbruch, *Science* **259,** 984 (1993).
[56] L. Ferradini, H. Gu, A. De Smet, K. Rejewsky, C. Reynaud, and J. Weill, *Science* **271,** 1416 (1996).

protein such as CD20.[57] To avoid cotransfections, the recombinase and marker gene can be expressed from the same plasmid. We have used FACS selection to enrich for Flp-modified cells that were cotransfected with either Flp and β-Gal expression plasmids or a plasmid that expresses both Flp and β-Gal from a single message containing an internal ribosome entry site (IRES).[27,29] Cotransfection with GFP has also been used to enrich for deletion of a selectable marker by Flp.[58] Although we have successfully used a transient expression cotransfection with β-Gal and FACS sorting to enrich for ES cells with a Flp-mediated deletion,[29] we do not favor this approach for ES cells due to concerns of interfering with the cells' potential to contribute to the germ line. We currently utilize transient expression of Cre without enrichment for ES cell studies *in vitro,* but for mutations we want to introduce into mice we rely on *in vivo* Cre-mediated deletion of selectable markers as this approach has the benefit of minimizing the time in culture of ES cells.

Selectable markers can also be deleted efficiently after stable integration of a recombinase.[27,30,37] Although simple and avoiding the need for FACS or panning antibodies, the theoretical disadvantage is that long-term expression of the recombinase in cells containing a recognition site could result in unwanted recombinase-mediated events via cryptic recombination sites. However, such events are exceedingly rare and have only been seen when experiments are designed specifically to select for them.[59] Another theoretical concern is that recombinase protein binding to the recombinase site(s) in the locus may influence gene regulation or function. Although no data support this possibility, it has not been addressed thoroughly. Deletion of selectable markers via stably integrated recombinase expression vectors can be achieved either by linking the recombinase and selectable marker expression cassettes on a single vector[30,37] or by cotransfecting the two vectors.[27] When cotransfection is used, a greater molar ratio of the recombinase expression cassette is used to increase the probability that the cells that express the selectable marker will also express the recombinase.

loxP-flanked selectable markers can be excised efficiently from transgenic mice after stable exposure to Cre[35,44,51–53,60] or after transient expression of Cre in fertilized eggs.[14,61] Deletion of sequences by expression of Flp in mice has been accomplished, but has not been as efficient as Cre.[50] Deletion of markers flanked by *loxP* sites in mice is done most frequently

[57] S. van den Heuvel and E. Harlow, *Science* **262,** 2050 (1993).

[58] E. Bouhassira, K. Westerman, and P. Leboulch, *Blood* **90,** 3332 (1997).

[59] B. Sauer, *J. Mol. Biol.* **223,** 911 (1992).

[60] A. Nagy, C. Moens, E. Ivanyi, P. J, M. Gertsenstein, A. Hadjantonakis, M. Pirity, and J. Rossant, *Curr. Biol.* **8,** 661 (1998).

[61] K. Araki, M. Araki, J. Miyazaki, and P. Vassalli, *Proc. Natl. Acad. Sci. U.S.A.* **92,** 160 (1995).

by breeding with a Cre recombinase-expressing transgenic line. Mice carrying both transgenes are identified by Southern blot or PCR analysis of tail DNA. The likelihood of germ line deletion of the marker in a particular mouse can be estimated by assaying for deletion of the *loxP*-flanked selectable marker in the somatic tail tissue. Although the marker may be excised efficiently in the tail biopsy, in our experience these F_1 mice are sometimes chimeric for recombinase-mediated deletion. To assure that the deletion has passed through the germ line and that there is no somatic tissue chimerism, these transgenics are bred to wild-type mice and offspring are identified that have not inherited the recombinase transgene but have inherited the HR-modified locus with the selectable marker deleted. These second-generation mice that lack the selectable marker but do not express recombinase must have inherited the modified allele through the germ line and therefore are not chimeric. Transgenic mouse lines that express the Cre recombinase efficiently in the male germ line have been reported,[51,53] and several investigators have reported the efficient deletion of *loxP*-flanked DNA when Cre-expressing mice are bred to transgenics carrying the appropriate *loxP* sites.[44,51,53,60]

Alternatively, excision of the selectable marker *in vivo* can be accomplished after the transient expression of Cre in fertilized oocytes. Oocytes are microinjected with a Cre-expressing plasmid and are transferred to a foster mother. The *in vivo* transient expression of Cre in fertilized oocytes has been shown to result in the deletion of a marker early in development without integration of the expression plasmid.[14,61]

For experiments in which the ultimate goal is a mouse line, these *in vivo* approaches for recombinase-mediated deletions offer two major advantages. First, mouse lines with and without the selectable marker will be derived from the same clone of ES cells, allowing assessment of how markers affect gene regulation. In addition, the ability of ES cells to contribute to the germ line often diminishes with passage number. Thus, after transfection, subcloning, and more time in culture, these higher passage number cells carrying the targeted mutation with the marker deleted may be less likely to contribute to the germ line than the parental ES cells from which they are derived. It is important to keep in mind that an investigator will not know that a clone does not have the potential for germ line transmission for several months, and only after the cost and effort of blastocyst injections and initial chimera breedings. The disadvantage of the *in vivo* approach is that it entails more time, mice, and animal space. In addition, if the Cre-expressing transgenic mouse strain and the ES cells differ in genetic background, the genetics of the ES-derived offspring will be modified. This latter problem is being addressed in our and other laboratories by introduction of the Cre transgene into the 129 mouse strain from which most ES cells are derived, as well as into the commonly used C57b16 strain.

Insertion/Excision and Replacement Strategies for Generating HRs without Selectable Markers

In order to generate targeted mutations lacking selectable markers and recombinase sites, several groups have utilized two-step systems, making use of an initial positive selection step, followed by a negative selection step. While some groups utilize an HPRT minigene that allows both positive and negative selection,[62] others have used neomycin phosphotransferase (Neo) for positive selection and HSV-TK for negative selection.[63] The HPRT minigene is versatile in its ability to be used as a positive or negative selectable marker in cultured ES cells, but this system requires an HPRT(−) ES cell line (of which there are several). The advantage of the Neo/HSV-TK systems is that by using a dominant selectable marker (Neo), there are no restrictions on the cell types used (see later).

Insertion/deletion strategies rely on an initial targeting event that occurs by a single recombination event leading to an *insertion,* which accomplishes two things: (1) insertion or activation of the positive and negative marker(s) and (2) duplication of the region of homology used for targeting, of which the copy arising from the targeting construct has the mutation of interest (Fig. 2). The second step entails negative selection to isolate subclones that have lost the marker(s) due to an HR event between the duplicated regions. Depending on the crossover point, the introduced mutation will be excised, leaving an intact wild-type locus, or the corresponding wild-type region will be excised, generating an intact locus except for the presence of the mutation. This approach has been used to make a subtle mutation in a gene-coding sequence[64] and a 1-kb deletion of an immunoglobulin regulatory sequence.[65]

To study the effects of a regulatory region, one would like to target the same region with different mutations repeatedly. Although the approaches, discussed here work, each new mutation requires repeating the initial and secondary steps. A variation of this strategy allows multiple mutations to

[62] V. Valancius and O. Smithies, *Mol. Cell. Biol.* **11,** 1402 (1991).
[63] P. Hasty, R. Ramiriez-Solis, R. Krumlauf, and A. Bradley, *Nature* **350,** 243 (1991).
[64] R. Ramirez-solis, J. Zheng, J. Whiting, R. Krumlauf, and A. Bradley, *Cell* **73,** 279 (1991).
[65] M. Serwe and F. Sablitzky, *EMBO J.* **12,** 2321 (1993).

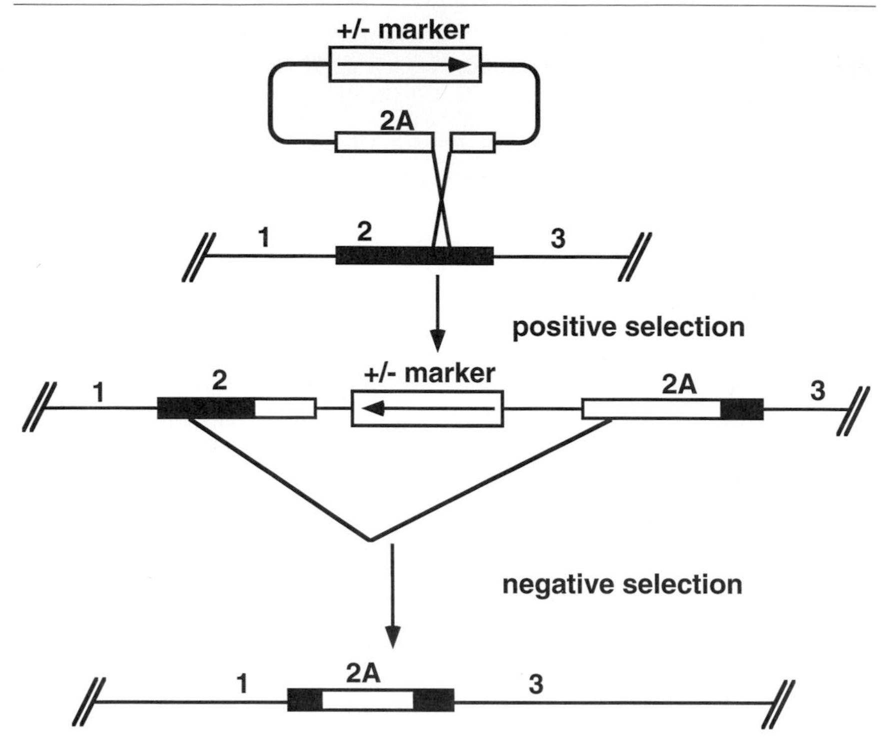

FIG. 2. Insertion/deletion reaction. The first step is an insertion type HR reaction that is selected by positive selection, which produces an insertion of the positive/negative selectable marker and a duplication of the homology region. A second recombination step occurs spontaneously due to the duplication and is selected by negative marker selection. Depending on the site within the duplication that this recombination occurs, the mutation introduced with the insertion (2A represents a mutation in the 2 region) will either replace the endogenous copy, as shown here, or be deleted.

be introduced after a common initial step.[66–68] The initial step involves targeting the locus of interest with a *replacement* construct using one of the positive/negative marker strategies described earlier (Fig. 3). Note that no duplication is generated. Once this targeting event has been achieved, a second double crossover homologous recombination event leads to the replacement of the selectable marker(s) with the the mutant region of interest. Because the second HR event can be isolated by negative selection

[66] G. Askew, T. Doetchman, and J. Lingrel, *Mol. Cell. Biol.* **13,** 4115 (1993).

[67] H. Wu, X. Liu, and R. Jaenisch, *Proc. Natl. Acad. Sci. U.S.A.* **91,** 2819 (1994).

[68] A. Stacey, A. Schnieke, J. McWhir, J. Cooper, A. Colman, and D. Melton, *Mol. Cell. Biol.* **14,** 1009 (1994).

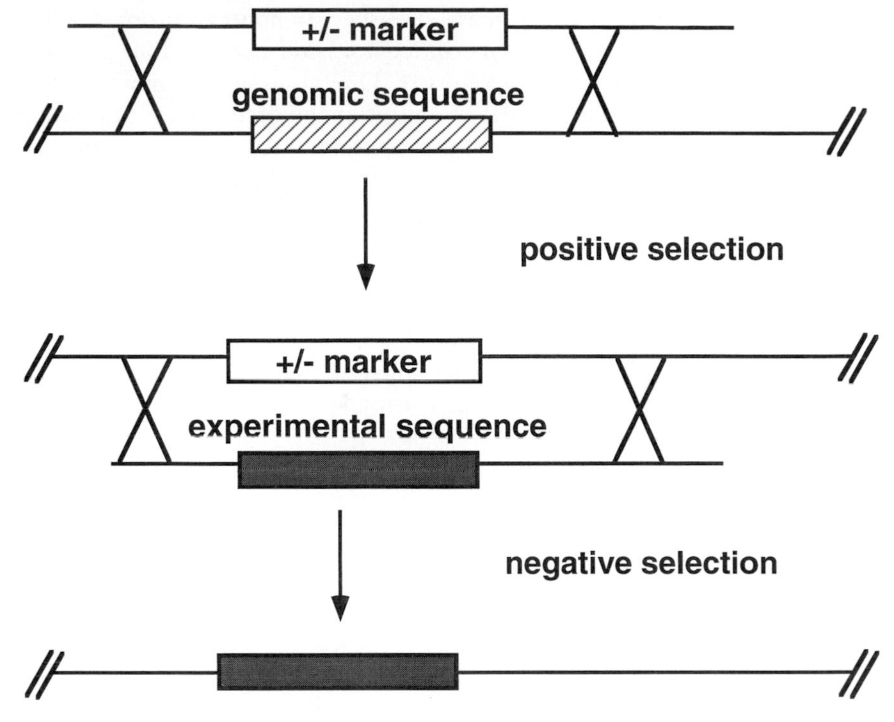

FIG. 3. Double-replacement reaction. The first step is a replacement HR that replaces a sequence of interest (hatched box) with a positive/negative selectable marker. A second homologous recombination reaction, selected for by negative selection, replaces the positive/negative marker with a sequence of experimental interest (solid box).

only, there is no need for selectable markers on the second targeting construct, so the locus is left intact with the exception of the new mutated regulatory region.

Although these strategies result in a mutant locus without markers or recombinase sites, there are some potential disadvantages. Approaches that rely on the loss of a marker (negative selection) have a background of clones that resist the negative selection, but have not replaced the negative selectable marker. This inactivation of the negative marker can arise from epigenetic gene silencing or chromosomal loss rather than the desired inactivation via replacement with secondary modifications.[66,67,69] In addition, the strategies discussed in this section all require two rounds of subcloning, each of which can be associated with a decrease in the germ line

[69] J. Seibler, D. Schubeler, S. Fiering, M. Groudine, and J. Bode, *Biochemistry* **37,** 6229 (1998).

transmissibility of ES cells, as discussed earlier. In contrast, if *loxP* or Frt sites are used, the recombinase-mediated excision can be done *in vivo*, thus diminishing manipulations in culture. One approach to minimize this problem is to diligently screen HR clones after the first round of targeting and to only use clones that maintain their potential for germ line transmission for second targeting events. Although this is tedious, time-consuming, and does not always generate mice of interest, this screening helps assure ultimate success. In addition, the choice of markers used presents potential disadvantages. Expression of the HSV-TK gene in the testes from a cryptic promoter within the gene itself has been observed to cause male mouse sterility.[70] Thus, assessing the clones for germ line transmission after the insertion of the HSV-TK gene in the first HR may not be feasible. This caveat is also applicable to any strategy using HSV-TK as a negative selection in ES cells.

Systems to Facilitate Multiple Modifications at Single Locus

As stated previously, to analyze regulatory regions, one would like to generate many mutations in the same region by HR rapidly without the effects of selectable markers. Ideally, systems for multiple modifications of a targeted site should (1) facilitate efficient isolation of modified clones; (2) facilitate the removal of, or avoid the use of, selectable markers; (3) have the flexibility to make large deletions, insert nonhomologous sequences, or make subtle mutations in protein-binding sites; and (4) minimize the time in culture when ES cells are used. When accessing new strategies to multiply target a region, one must compare it critically with the approach of performing independent HRs for each mutation, which has the advantage of minimizing time in culture (Table I). The latter is appropriate when there is a high rate of HR at the site of interest (e.g., 1/10). However, at many sites the HR frequency is 10- to 100-fold lower, making the identification of HR-modified clones difficult, and if the particular HR of interest has not been generated previously, the frequency cannot be predicted.

The double-replacement strategy described earlier can be used to facilitate multiple modifications of a targeted locus (Table I). These systems depend on selection against a negative selectable marker, which, as discussed earlier, suffer from a background of clones that escape selection spontaneously. Several groups using this approach reported replacement efficiencies of 10 to 88%.[66–68] However, the frequency of double replacement

[70] R. Al-Shawi, J. Burke, H. Wallace, C. Jones, S. Harrison, D. Buxton, S. Maley, A. Chandley, and J. Bishop, *Mol. Cell. Biol.* **11,** 4207 (1991).

TABLE I
CHARACTERISTICS OF VARIOUS STRATEGIES FOR REPEATEDLY TARGETING DNA REGION OF INTEREST[a]

System	Cell subcloning steps for marker removal		Efficiency of repeat targeting	Potential advantages	Potential disadvantages
	In mice	In cells			
Standard HR with recombinase sites	1	2	HR frequency	Simple; minimal time in culture	Dependent on HR frequency for repeat targeting
Double replacement	No	2	10–88%	No residual recombinase site	Dependent on negative selection for marker removal and repeat targeting
Plug and socket with recombinase sites	2	3	80–100%	Can be used to make larger deletions on one side of the socket; highly efficient	More time in culture
RMCE					
Negative selection only	2	2	21–38%	Minimal time in culture	Not fully developed; dependent on negative selection for repeat targeting
Positive plus negative selection	2	3	54–100%	Highly efficient	Not fully developed; more time in culture

[a] References for the frequency of repeat targeting are in the text.

may vary according to the frequency of HR at the locus, a point that has not been investigated.

The following strategies are similar to double replacement in that they are two-step processes in which the initial step is a standard HR event that introduces sequences into the targeted locus, which will then allow multiple efficient second targeting events. The "plug and socket" strategy[71] achieves this by selecting for HR events leading to the formation of an intact selectable marker, whereas recombinase-mediated cassette exchange (RMCE) utilizes recombinases to facilitate targeting in association with positive selection, negative selection, or a combination thereof.

"Plug and Socket" Strategy

In this approach, the first step is a standard HR event replacing genomic sequences with a positive selectable marker (M1) and a "socket" consisting of a second truncated positive selectable marker gene (M2). This site can now be retargeted with multiple "plug" vectors whose correct targeting is selected because they complete and activate the inactive selectable marker. The plugs can be used to introduce the desired locus modifications efficiently, e.g., replacement with a mutated regulatory region (see detailed description later). After generation of the initial socket-containing cell line, different plugs were used to generate deletions or add heterologous sequences back with an efficiency of nearly 100%.[71] Although an extremely powerful demonstration, the versatility of this system is limited by two points: (1) HPRT was used for M2, necessitating the use of HPRT(−) cells and (2) no provision was made to remove the selectable markers.

To address these issues, we have modified this system by (1) using dominant selectable markers for both M1 and M2 and (2) flanking all markers with Lox sites as proposed previously[72] (Fig. 4). Complementary fragments of neomycin phosphotransferase were chosen for use as M2, which have been shown previously to result in efficient HR and not to result in any background drug resistance when either single fragment is used.[73,74] Figure 4 diagrams the use of the modified plug and socket approach combined with a site-specific recombinase-mediated deletion to make multiple modifications of a putative regulatory region. The reaction shown in

[71] P. Detloff, J. Lewis, S. John, W. R. Shehee, R. Langenbach, N. Maeda, and O. Smithies, *Mol. Cell. Biol.* **14,** 6936 (1994).

[72] J. Lewis, B. Yang, P. Detloff, and O. Smithies, *J. Clin. Invest.* **97,** 3 (1996).

[73] R. Kucherlapati, E. Eves, K. Song, B. Morse, and O. Smithies, *Proc. Natl. Acad. Sci. U.S.A.* **81,** 3153 (1984).

[74] O. Smithies, M. Koralewski, K. Song, and R. Kucherlapati, *Cold Spring Harb. Symp. Quant. Biol.* **49,** 161 (1984).

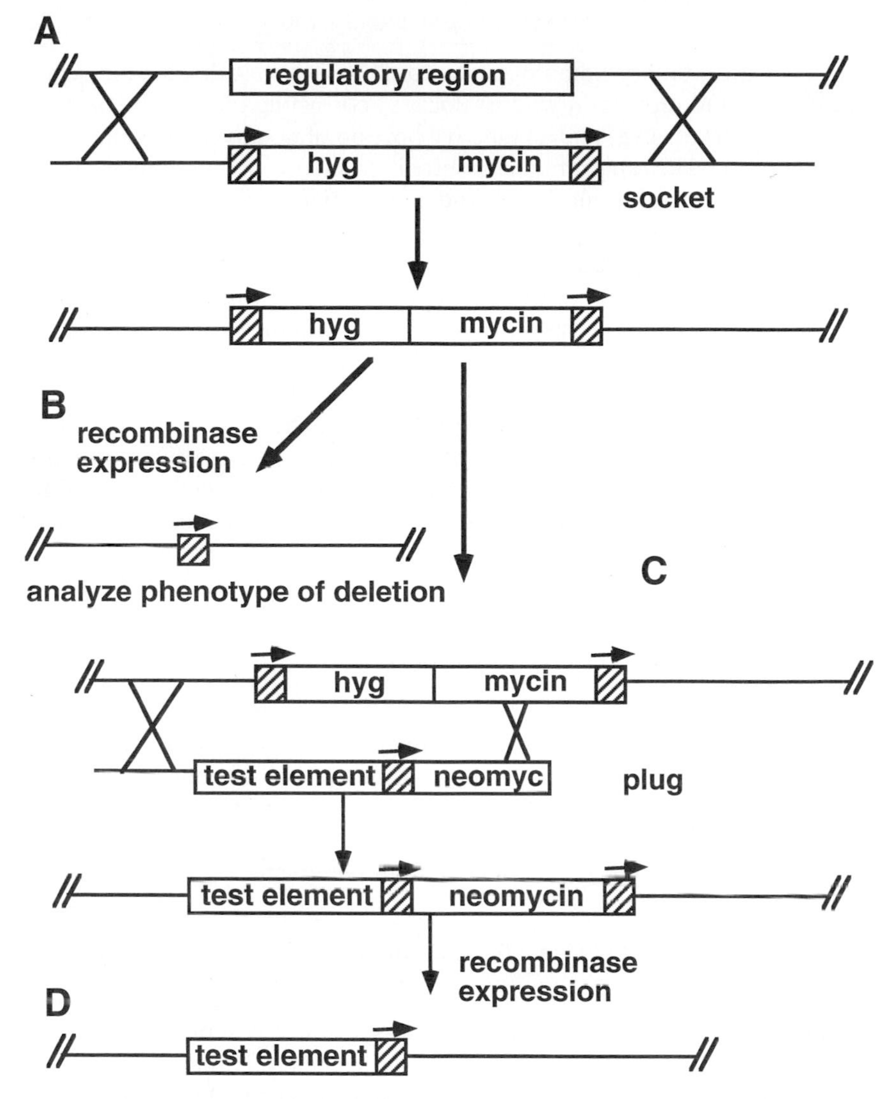

FIG. 4. Strategy for multiple modifications of a locus using a plug and socket. (A) The HR reaction replaces the region of interest with an active selectable marker (hyg) and a partial (inactive) selectable marker (mycin, a portion of the neomycin gene). This is the "socket" cell line genotype. Striped boxes represent recombinase target sites as in Fig. 1. (B) Expression of the recombinase deletes the selectable marker cassette and allows analysis of the phenotype of the simple deletion. (C) Replacement with a "plug." A modified regulatory element (test element) is added to the locus by homologous recombination that completes and activates the second selectable marker (neomycin). The homology on the left is genomic and the homology on the right is the partial marker, some of which is found in both the genomic portion (MYCin) and the incoming portion (neoMYC) needed for marker complementation. (D). Recombinase target sites flank the restored marker so that the recombinase can be expressed to delete the neomycin marker and the phenotype of the deletion plus test element can be assayed.

Fig. 4A is the primary homologous recombination that deletes the region under study and replaces it with a "socket" consisting of an active positive selectable marker (hyg) and an inactive portion of neomycin phosphotransferase (mycin). The replacement cassette is flanked with recombinase target sites so that selectable marker 1 and the partial marker (mycin) can be deleted, allowing phenotypic assessment of the deletion of the region under study (Fig. 4B). To introduce a test element and assess the activity of this element in the locus, the "socket" is complemented with the "plug" (Fig. 4C) by HR. A "plug" consists of a region of flanking homology from the locus, the element to be inserted (test element), a recombinase target site, and the complementing portion of the partial socket marker (neomyc). The homology between the socket (*myc*in) and the plug (neo*myc*) forms the second region of homology needed for a homologous recombination replacement reaction. The correctly targeted cells are directly selected for after complementation of selectable marker 2. The final step in producing the desired configuration is to express the recombinase and delete marker 2 (Fig. 4D), allowing the phenotypic assessment of the locus with the regulatory region replaced cleanly by the test element plus a recombinase target site.

Using this approach, we have been able to add plugs back to a socket line, generating 100% homologous recombinants, and then remove the selectable marker after the transient expression of Cre. This has been done both in ES cells and on a human chromosome in DT-40 cells. Once an ES cell socket line capable of giving germ line transmission efficiently has been generated, making multiple modifications is straightforward. In fact, multiple "plugs" could be transfected simultaneously.[75]

Recombinase-Mediated Cassette Exchange

In contrast to the "plug and socket," which relies on endogenous recombination machinery, RMCE exploits site-specific recombinases to attain specificity of the recombination event. Site-specific recombinases mediate recombination between any two appropriate target sites (Fig. 1). When the sites are oriented in the same direction and linked in cis, the result of such an excision reaction is deletion of the sequence between the two sites (leaving a single target site in the genome) and production of a closed circle that contains a single target site (Fig. 1A). This reaction is reversible, the recombinase can mediate a recombination between a closed circle (e.g., plasmid) containing a single target, and a target site in the genome. This insertion reaction occurs infrequently, in part due to it being a bimolecular

[75] J. Rabinowitz and T. Magnuson, *Anal. Biochem.* **228,** 180 (1995).

rather than an intramolecular reaction. To detect these rare insertion events, most insertion schemes mediated by site-specific recombinases have depended on a "trap," a system in which the insertion complements a previously inactive selectable marker.[76]

The development of RMCE has improved the efficiency of recombinase-mediated insertion geatly.[58,69,77–79] RMCE is based on the idea that the bimolecular insertion reaction occurs more frequently than it is observed, but that after the insertion occurs, it is reversed by the intramolecular excision reaction. RMCE depends on the decreased recombination frequency between two nonidentical recombinase sites to maintain the rate of the recombinase-mediated insertion while decreasing the chance of the normally favored excision reaction. Target sites for Cre and Flp recombinases have been mutated in such a way that their recognition by the recombinase is unaltered, but the enzymes exhibit a preference for recombination between two target sites with similar mutations. Both Flp and Cre recognition sequences consist of inverted repeats with the actual recombination reaction taking place within a short internal spacer (6 bp for Cre and 8 bp for Flp). Although recombinases cannot tolerate significant variation in their binding sites, they can tolerate changes in the spacer regions. In general, recombinase sites with identical spacer sequences recombine far more efficiently than sites with significant differences between their spacer sequences (heterospecific sites).[77,79–81]

As with the plug and socket technique, the first step in RMCE involves the replacement of genomic sequences with a selectable marker(s) using standard HR techniques. The marker(s) is flanked by a pair of heterospecific recombinase sites (site 1 and site 2 in Fig. 5A). Tissue culture cells carrying this modification are then cotransfected with two plasmids: one containing site 1 and site 2 flanking the region to replace the markers and a second one expressing the recombinase. RMCE can occur by either of two mechanisms, both of which entail two recombinase-dependent recombination events: one between the two site 1s and the other between the two site 2s. This results in the replacement or "exchange" of genomic sequences between the site 1 and site 2 sites with the incoming site 1/site 2 flanked sequences. In the first mechanism, a simultaneous double recombination event occurs (Fig. 5A). In the second mechanism the recombination events occur sequentially, with the first bimolecular recombination event leading to the insertion

[76] B. Sauer and N. Henderson, *New Biol.* **2,** 441 (1990).
[77] T. Schlake and J. Bode, *Biochemistry* **33,** 12746 (1994).
[78] J. Seibler and J. Bode, *Biochemistry* **36,** 1740 (1997).
[79] B. Bethke and B. Sauer, *Nucleic Acids Res.* **25,** 2828 (1997).
[80] J. Senecoff and M. Cox, *J. Biol. Chem.* **261,** 7380 (1986).
[81] R. Hoess, A. Wierzbicki, and K. Abremski, *Nucleic Acids Res.* **14,** 2287 (1986).

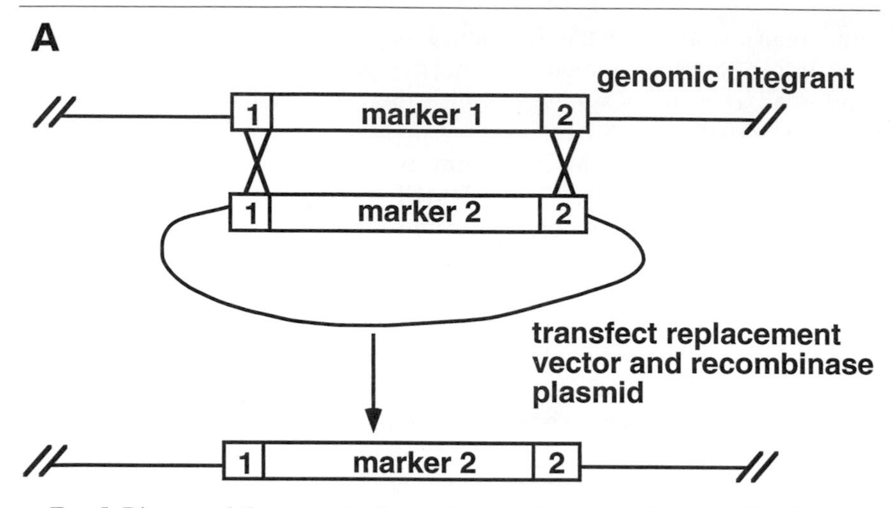

FIG. 5. Diagram of the two potential mechanisms for a recombinase-mediated cassette exchange. Recombinase sites 1 and 2 (small numbered boxes) are heterospecific; they recombine efficiently with their identical partner site but poorly or not at all with the other site. (A) Simultaneous double crossover. If both pairs of sites recombine simultaneously, a replacement results. Selection for clones that are resistant to marker 2 and sensitive to marker 1 will produce clones with the desired replacement. (B) Single crossover and resolution of the produced array. Another potential mechanism begins with a single crossover at one of the pairs of heterospecific recombinase target sites (site 2 in this case). This recombination inserts the replacement vector and produces an array containing two copies of each of the heterospecific target sites. A second recombinase-mediated event reduces the array by recombining between either pair of sites and deleting the intervening region. If the recombination occurs between the sites not used for integration, the replacement event occurs (left). If the recombination occurs between the sites used for integration, there is no net change in the locus (right). Once again, selection of clones resistant to marker 2 and sensitive to marker 1 will produce the desired clones with the exchange. In ES cells we have seen evidence of this mechanism in clones that have been resistance mediated by both markers and an unresolved array as diagrammed. This reaction can also be done with marker 1 as a positive/negative selectable marker and no marker 2, with the exchange selected by negative selection against marker 1.

of the entire targeting plasmid into the locus (Fig. 5B, recombination between site 2 and site 2). If the second intramolecular event occurs between the recombinase sites used to generate the insertion (e.g., site 2), the resulting excision leads to an unmodified locus and the event will not be detected. If, however, the second recombination occurs between the two site 1s, excision will result in the replacement of the marker with the targeting sequences. Heterospecific sites have been used to achieve efficient Flp-mediated RMCE in ES cells utilizing positive selection (6–50%), negative selection against an HSV-TK negative selectable marker (21–38%), or combined positive and negative selection (54–100%).[69] When using negative

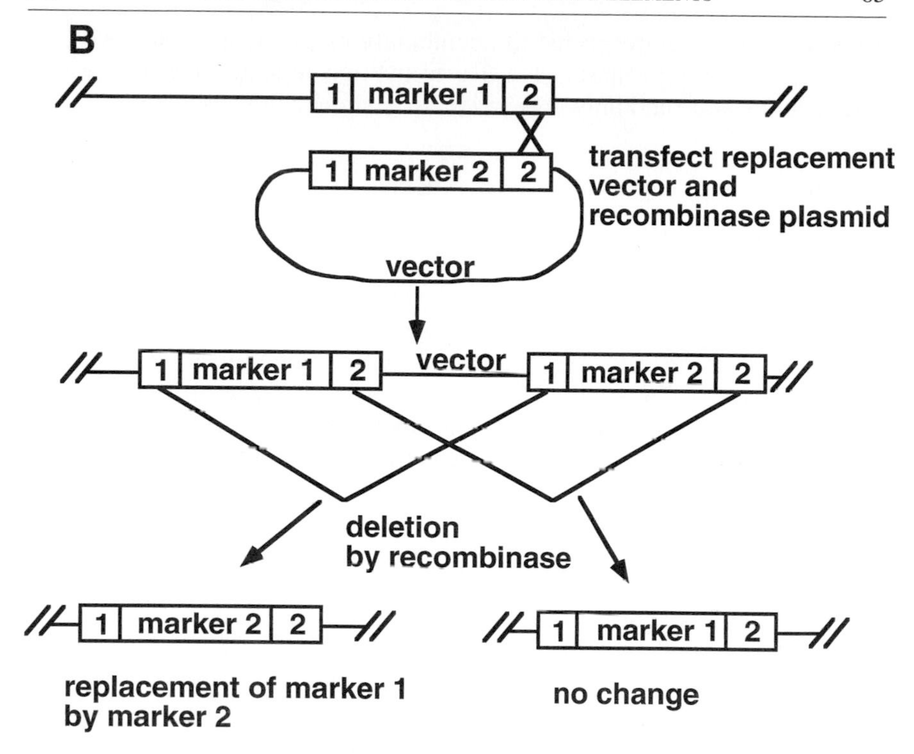

FIG. 5. (*continued*)

selection only, this system has the potential of allowing multiple modifications of a particular region without inserting a second selectable marker, thus avoiding the need for a further recombinase reaction to remove the selectable marker. If positive selection is needed to isolate the secondary modification, the marker could be flanked by *loxP* sites and subsequently deleted with Cre.

As with the prior techniques, there are several difficulties with the RMCE system: (1) recombinase sites must be paired carefully such that the efficiency of recombination is similar with both pairs; (2) recombination between site 1 and site 2 can potentially occur (the frequency can be negligible[78] and varies between heterospecific site combinations), leading to the excision of any negative selectable markers, thus increasing the background of nonexchanged clones when using only negative selection; (3) problems of finding an ideal negative selectable marker mentioned earlier exist; and (4) as with the double replacement and plug and socket, RMCE requires two rounds of cloning *in vitro*. The development of RMCE technology is in its infancy, and additional work, including the development of new efficient heterospecific site combinations, is ongoing. However, the approach is quite promising for use in making repeated modification of endogenous and randomly tagged loci in mammalian cells.

Summary

The use of homologous recombination to modify and thereby functionally analyze cis-regulatory DNA elements in mammalian cells has become an important approach in mammalian gene expression research. We have emphasized the necessity of designing a system that allows the removal of selectable markers used in targeting and facilitates the further modification of the region under study. To perform these tasks, we presently favor making an initial HR-mediated replacement of the entire element under study with an active positive selectable marker in combination with either an inactive second positive selectable marker or an active negative selectable marker. The plug and socket system, in which an inactive selectable marker is complemented by HR, is the most dependable and well-characterized option for making secondary modifications. However, the double-replacement system has certain advantages, and the recently developed RMCE approach, which allows replacement of a negative selectable marker by site-specific recombinase-mediated insertion without using a positive selectable marker, will likely prove very valuable in future experiments. Each of the systems, or combinations thereof, should be considered in light of the specifics of any given experiment to select the most appropriate option. Although the emphasis of this article has been the analysis of cis-acting regulatory elements involved in transcription, these same approaches can be used to analyze other regulatory elements (e.g., origins of replication) and to make multiple subtle mutations in polypeptides.

Acknowledgments

We acknowledge Jianzhu Chen for helpful discussions and for providing unpublished data and Andras Nagy for providing unpublished Cre-expressing mice. This work was supported by the American Society of Hematology ASH Scholars Award (SF), the Burroughs-Wellcome Career Development Award (SF), and the National Institutes of Health Grant P30 HD28834, through the University of Washington Child Health Research Center (MAB). MAB is a Howard Hughes Medical Institute Physician Postdoctoral Fellow.

[4] Genetically Directed Representational Difference Analysis-Based Positional Cloning and Snorthern Analysis of Candidate Genes

By MARK W. SORNSON and MICHAEL G. ROSENFELD

Introduction

Positional cloning allows the identification of genes through knowledge of their location within the genome. The positional cloning strategy in mice typically involves four distinct steps: (1) the locus is mapped genetically using an intercross or backcross to a genetically divergent strain, (2) DNA probes located close to the locus are obtained, (3) the probes and/or clone walking is used to identify cloned DNA segments that contain the locus, and (4) genes within the cloned segments are identified, and a candidate gene is proposed as the source of the phenotypic change based on expression and functional studies.[1] In mice, positional cloning has had several recent successes in identifying the basis of phenotypes such as *iv* and *lgl*,[2] *sm*,[3] *lurcher*,[4] and *df*.[5]

This article describes methods used in the positional cloning of *df* that have general applications, including aspects of intercross development, genetically directed representational difference analysis (GDRDA), candidate gene identification, and the technique of Snorthern analysis, a novel method for examining gene expression in restricted tissue samples.

Genetic Crosses for Linkage Analysis

Genetic crosses were widely used for establishing linkage analysis between phenotypic markers prior to the advent of molecular technologies for examining linkage analysis between specific DNA segments isolated

[1] F. S. Collins, *Nature Genet.* **9,** 347 (1995).

[2] D. M. Supp, D. P. Witte, S. S. Potter, and M. Brueckner, *Nature* (*London*) **389,** 963 (1997).

[3] A. Sidow, M. S. Bulotskj, A. W. Kerrebrock, R. T. Bronson, M. J. Daly, M. P. Reeve, T. L. Hawkins, B. W. Birren, R. Jaenisch, and E. Lander, *Nature* (*London*) **389,** 722 (1997).

[4] J. Zuo, P. L. DeJager, K. A. Takahashi, W. Jiang, D. J. Linden, and N. Heintz, *Nature* (*London*) **388,** 769 (1997).

[5] M. W. Sornson, W. Wu, J. S. Dasen, S. E. Flynn, D. J. Norman, S. M. O'Connell, I. Gukovsky, C. Carriere, A. K. Ryan, A. P. Miller, L. Zuo, A. S. Gleiberman, B. Andersen, W. G. Beamer, and M. G. Rosenfeld, *Nature* (*London*) **384,** 327 (1996).

0076-6879/99 $30.00

as probes.[6] In the early 1980s, the continuing development of molecular technologies allowed linkage experiments to be undertaken utilizing polymorphic DNA markers instead of polymorphisms that produce visible phenotypic effects. Crosses between divergent species or subspecies of mice facilitated the analysis of multiple markers within the same cross, as the entire genomes of the parental mice had polymorphisms that could be exploited for linkage analysis.[7,8]

Intercross and Backcross Strategies

For investigating recessive mutations, intercross and backcross breeding strategies allow recombination events to be recorded in the second generation. In both crosses, an initial cross to generate F_1 animals is performed between a strain homozygous for the mutation of interest and a genetically divergent strain. The intercross then uses a cross between two F_1 progeny, whereas a backcross uses an F_1 to parental homozygote cross. Should the homozygote be infertile or not viable, an intercross can be developed by a cross between a heterozygote and a genetically divergent strain, with the alleles within the F_1 progeny determined through test matings. For investigating mutations within *Mus musculus domesticus,* commonly used divergent strains are the *Mus* subspecies *Mus musculus castaneus* and *Mus musculus molossinus* and the separate *Mus* species *Mus spretus*. The backcross requires large numbers of matings to homozygous parentals, which may not be possible if the homozygote has low viability or fertility. In an intercross, homozygote breedings are not necessary; however, progeny may have zero, one, or two recessive alleles. Animals carrying zero or one allele may be indistinguishable phenotypically and may require additional breedings to determine their genotype.

An intercross can be developed more quickly from small numbers of founder animals if the F_2 animals are characterized by genetic testing to determine animals heterozygous at the locus and its surrounding region. These animals can then be used as equivalents to F_1 animals. The development of simple sequence length polymorphism (SSLP) markers by the

[6] J. F. Gusella, N. S. Wexler, P. M. Conneally, S. L. Naylor, M. A. Anderson, R. E. Tanzi, P. C. Watkins, K. Ottina, M. R. Wallace, A. Y. Sakaguchi *et. al., Nature (London)* **306,** 234 (1983).

[7] R. Robert, P. Barton, A. Minty, P. Daubas, A. Weydert, F. Bonhomme, J. Catalan, D. Chazottes, J. L. Guenet, and M. Buckingham, *Nature (London)* **314,** 181 (1985).

[8] L. Farrer, *Otolar Clin. N. Am.* **25,** 907 (1992).

MIT/Whitehead group[9] and other groups[10] has provided a fast method of genotyping animals, as well as high probe density throughout the genome. The use of at least two flanking markers on each side can prevent analysis errors or ambiguities due to polymerase chain reaction (PCR) artifacts. Figure 1 shows a diagram of these intercross and backcross strategies.

Small-Scale Animal DNA Preparation

Preparation of samples for analysis and genotyping can be performed as follows. For tail preparations, mice are anesthetized using halothane anesthesia in groups of two to six animals. Mouse tail pieces (0.5–1 cm) are removed using a cauterizing blade and are placed in 0.5 ml tissue digestion buffer [25 m*M* EDTA, 10 m*M* Tris–HCl, pH 8, 100 m*M* NaCl, 0.5% sodium dodecyl sulfate (SDS)], supplemented with 0.6 mg/ml proteinase K (Boehringer-Mannheim, Indianapolis, IN) at 50° for 8–16 hr. The samples are then extracted once with buffer-saturated phenol : chloroform : isoamyl alcohol (25 : 24 : 1), pH 8. Next, 0.35 ml of the aqueous supernatant is removed with a large-bore 1-ml pipette tip, with care taken not to disturb the interface. The DNA is then precipitated by the addition of 150 μl of 10 *M* ammonium acetate and 950 μl 100% ethanol and is recovered by microcentrifugation (Eppendorf microcentrifuge, 13,000*g* at 4° for 15 min), followed by washing with 80% ethanol and air drying to near completion. Resuspend in 100 μl $T_{10}E_{0.1}$ (10 m*M* Tris–HCl, pH 8, 0.1 m*M* EDTA) and incubate for 4 hr at 50° to solubilize.

Note: It is also possible to analyze the tail preparations as follows: Take 2 μl of the digested tail preparation prior to phenol–chloroform extraction and dilute to 100 μl with 10 m*M* Tris–HCl, pH 8.5. Next, place the diluted sample at 98° for 5 min to inactivate proteinase K. Use 1 μl as template for 15 μl PCR reactions, with other PCR conditions as described next.

Simple Sequence Length Polymorphism Genotyping

Genotyping of the animals through the use of SSLP markers[9] can be performed. Ten-microliter PCR reactions include 1.5 μl of a 1 : 10 dilution of the tail preparations in 10 m*M* Tris–HCl, pH 8, 1.0 μl of 10× PCR

[9] W. F. Dietrich, J. Miller, R. Steen, M. A. Merchant, D. Damron-Boles, Z. Husain, R. Dredge, M. J. Daly, K. A. Ingalls, T. J. O'Connor, C. A. Evans, M. M. DeAngelis, D. M. Levinson, L. Kruglyak, N. Goodman, N. G. Copeland, N. A. Jenkins, T. L. Hawkins, L. Stein, D. C. Page, and E. S. Lander, *Nature (London)* **380,** 149 (1996).

[10] C. M. Hearne, M. A. McAleer, J. M. Love, T. J. Aitman, R. J. Cornall, S. Ghosh, A. M. Knight, J.-B. Prins, and J. A. Todd, *Mamm. Genome* **1,** 273 (1991).

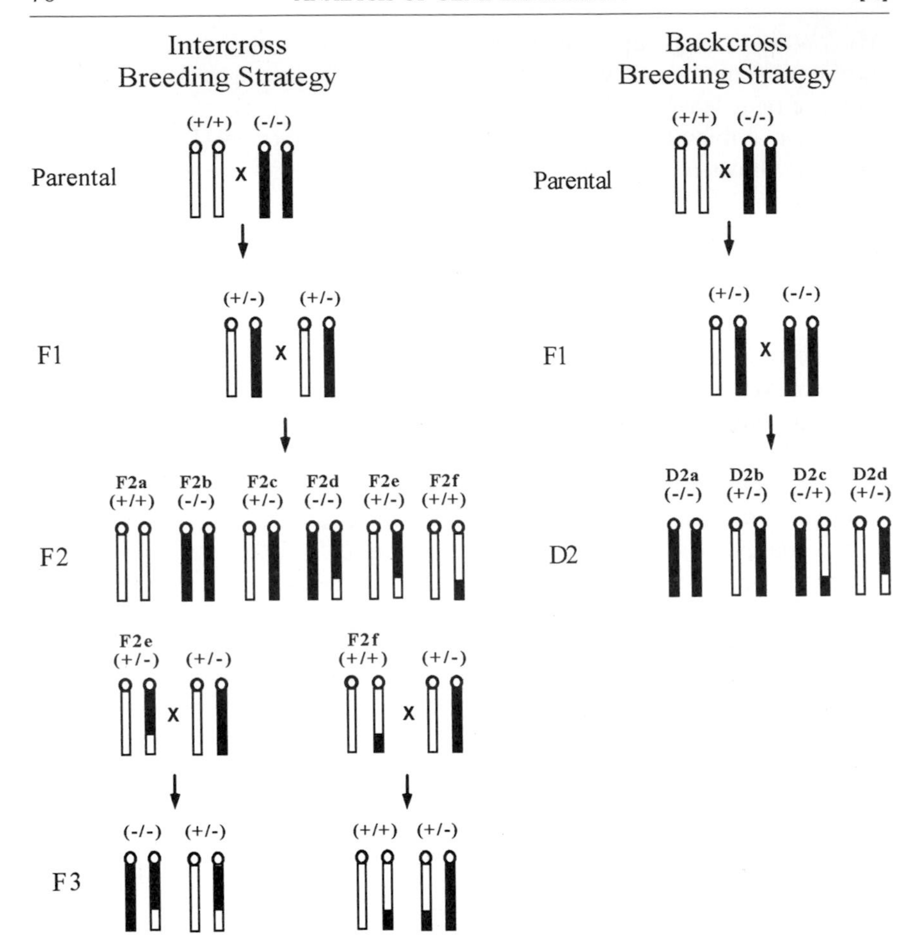

FIG. 1. Intercross and backcross breeding strategies for analyzing recessive mutations. The wild-type strain is shown in white and the strain containing the phenotypic mutation is shaded black. The intercross uses breeding between the divergent parental strains to generate F_1 animals, followed by an intercross to generate F_2 animals. Each F_2 animal records recombination events occurring during two parental meioses. Animals may have zero, one, or two mutant alleles; the presence of zero or one mutant allele is indistinguishable by phenotype, necessitating additional breedings to determine the exact genotype. As shown, animal F_{2e} has one mutant allele, proven by an intercross to an F_1 animal producing homozygous mutant phenotype progeny. Animal F_{2f} has no mutant alleles and all intercross progeny are wild-type; however, the presence of homozygous mutant strain DNA, as shown at the telomeric region, can be used to localize the recombination event. Heterozygous F_2 progeny (F_{2c}, F_{2e}) that are found to be nonrecombinant near the phenotypic locus can be considered equivalent to F_1 animals and may be used for additional intercrossing. In the backcross breeding strategy, parental mice are bred to produce F_1 animals, then backcrossed to the homozygous mutant parental strain to generate D_2 animals possessing one or two mutant alleles. Each D_2 animal records recombination events occurring in a single meiosis.

buffer (200 m*M* Tris–HCl, pH 8.4, 500 m*M* KCl; GIBCO/BRL, Gaithersburg, MD), 0.6 μl of each oligonucleotide (100 ng/μl), 0.3 μl of 50 m*M* $MgCl_2$, 1.6 μl of dNTP mix (1.25 m*M* of each dNTP), 0.1 μl (0.5 U) of *Taq* polymerase (GIBCO/BRL), and 4.3 μl of doubly distilled H_2O. Dilutions of the tail samples are performed in 96-well microtiter plates, allowing large numbers of PCR reactions with multiple oligonucleotide pairs to be performed. Twelve channel pipettors (Brinkman) are used to allow rapid sample transfer. Thermal cycling reactions are performed (Perkin Elmer Cetus, Norwalk, CT, Model 9600) with cycle parameters of denaturation at 94° for 20 sec, annealing at 55° for 30 sec, and extension at 72° for 30 sec, for 40 cycles, with a 5-min terminal extension period at 72°. For thermocyclers measuring block temperature instead of tube temperature, times should be increased by 25 sec for each cycling step. The annealing temperature of 55° assumes oligonucleotides have been optimized for that annealing temperature.[9] Gel analysis of products can be performed using gels made from a mixture of 2.5% MetaPhor agarose (FMC Bioproducts, Rockland, ME) and 1% UltraPure agarose (GIBCO/BRL) dissolved in 0.5× TBE. This gel mix provides optimum resolution and handling characteristics for examining polymorphic fragments in the range of 70 to 250 bp. After analysis, gels can be reheated and repoured up to four times without loss of resolution or clarity.

Large-Scale Animal DNA Preparation

DNA from animals with local recombination events can be prepared on a larger scale as follows. Mice are euthanized in accordance with mouse husbandry protocols. Dissected liver, heart, spleen, lungs, and brain are crushed to powder in liquid nitrogen using a mortar and pestle. Following evaporation of the liquid nitrogen, the tissue powder is resuspended in 4 volumes of tissue digestion buffer [25 m*M* EDTA, 10 m*M* Tris–HCl, pH 8, 100 m*M* NaCl, 0.5% sodium dodecyl sulfate (SDS)], supplemented with 0.2 mg/ml proteinase K (Boehringer-Mannheim). Samples are incubated for 8–16 hr at 50° with gentle rotation. Buffer-saturated phenol : chloroform : isoamyl alcohol, pH 8 (25 : 24 : 1), is used for two successive extractions: add an equal volume, mix gently by inversion 10–15 times, spin for 10 min at 4000*g*, and then remove the supernatants carefully with a blunted 5-ml pipette, being careful not to disturb the interface. The DNA is then precipitated from the aqueous phase by adding 1/2 volume 7.5 *M* ammonium acetate, mixing, adding 2 volumes 100% ethanol, mixing, and removal of the DNA clump using a 200-μl pipette tip. Wash the DNA with 80% ethanol and allow it to air dry to near completion, then resuspend in 1.5 ml $T_{10}E_{0.1}$, pH 8, by a 4-hr incubation at 50°.

Representational Difference Analysis

Representational difference analysis (RDA) is a PCR-based approach for subtracting complex genomes.[11] In RDA, PCR is used to amplify digests of genomic DNA that have been adaptor ligated. The PCR-amplified, adaptor-ligated digests are termed "amplicons" and do not contain all of the sequence information of the genomic DNA, as digest fragments greater than 1.5 kb are not amplified. If two strains have an RFLP such that one strain has a fragment significantly less than 1.5 kb and another strain has a fragment greater than 1.5 kb, then the amplicons will differ in possessing that locus. These differences can then be exploited through the use of subtractive hybridization, kinetic enrichment during reassociation, and PCR amplification. For terminology, an excess of "driver" amplicon is subtracted from the "tester" amplicon. The driver has its adaptors digested away before hybridization, whereas the tester has its adaptors replaced with adaptors of different sequences. During subtractive hybridization, tester : tester, tester : driver, and driver : driver hybrids can be formed. The subsequent PCR amplification step allows substantial enrichment of sequences present solely in the tester, as only tester : tester hybrids amplify exponentially. The amplification product is used as a tester amplicon in a new round of the RDA procedure, with a decreased amount of tester used in the subtractive hybridization. Sequences present only in the tester are enriched compared to nonselected sequences, allowing them to continue to reanneal, whereas nonselected sequences are at a lower concentration and their hybridization is disfavored kinetically.

Representational Difference Analysis: Method

A detailed protocol for representational difference analysis has been reported previously.[12] Minor modifications that we have introduced to the technique are as follows.

1. PCR Amplification (Step B.4 of Detailed Protocol in Ref. 12)

a. After the tester and driver ligations have been performed, dilute each ligation with 940 μl of TE (10 m*M* Tris–HCl, 1 m*M* EDTA, pH 8) plus tRNA (20 μg/ml) buffer. Make two tubes of PCR mix for the preparation of the tester amplicon and six tubes for the preparation of the driver amplicon, each containing 9 μl of 10× PCR buffer (200 m*M* Tris–HCl, pH 8.4, 500 m*M* KCl; GIBCO/BRL), 2 μl of 24-mer oligonucleotide, 8 μl of dNTP mix (2.5 m*M* each dATP,

[11] N. A. Lisitsyn, N. M. Lisitsyn, and M. Wigler, *Science* **259,** 946 (1993).

[12] N. A. Lisitsyn and M. Wigler, *Methods Enzymol* **254,** 291 (1995).

dCTP, dGTP, dTTP), 4 μl of 50 m*M* $MgCl_2$, 10 μl of diluted ligate, and 57 μl of doubly distilled H_2O.

At the same time, prepare a *Taq* DNA polymerase dilution cocktail in 1× PCR buffer, multiplying volumes by the number of tubes, allowing an additional 10% for wastage: 0.75 μl (3.75 U) of *Taq* polymerase (GIBCO/BRL), 1 μl of 10× PCR buffer (GIBCO/BRL), and 9.25 μl of doubly distilled H_2O.

b. Hot start: Place tubes in thermocycler (Perkin-Elmer Cetus, Model 9600) at 72° for 30 sec.
c. Add 10 μl *Taq* polymerase dilution and continue 72° incubation for 5 min to fill in the overhang present after the 12-mer disassociates.
d. Amplify for 15 cycles (30 sec at 95° and 3 min at 72°), with the final cycle followed by extension at 72° for 10 min. Store the tubes on ice.
e. Check concentration of amplicons by gel electrophoresis of 10-μl samples, loading several dilutions of standard DNA (*Sau*3A digest of human DNA, *Dpn*II digest of mouse DNA, or amplicons from previous experiments).
f. Set up secondary PCR amplification reactions.

Make two tubes of PCR mix for the preparation of the tester amplicon and six tubes for the preparation of the driver amplicon, each containing 30 μl of 10× PCR buffer (GIBCO/BRL), 6 μl of 24-mer oligonucleotide, 24 μl dNTP mix (2.5 m*M* each dATP, dCTP, dGTP, dTTP), 12 μl 50 m*M* $MgCl_2$, 90 μl initial PCR reaction, 2.5 μl *Taq* polymerase, and 235.5 μl doubly distilled H_2O for a total volume of 400 μl.

Cover with 110 μl mineral oil. If the concentration of the initial PCR reaction was more than 0.3 μg/10 μl, then amplify for five cycles (Perkin Elmer Cetus, Model 480). If the concentration was between 0.1 and 0.3 μg/10 μl, amplify for seven cycles. If the concentration is less than 0.1 μg/10 μl, amplify for nine cycles. Cycle parameters are 1 min at 95° and 3 min at 72°, with the final cycle followed by extension at 72° for 10 min.

Note: Use of the two-part PCR reactions allows easier setup of the hot starts, more rapid PCR reactions, and the ability to identify failed reactions before performing the expensive large-scale reactions.

2. Changing of Adaptors on Tester Amplicon (Step B.6 of Detailed Protocol in Ref. 12)

a. Electrophorese 5–10 μg of tester amplicon DNA digest on a 2% NuSieve (low melting point) agarose gel (FMC Bioproducts), until the bromphenol blue dye marker has moved 5 cm.

Note: For optimum subsequent PCR amplification, it is important not to expose the tester to UV irradiation. Prepare a 1-kb ladder DNA standard in 1× PCR buffer and electrophorese with an intervening empty lane. Gels are run with ethidium bromide at 4 V/cm. The marker lane can be removed, UV visualized, and replaced to check migration.

b. Using the lengths of migration in the marker lane as a guide, cut the agarose slice (0.2–0.8 g) containing fragments 150–1500 bp in length and place in a 5-ml Falcon tube.

c. Spin column purify DNA from slice (QIAquick Gel Extraction Kit; Qiagen, Santa Clarita, CA) and elute in 30 μl TE.

d. Check DNA concentration by gel electrophoresis, adjust to 0.1 mg/ml.

Hybridization/Amplification Steps (Step C.1-2 of Detailed Protocol in Ref. 12). All volumes in the hybridization/amplification steps were reduced by 50% in order to conserve reagents. Times of hybridization and hybridization conditions were unchanged. After hybridization, prepare the DNA dilution by adding 4 μl of tRNA solution (5 mg/ml) to the reannealed DNA, mix thoroughly by pipetting, add 195 μl of $T_{10}E_{0.1}$ buffer, and mix again. Continue with remainder of hybridization/amplification steps as described.

Genetically Directed Representational Difference Analysis

For GDRDA, the "genetic direction" is provided by generating the driver amplicon from pooled genomic DNA of recombinant animals with the phenotype such that the driver pool has alleles corresponding to the tester throughout the genome except for the region around the phenotypic locus.[13] The tester amplicon is produced from the divergent strain used to develop the intercross or backcross.

In the driver amplicon, a small region surrounding the phenotypic locus lacks tester alleles. A fraction of the fragments within this interval are polymorphic with shorter tester fragments and large driver fragments; these will be enriched by RDA. The prevalence of selectable polymorphic fragments is determined by the variation between the two strains involved; an estimate has been given for an intersubspecific cross as one to two selectable polymorphisms per megabase (Mb) per enzyme used to make amplicons.[13]

GDRDA Strategy

The GDRDA strategy used to positionally clone the Ames dwarf locus (*df*) serves as a generally applicable model for designing GDRDA tester

[13] N. A. Lisitsyn, J. A. Segre, K. Kusumi, N. M. Lisitsyn, J. H. Nadeau, W. N. Frankel, M. H. Wigler and E. S. Lander, *Nature Genet.* **6,** 57 (1994).

and driver pools for studying recessive phenotypic mutations. For the cloning of *df*, the intercross [(DF/B *df/df* × Cast/Ei) × (DF/B *df/df* × Cast/Ei)] was developed. The subset of *df/df* animals from the first 2530 meioses (1265 animals) was used for developing the driver pools for GDRDA, and Cast/Ei DNA was used to develop the tester. The number of homozygous animals used for developing driver pools has been examined in detail,[13] with mathematical models suggesting the use of a minimum of 10 F_2 homozygous intercross progeny. With large intercrosses, it is possible that too small a region could be targeted and that the GDRDA strategy could fail due to a lack of selectable polymorphisms within the nonrecombinant interval. In order to prevent this, two driver pools can be developed, with one targeting a larger interval through the use of a subset of the available animals with recombination events close to the phenotypic interval. A diagram of the driver pools used in the GDRDA of *df* is presented as Fig. 2, illustrating this strategic point.

GDRDA experiments are conveniently performed using groups of four to six enzymes. The GDRDA technique requires 12- and 24-mer oligonucleotides for each amplification round. Four 24-mer oligonucliotides were

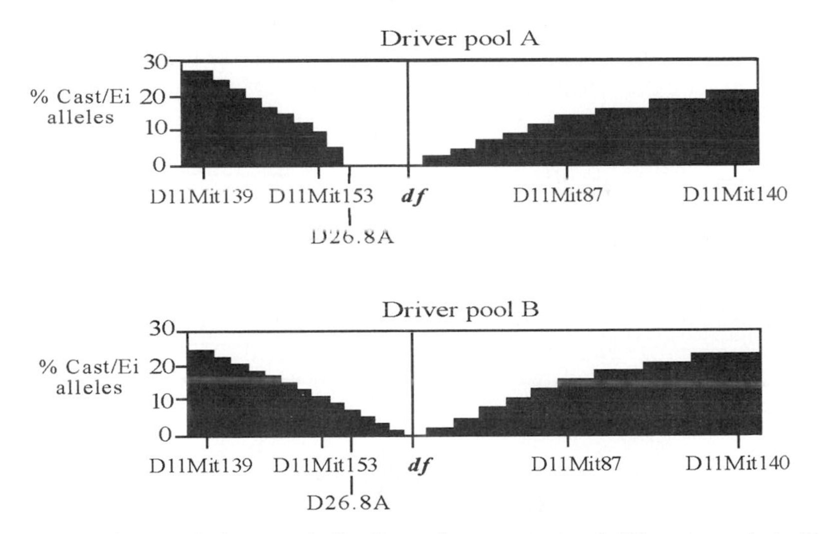

FIG. 2. Driver pools for genetically directed representational difference analysis. The percentage of Cast/Ei alleles in the driver pools is plotted against position along mouse chromosome 11. Driver pool B used all DF/B *df/df* animals with recombination events located close to *df* to subtract the Cast/Ei tester through RDA, whereas driver pool A used a subset of *df/df* animals. As the physical distance between the closest recombination events in driver pool B was unknown, the omission in driver pool A of animals with recombination events between D26.8A and *df* provided a larger region for the recovery of polymorphisms in the initial experiments.

made for each of the R, J, and N rounds. The enzymes used and the sequences of the 24- and 12-mer oligonucleotides used with each enzyme are provided in Table I.

Not shown in Table I are *Afl*II and *Ppu*10I, which did not produce amplicons when used. *Bsp*120I produced a low-yield amplicon with a size distribution primarily above 600 bp. *Acc*65I, *Apa*LI, *Ban*I, *Mfe*I, and *Nhe*I required two to four additional cycles of PCR for the production of amplicons. Of the enzymes, *Bsp*HI, *Bgl*II, *Eco*RI, and *Nhe*I produced probes located within the nonrecombinant region after three rounds of RDA; *Bcl*I and *Bsr*GI produced nonrecombinant probes after four rounds of RDA.[5] In addition to the previously reported *Bam*HI and *Hin*dIII,[13] these enzymes have thus been proven successful in targeting GDRDA.

Note: Important control experiments that should be considered are parallel experiments using tester DNA supplemented with an equimolar amount of foreign DNA fragments recoverable through the RDA process. DNA used for supplementation should possess a fragment of between 200 and 800 bp when digested with the enzyme. An additional control that should be performed with each round of the RDA procedure is to conduct in parallel a previously successful RDA subtraction/hybridization and amplification. Used together, these controls provide a basis for troubleshooting should a given enzyme or round of RDA not produce a product.

For investigating the location of probes, cloned fragments are first tested against digests of genomic DNA to determine if they are single-copy sequences. Subsequently, restriction fragment length polymorphism (RFLP) can be determined through hybridization to a Southern blot of multiple enzyme digests of parental strain DNAs. One of the enzymes should be the enzyme used to obtain the probe, as an RFLP should be evident for all nonartifactual RDA products. Products can then be mapped genetically using Southern blots of polymorphic enzyme digests of intercross animal DNA.

Snorthern Analysis

Historically, one of the most difficult steps in positional cloning has been the identification of candidate genes once the physical region known to contain the gene has been isolated. There are currently several gene identification techniques available, most of which provide large numbers of clones, for which adequate secondary screening strategies are often unavailable. The Snorthern technique provides readily made small-sized blots that can be used with large numbers of probes to investigate gene expression in restricted tissue sources. The following steps are used: (1) cDNA is produced from the tissue(s) of interest. (2) The cDNA is digested with three four-base restriction enzymes in separate reactions. (3) Adaptors

TABLE I

SEQUENCES OF OLIGONUCLEOTIDES USED FOR GDRDA[a]

Enzyme[b]	24-mers[c]	R 12-mers[d]	J 12-mers[d]	N 12-mers[d]
*Acc*65I	G	5′-GTACCTCGGTGA-3′	5′-GTACCGTTCATG-3′	5′-GTACCTCCCTCG-3′
*Afl*III	A	5′-C(AG)(CT)GTGCGGTGA-3′	5′-C(AG)(CT)GTGTTCATG-3′	5′-C(AG)(CT)GTTCCCTCG-3′
*Apa*LI	G	5′-TGCACTCGGTGA-3′	5′-TGCACGTTCATG-3′	5′-TGCACTCCCTCG-3′
*Ava*II	G	5′-G(AT)CCTCGGTGA-3′	5′-G(AT)CCGTTCATG-3′	5′-G(AT)CCTCCCTCG-3′
*Ban*I	G	5′-G(CT)(AG)CCTCGGTGA-3′	5′-G(CT)(AG)CCGTTCATG-3′	5′-G(CT)(AG)CCTCCCTCG-3′
*Bcl*I	T	5′-GATCAGCGGTGA-3′	5′-GATCAGTTCATG-3′	5′-GATCATCCCTCG-3′
*Bsp*120I	G	5′-GGCCCTCGGTGA-3′	5′-GGCCCGTTCATG-3′	5′-GGCCCTCCCTCG-3′
*Bsp*HI	T	5′-CATGAGCGGTGA-3′	5′-CATGAGTTCATG-3′	5′-CATGATCCCTCG-3′
*Bsr*GI	T	5′-GTACAGCGGTGA-3′	5′-GTACAGTTCATG-3′	5′-GTACATCCCTCG-3′
*Eco*RI	G	5′-AATTCTCGGTGA-3′	5′-AATTCGTTCATG-3′	5′-AATTCTCCCTCG-3′
*Mfe*I	C	5′-AATTGTCGGTGA-3′	5′-AATTGGTTCATG-3′	5′-AATTGTCCCTCG-3′
*Nco*I	C	5′-CATGGTCGGTGA-3′	5′-CATGGGTTCATG-3′	5′-CATGGTCCCTCG-3′
*Nhe*I	G	5′-CTAGCTCGGTGA-3′	5′-CTAGCGTTCATG-3′	5′-CTAGCTCCCTCG-3′
*Sty*I	C	5′-C(AT)(AT)GGTCCGTGA-3′	5′-C(AT)(AT)GGGTTCATG-3′	5′-C(AT)(AT)GGTCCCTCG-3′
*Xba*I	T	5′-CTAGAGCGGTGA-3′	5′-CTAGAGTTCATG-3′	5′-CTAGATCCCTCG-3′

A 24-mers		C 24-mers	
R 24 A	5′-AGCACTCTCCAGCCTCTCACCGCA-3′	R 24 C	5′-AGCACTCTCCAGCCTCTCACCGAC-3′
J 24 A	5′-ACCGACGTCGACTATCCATGAACA-3′	J 24 C	5′-ACCGACGTCGACTATCCATGAACC-3′
N 24 A	5′-AGGCAACTGTGCTATCCGAGGGAA-3′	N 24 C	5′-AGGCAACTGTGCTATCCGAGGGAC-3′
G 24-mers		T 24-mers	
R 24 G	5′-AGCACTCTCCAGCCTCTCACCGAG-3′	R 24 T	5′-AGCACTCTCCAGCCTCTCACCGCT-3′
J 24 G	5′-ACCGACGTCGACTATCCATGAACG-3′	J 24 T	5′-ACCGACGTCGACTATCCATGAACT-3′
N 24 G	5′-AGGCAACTGTGCTATCCGAGGGAG-3′	N 24 T	5′-AGGCAACTGTGCTATCCGAGGGAT-3′

[a] The R series is used for the representation stage, and the J series and N series are used for odd and even hybridization/amplification steps, respectively.

[b] For *Bam*HI, *Bgl*II, and *Hin*dIII, 12- and 24-mer oligonucleotides were used as described previously.[13]

[c] The 24-mer column refers to the set of 24-mers that should be used. The nucleotide sequences of the 24-mers are provided.

[d] Nucleotides in parentheses represent equimolar mixtures of the two at that position.

are attached. (4) The adaptor-ligated cDNA digests are amplified by PCR and combined. (5) The amplified products are gel electrophoresed and Southern blotted.

The result is a Southern blot that provides expression information, hence the name "Snorthern" blot. The technique provides a method for investigating expression in a restricted tissue source, with less than 100 ng of starting cDNA having been shown to work well for amplification.[5] Additionally, the amplified material can be used for the production of large numbers of blots, allowing the expression pattern of hundreds of probes to be investigated in multiple tissue sources with relatively little effort. The technique does not give information on the size of the transcript from the Snorthern results, as the Snorthern represents a digest of the cDNA. Because of PCR amplification, there is also a chance for some genes and/or gene regions not to be represented in the amplified digested cDNA. The gene regions expected to be absent would be the 5′ end of the gene up to the first occurrence of one of the three four-base restriction sites and gene regions that do not have a pair of the four-base restriction enzyme sites producing an amplifiable fragment, which should happen very rarely. Difficult to amplify GC-rich regions would also be expected to be underrepresented.

mRNA Preparation

Poly(A) purified mRNA is prepared directly from tissue using standard microscale protocols (Micro FastTrack, Invitrogen; Oligotex, Qiagen). The yield of the small-scale mRNA yield is quantitated (DNA Dipstick; Invitrogen, Carlsbad, CA).

cDNA Preparation

First-strand synthesis is performed using the SuperScript Choice system (GIBCO/BRL), described briefly as follows. Dissolve mRNA (≤ 1 μg) in 9 μl (diethyl pyrocarbonate (DEPC)-treated doubly distilled H_2O; add 1 μl (50 ng) of random hexamers and 2 μl (1700 ng) of oligo$(dT)_{18} \cdot$ *Mbo*I primer (sequence: 5′-GACAGAGATCTTTTTTTTTTTTTTTTTT-3′). Heat mixture to 70° for 10 min and quick chill on ice. Centrifuge briefly at 4° to collect contents of tube, then add 4 μl first-strand buffer (250 m*M* Tris–HCl, pH 8.3, 375 m*M* KCl, 15 m*M* $MgCl_2$), 2 μl 0.1 *M* dithiothreitol (DTT), and 1 μl 10 m*M* dNTP mix for a total volume of 19 μl. Mix by gentle vortexing and spin briefly to collect. Place tube at 37° for 2 min for temperature equilibration. Prepare a tube for a labeled side reaction containing 0.1 μl [α-^{32}P]dCTP (1 μCi/μl). Add 1 μl SuperScript II reverse transcriptase (10000 U/μl, GIBCO/BRL) to the main reaction, mix gently, and immediately add 2 μl of the mixed reaction to the side reaction tube. Incubate reactions for 1 hr at 37°. Place on ice to terminate. Using the labeled side reaction, calculate yield using the activity of acid-precipitable counts and electrophorese on an alkaline agarose gel to determine the size distribution of the cDNA in accordance with manufacturer's instructions. Use the 18-μl unlabeled reaction for second-strand synthesis, according to manufacturer's instructions, adding the following: 93 μl of DEPC-treated water, 30 μl of 5× second-strand buffer [100 m*M* Tris–HCl, pH 6.9, 450 m*M* KCl, 23 m*M* $MgCl_2$, 0.75 m*M* β-NAD^+, 50 m*M* $(NH_4)_2SO_4$], 3 μl of 10 m*M* dNTP mix, 1 μl of *Escherichia coli* DNA ligase (10 U/μl) 4 μl of *E. coli* DNA polymerase (10 U/μl), and 1 μl of *E. coli* RNase H (2 U/μl) for a total volume of 150 μl.

Vortex gently to mix and incubate for 2 hr at 16°. Add 2 μl (10 U) of T4 DNA polymerase and continue incubation at 16° for 5 min. Place reaction on ice and add 10 μl 0.5 *M* EDTA. Isolate the ds-cDNA using spin column purification (QIAquick PCR purification kit, Qiagen), using 50 μl of 10 m*M* Tris–HCl, pH 8.3, for elution.

cDNA Digestion

Digestions are performed separately with *Dde*I, *Dpn*II (or *Mbo*I), and *Nla*III. If other restriction enzymes are used, they should be heat labile, as there are no purification steps prior to the adaptor ligation reactions. Each digestion includes 16 μl of cDNA eluate, 4.5 μl of 10× restriction buffer (New England Biolabs, Beverly, MA), 0.5 μl (5 U) of restriction endonuclease, and 24 μl of H_2O for a total volume of 45 μl. Add 0.5 μl of bovine serum albumin (BSA, 100 mg/ml) to the *Nla*III digestion only. Incubate for 1 hr at 37° and then for 20 min at 65° to inactivate the restriction endonucleases.

Adaptor Ligation

The following adaptor primer sets are used:

Mbo-24: 5′-AGCACTCTCCAGCCTCTCACCGGT-3′
Mbo-12: 5′-GATCACCGGTGA-3′
Dde-25: 5′-AGCACTCTCCAGCCTCTCACCGGTC-3′
Dde-11: 5′-T(AGTC)AGACCGGTG-3′
Nla-28: 5′-AGCACTCTCCAGCCTCTCACCGGTCATG-3′
Nla-10: 5′-ACCGGTGAGA-3′

Each ligation reaction includes 45 μl of cDNA digest, 1.5 μl of 10× T4 DNA ligase buffer (without rATP), 4.5 μl of rATP (10 m*M*), 4.5 μl of

12-mer at 0.5 μg/μl, and 4.5 μl of 24-mer at 1 μg/μl for a total volume of 60 μl.

To anneal the oligonuceotides, place the tubes in a heating block (Thermoline DriBath, holes filled with glycerol) at 50–55° and place the block in a cold room for 1–2 hr until the temperature drops to 10–15°. Alternatively, a 250-ml beaker with 200 ml H_2O can be used. Next, add 2 μl of T4 DNA ligase and incubate overnight at 12–16°. Dilute the adaptor ligated product with 240 μl doubly distilled H_2O.

Hot-Start PCRs to Determine Optimum Number of Cycles for Amplification

Prepare 12 tubes per enzyme and tissue source: 1.5 μl of adaptor-ligated cDNA, 2 μl of 10× PCR buffer (200 m*M* Tris–HCl, pH 8.4, 500 m*M* KCl, GIBCO/BRL), 4 μl of dNTP mix (1.25 m*M* each), 1 μl of 50 m*M* $MgCl_2$, 0.15 μl of 24-mer (1 μg/μl), and 11.35 μl of doubly distilled H_2O for a total volume of 20 μl each. In a thermocycler, heat to 72° for 30 sec, then pause machine and add 5 μl *Taq* DNA polymerase dilution to each tube and mix. *Taq* DNA polymerase dilution includes 0.2 μl of *Taq* polymerase, 0.5 μl of 10× PCR buffer, and 4.3 μl doubly distilled H_2O. Following addition, the total volume is 25 μl. Use PCR conditions (Perkin Elmer Cetus, Model 9600) of an initial extension at 72° for 5 min to extend oligonucleotide sequences, followed by 30 cycles of denaturation at 94° for 30 sec and annealing/extension at 72° for 3 min.

Tubes are removed after every two cycles from 10 to 30. PCR products are then run on a 1.4% agarose gel with *Dpn*II-digested mouse genomic DNA used as a concentration standard curve. The number of cycles where the yield appears to plateau is chosen as the optimum number for the large-scale reactions to produce material for the cDNA selection. Large-scale reactions are performed with eight tubes of 100 μl volume, using scaled-up reagents and the same cycle parameters. An aliquot of the final reaction should be run on a 1.4% gel and compared to a concentration standard curve. Mixtures by visualized concentration (1 : 1 : 1) are prepared and used for the final Snorthern blot gel. If necessary, PCR products can be isopropanol precipitated to concentrate, using 0.3 *M* sodium acetate as the precipitating salt.

Southern Blotting

For Snorthern blots, products are run at 6 V/cm with ethidium bromide on 1.2–1.4% agarose gels. Two micrograms of PCR product is used per lane, with lane sizes of 3.0 × 0.75 mm. Gels are then Southern blotted to a ICN (Costa Mesa, CA) Biotrans(+) membrane. A convenient approach to setting up large numbers of small-sized blot hybridizations is to perform the hybridizations in 15-ml screw-cap tubes containing 2–4 ml of hybridization solution (Church Buffer, from stock solutions: 35 ml 20% SDS, 0.2 ml 0.5 *M* EDTA, 34.2 ml 1 *M* Na_2HPO_4, 15.8 ml 1 *M* NaH_2PO_4, 14.8 ml doubly distilled H_2O; final pH 7.4) with 100 μg/ml denatured sheared salmon sperm DNA and two- to five-million cpm/ml probe), with the tubes held in a rotator at 65°. Figure 3 shows an example of Snorthern blot hybridization results.

Gene Identification Using Modified Form of cDNA Selection

Some of the currently available gene identification techniques include exon amplification/exon trapping,[14] cDNA selection,[15,16] direct DNA sequencing,[17] CpG island rescue,[18] and transgenic analysis via phenotypic rescue using large insert clones.[18] This section focuses on a method of cDNA selection used to ensure redundant coverage of expressed transcripts.

cDNA selection uses hybridization of cDNA to a genomic template to capture expressed sequences, which are then amplified by PCR.[15,16] A series of protocols utilize solution-based hybridization of biotinylated genomic DNAs to adaptor-ligated cDNA fragments, followed by recovery with streptavidin-coated magnetic beads, washing and elution steps, and then PCR amplification.[15,19–23] The following modified protocol introduces changes to allow the use of smaller tissue samples, to give more comprehensive coverage of sequences from the entire length of transcripts, and to provide RDA style kinetic enrichment at the second round of cDNA selection.

[14] A. J. Buckler, D. D. Chang, S. I. Graw, J. D. Brook, D. A. Haber, P. A. Sharp, and D. E. Housman, *Proc. Natl. Acad. Sci. U.S.A.* **88,** 4005 (1991).

[15] M. Lovett, J. Kere, and L. M. Hinton, *Proc. Natl. Acad. Sci. U.S.A.* **88,** 9628 (1991).

[16] S. Parimoo, S. R. Patanjali, H. Shulkla, D. D. Chaplin, and S. M. Weissman, *Proc. Natl. Acad. Sci. U.S.A.* **88,** 9623 (1991).

[17] A. Kamb, C. Wang, A. Thomas, B. S. DeHoff, F. H. Norris, K. Richardson, J. Rine, M. H. Skolnick, and P.R. Rosteck, *Comp. Biomed. Res.* **28,** 140 (1995).

[18] J. M. Valdes, D. A. Tagle, and F. S. Collins, *Proc. Natl. Acad. Sci. U.S.A.* **91,** 5377 (1994).

[19] J. G. Morgan, G. M. Dolganov, S. E. Robbins, L. M. Hinton, and M. Lovett, *Nucleic Acids Res.* **20,** 5173 (1992).

[20] D. A. Tagle, M. Swaroop, M. Lovett, and F. S. Collins, *Nature (London)* **361,** 751 (1993).

[21] N. Dracopoli, J. Haines, B. Korf, D. Moir, C. Morton, C. Seidman, J. Seidman, and D. Smith, eds., "Current Protocols in Human Genetics." Greene Publishing Associates and Wiley (Interscience), New York, 1994.

[22] M. Lovett, *Trends Genet.* **10,** 352 (1994).

[23] A. Simmons, S. Goodart, T. Gallardo, J. Overhauser, and M. Lovett, *Hum. Mol. Gen.* **4,** 295 (1995).

FIG. 3. Snorthern blot of multiple tissue sources probed with *P-Lim.* Digests of cDNA with *Dde*I, *Mbo*I, and *Nla*III were adaptor ligated, amplified, mixed, agarose gel electrophoresed, and Southern blotted. Tissues present are 13.5 and e15.5 embryonic pituitaries (mixed after amplification), adult liver, adult skeletal muscle, and an e14.5 animal from which the pituitary had been dissected away (Emb. no Pit). Genomic c57BL/6J DNA (4 μg) was added to the blots to be used as a positive control and to check for repetitive sequences. Probing with *P-Lim* reveals strong signals specific to the embryonic pituitary lane, as expected, given previous RNase protection and *in situ* hybridization results.[26] Multiple bands are the result of the multiple fragments present in the three digests.

To improve the coverage of sequences from the length of the transcripts, cDNA is prepared using an oligo(dT) primer with a four-base restriction enzyme site at the 5′ end, allowing sequences close to the poly(A) tail to be amplified. Three different four-base restriction enzymes are used to prepare digests for adaptor ligation, allowing small fragments to be generated that cover the length of the transcript. These adaptor-ligated digests are amplified separately by PCR, then mixed to provide a source of cDNA for genomic selection.

Amplified cDNA Preparation

The mRNA preparation, cDNA preparation, digestion, and adaptor ligation steps are performed essentially as described earlier for Snorthern analysis, with minor modifications. Random primers should be omitted from the cDNA preparation step if the genomic template to be used for selection is contaminated by rRNA sequences [e.g., yeast artificial chromosome (YAC) DNA preparations]. Hot-start PCRs are performed with 25-μl reactions for the three amplified cDNA digests to determine the optimum number of cycles. Shorter oligonucleotides (17-mers) are used to prevent reannealing during the cDNA hybridizations. The oligonucleotide sequences are

Mbo-17: 5′-GCCTCTCACCGGTGATC-3′
Dde-17: 5′-CAGCCTCTCACCGGTCT-3′
Nla-17: 5′-GCCTCTCACCGGTCATG-3′

The PCR reaction reagent concentrations are identical to the conditions described for Snorthern analysis. The PCR cycle parameters (Perkin Elmer Cetus, Model 9600) are 72° for 5 min to extend the adaptor oligonucleotide sequences, followed by 10–30 cycles of denaturation at 94° for 30 sec, annealing at 52° for 45 sec, and extension at 72° for 2 min and 30 sec. Tubes are removed every two cycles from 10 to 30, and then the tubes are replaced in the machine for a final extension at 72°. The 17-mer oligonucleotides typically require four additional PCR cycles to reach saturation, compared to the 24- to 28-mers. Scaled-up reactions are performed with 100-μl volumes and similar conditions, with the number of cycles given by the cycle number prior to the plateau in product concentration of the smaller scale reactions. Aliquots of the scaled-up reactions should be run on a 1.4% gel and compared to a standard curve to determine concentrations. 2-Propanol precipitate the amplified cDNA with 0.3 *M* sodium acetate as the salt, wash with 80% ethanol, and resuspend at a final concentration of 1 μg/μl.

Bacterial Artificial Chromosome DNA Preparation

Bacterial artificial chromosome (BAC) DNA is prepared by alkaline lysis, as described.[24] For cDNA selection experiments, BAC DNA is purified by CsCl banding, according to standard protocols.[25] At all steps, care should be taken to minimize shearing of the DNA. Yields are typically low, with 20–50 μg obtained per liter of culture. DNA should be ethanol precipitated and resuspended at a concentration of approximately 1 μg/μl. *Bam*HI and *Bgl*II digests should be electrophoresed to check for the presence of contaminating bacterial chromosomal DNA, which appears as a smear behind the bands of the digest. The best results in the cDNA selection protocol are obtained using BAC DNA preparations that have very little apparent contaminating chromosomal DNA.

BAC DNA Digestion

The BAC DNA is digested to allow biotinylation by end filling. Each reaction includes 5.0 μl of BAC DNA (1 μg/μl), 3 μl of 10× restriction enzyme buffer (1.5 *M* NaCl, 10 m*M* DTT, 100 m*M* Tris–HCl, pH 7.9, New England Biolabs), 1 μl of *Bam*HI (20 U/μl, New England Biolabs), 2 μl of *Bgl*II (10 U/μl), 0.3 μl of bovine serum albumin (100 μg/μl), and 18.7 μl of doubly distilled H_2O for a total volume of 30 μl. Digestions are performed for 4 hr at 37°. Digests are then incubated at 65° for 20 min to inactivate the restriction endonucleases.

BAC DNA Digest Biotinylation

The following components are added to the heat-inactivated digestion: 2.0 μl of 10× restriction enzyme buffer (500 m*M* NaCl, 10 m*M* DTT, 100 m*M* Tris–HCl, pH 7.9, New England Biolabs), 2 μl of dNTP mix (1.25 m*M*), 2.5 μl of biotin-16-dUTP (1 m*M*, Boehringer-Mannheim), 1 μl of Klenow fragment DNA polymerase (5 U/μl), and 12.5 μl of doubly distilled H_2O for a total volume of 50 μl.

Incubate for 30 min at room temperature and stop reaction with 2.5 μl 0.5 *M* EDTA. DNA fragments are then separated from unincorporated biotin-16-dUTP and other nucleotides by two successive spin column purifications (QIAquick PCR purification kit, Qiagen), adding 3 μl 1 *M* Tris–

[24] H. Shizuya, B. Birren, U. J. Kim, V. Mancino, T. Slepak, Y. Tachiiri, and M. Simon, *Proc. Natl. Acad. Sci. U.S.A.* **89,** 8794 (1992).

[25] J. Sambrook, E. F. Fritsch, and T. Maniatis, "Molecular Cloning: A Laboratory Manual," 2nd Ed. Cold Spring Harbor Laboratory, Cold Spring Harbor, NY, 1989.

[26] I. Bach, S. J. Rhodes, R. V. I. Pearse, T. Heinzel, B. Gloss, K. M. Scully, P. E. Sawchenko, and M. G. Rosenfeld, *Proc. Natl. Acad. Sci. U.S.A.* **92,** 2720 (1995).

HCl, pH 7.2, prior to the spin column purification to promote DNA binding to the spin column through lowered pH; elute with 30 μl 10 mM Tris–HCl, pH 8.0. Check concentration and dilute to 40 ng/μl. The concentration can be determined by making serial dilutions into doubly distilled H_2O, mixing 1 μl of each dilution with 1 μl of 1 μg/ml ethidium bromide on plastic wrap, and UV transilluminating with a standard curve of similarly prepared quantitated DNA.

cDNA Repeat Suppression

cDNA repeat suppression is performed with hybridization to Cot-1 selected DNA (GIBCO/BRL) that has been enriched for sequences hybridizing to a $C_0t_{1/2}$ value of 1. The 2× hybridization buffer[21] is made from stock solutions by combining 300 μl of 1.5 M NaCl, 60 μl of 60 mM (N-[2-hydroxyethyl]piperazine-N'-[3-propanesulfonic acid] (EPPS), pH 8, 20 μl of 10 mM EDTA, pH 7.6, 500 μl of 10× Denhardt's, 60 μl of 1.2% SDS, and 60 μl of doubly distilled H_2O for a total volume of 1 ml. The 5× EE buffer (50 mM EPPS, pH 8.0, 5 mM EDTA) and 10× EE buffer (100 mM EPPS, pH 8.0, 10 mM EDTA) should also be prepared as stock solutions. The hybridization solution should be prewarmed briefly at 65° to solubilize the SDS. For cDNA repeat suppression, the following are mixed in a 0.5-ml Eppendorf tube: 2 μl (2 μg) of cDNA amplicon mixture, 1 μl of Cot-1 DNA (1 μg/μl, GIBCO/BRL), and 1 μl 5× EE buffer for a total volume of 5 μl. Overlay with 50 μl of mineral oil and incubate at 95° for 5 min to denature. Next, add 5 μl 2× hybridization solution and mix by touching the pipette tip to the aqueous droplet under the surface and quickly pipetting once up and down. Incubate for 2 hr at 65°.

First Hybridization Step

Prepare a solution of biotinylated genomic DNA by adding 2.6 μl biotinylated genomic DNA (100 ng), 1 μl Cot-1 DNA (GIBCO/BRL), and 0.4 μl 10× EE buffer. Cover with 40 μl of mineral oil and denature at 95° for 5 min. Quickly add 10 μl prehybridized cDNA and 5 μl of prewarmed 2× hybridization solution. Incubate for 32 hr at 65°.

Binding, Washing, and Elution of cDNA Fragments

cDNA fragments are isolated by magnetic bead capture of the hybridized biotinylated genomic DNAs. Prior to terminating the hybridization, prepare the magnetic bead binding, wash, elution, and neutralization solutions. Prepare 100 μl magnetic beads (Dynabeads M280, Dynal) per hybridization reaction by capturing magnetically and washing beads three times with binding buffer (10 mM Tris–HCl, pH 7.5, 1 mM EDTA, 1 M NaCl).

Capture beads again and resuspend with cDNA hybridization reaction. Place on rotator or agitate gently for 30 min at room temperature. Capture beads and wash twice using 0.5 ml 1× SSC, 0.1% SDS at room temperature, allowing 15 min per wash for nonannealed sequences to be released. Capture beads and wash three times in 0.1× SSC, 0.1% SDS at 65°, allowing 15 min per wash. Elute cDNA from genomic DNA and beads by adding 50 μl 100 m*M* NaOH, allowing 10 min at room temperature. Capture beads magnetically and remove supernatant. Neutralize supernatant with 50 μl 1 *M* Tris–HCl, pH 7.2. Purify by spin column purification (QIAquick PCR purification kit, Qiagen) and elute from column using 50 μl 10 m*M* Tris–HCl, pH 8.8.

cDNA Fragment PCR Amplification

Prepare a 100-μl PCR reaction using 25 μl eluted cDNA template, 1.2 μl 17-mer oligonucleotide (0.5 μg/μl), 16 μl dNTP mix (1.25 m*M* each dNTP), 4 μl $MgCl_2$ (50 m*M*), 10 μl 10× PCR buffer (GIBCO/BRL), and 43 μl doubly distilled H_2O. Perform a hot-start PCR, heating tube to 72° for 1 min prior to adding 0.8 μl *Taq* DNA polymerase and mixing. Thermal cycling reactions are then performed (Perkin Elmer Cetus, Model 9600) for eight cycles with cycle parameters of denaturation at 94° for 30 sec, annealing at 52° for 45 sec, and extension at 72° for 2 min and 30 sec, followed by a terminal extension at 72° for 5 min.

The initial amplification is followed by a second round of amplification as follows. Prepare 12 25-μl PCR reactions containing 1 μl of the initial PCR product, 0.3 μl 17-mer oligonucleotide (0.5 μg/μl), 4 μl dNTP mix (1.25 m*M* each dNTP), 1 μl $MgCl_2$ (50 m*M*), 2.5 μl 10× PCR buffer (GIBCO/BRL), 0.2 μl *Taq* polymerase (1 U, GIBCO/BRL), and 16 μl doubly distilled H_2O. Thermal cycling reactions are performed for 2 to 24 cycles, with parameters of denaturation at 94° for 30 sec, annealing at 52° for 45 sec, and extension at 72° for 2 min and 30 sec per cycle, removing one tube every 2 cycles at the end of the extension. The tubes are replaced for a final extension at 72° for 5 min.

Prepare a 1.4% agarose gel (UltraPure, GIBCO/BRL) and electrophorese 10-μl aliquots of each sample to determine the cycle number at which the product concentration plateaus. Also include a concentration standard curve of *Dpn*II or *Sau*3A-digested genomic DNA and 0.5-μg samples of each amplicon. This gel can be Southern blotted and used to determine enrichment in the first round by probing with a cDNA fragment known to be located within the genomic DNA and the tissue of interest.

Prepare eight 100-μl PCR reactions using the same reagent ratios used for the 25-μl PCR reactions. Amplify for the number of cycles shown by the smaller scale PCRs to be just prior to plateau in product concentration. Combine and extract once with 600 μl phenol : chloroform : isoamyl alcohol. 2-Propanol precipitate using 0.3 *M* sodium acetate as the salt. Resuspend at an assumed concentration of 0.5 μg/μl and electrophorese on a 1.4% agarose gel with a standard curve of *Dpn*II or *Sau*3A-digested genomic DNA to determine the exact concentration.

Second Hybridization Step: Washing and Elution of cDNA Fragments

Prepare a dilution of the first-round selected cDNA at a concentration of 80 ng/μl. Prepare a dilution of the genomic DNA template at a concentration of 8 ng/μl. Combine 0.25 μl genomic DNA dilution (2 ng), 0.25 μl cDNA dilution (20 ng), 4 μl COT-1-selected DNA (GIBCO/BRL), and 0.5 μl 10× EE buffer, covered with 40 μl mineral oil. Incubate for 5 min at 95°, add 5 μl prewarmed 2× hybridization solution, and hybridize for 8 hr at 65°. Washing and elution of the cDNA fragments are performed in a manner identical to the protocol used for the first round of cDNA selection.

Note: The cDNA concentration has been reduced by 100-fold, the genomic DNA concentration has been reduced by 100-fold, the COT-1 DNA concentration has been doubled, and the hybridization time has been decreased 6-fold. The effect of the decreased cDNA concentration and decreased time is to provide a kinetic-style enrichment similar to that used in RDA. Those cDNA fragments that have been enriched during the initial hybridization will be able to reach a $C_0t_{1/2}$ value sufficient for reannealing. The decreased genomic template concentration acts to normalize the concentrations of the selected cDNA products. The much higher relative concentration of COT-1 DNA allows the repetitive sequences present within the COT-1 DNA to hybridize efficiently to the cDNA and genomic DNA, thus eliminating the need for a prehybridization step.

cDNA Fragment PCR Amplification

PCR amplification of the eluted cDNA fragments is performed in a manner identical to the first round of the cDNA selection procedure, including the two-step amplification and the gel analysis of the plateau in product concentration. The gel should include samples from the amplicons, the first-round selection products, and a digested genomic DNA standard curve for concentrations. This gel can be Southern blotted and used to analyze the enrichment achieved in the second round of hybridization. The product with the cycle number just prior to the plateau in product concentration should be prepared on a larger scale for subcloning, according to the same reaction conditions given for the larger scale PCRs for the first round.

Note: In parallel experiments performed with the original protocol,[21] the modified protocol showed a lower complexity, as determined by a marked increase in the intensity of discrete bands above the background smear (data not shown).

Product Subcloning

Subcloning of the products can be accomplished by shotgun subcloning of the large-scale PCR reaction, using standard techniques for blunting the ends of the PCR fragments and subcloning.[25] Alternatively, if a nonredundant set of clones for initial characterization of the products is desired, the PCR reaction can be size fractionated and subcloned as individual bands. This can be accomplished by electrophoresis of 7.5 μg of product in a 60-μl sample on a 1% MetaPhor agarose (FMC) plus 1% UltraPure agarose (GIBCO/BRL) gel with 0.5 μg/μl ethidium bromide in the gel and the 0.5× TBE buffer. Run the gel at 2 V/cm for 2 hr followed by 4 V/cm until the xylene cyanol marker has progressed two-thirds of the length of the gel (overnight for a 24-cm gel). The gel is then UV illuminated with an underlying protective dark sheet. Being careful to minimize the UV exposure, the gel is then moved off the end of the sheet and product bands are pierced quickly with a round, smooth toothpick and innoculated into 50-μl PCR reactions set up in a 96-well format for immediate PCR. Care is taken not to transfer pieces of agarose to the PCR reactions. PCR reaction conditions are 5 μl 10× Pfu PCR buffer (Stratagene, La Jolla, CA), 0.5 μl *Pfu* DNA polymerase (Stratagene), 8 μl dNTP mix (1.25 m*M* each dNTP, Pharmacia), 0.6 μl 17-mer oligonucleotide (0.5 μg/μl), and 35.9 μl doubly distilled H_2O. Thermocycler parameters (Perkin Elmer Cetus, Model 9600) are denaturation at 94° for 30 sec, annealing at 50° for 45 sec, and extension at 75° for 3 min, for 35 cycles, followed by extension at 75° for 5 min. Ten microliters of each PCR product should be run on a 1.4% agarose gel (UltraPure, GIBCO/BRL) for analysis. The remainder (40 μl) of the reaction can be kinased directly in the same microtiter dish by the addition of 10 μl T4 polynucleotide kinase buffer (NEB), 1 μl T4 polynucleotide kinase, 1 μl of 100 m*M* rATP, and 48 μl of doubly distilled H_2O, incubating at 37° for 30 min. Next, 10 μl of 1 *M* Tris–HCl, pH 7.2, is added to decrease the pH and the products are spin column purified (QIAquick PCR purification system, Qiagen) and eluted in 30 μl $T_{10}E_{0.1}$. Kinased products are then ligated into *Sma*-cut dephosphorylated vector according to standard protocols.[25]

Note: Snorthern blot expression analysis provides a very useful secondary screen for expression information when large numbers of individual clones need to be analyzed. Partial sequencing of subcloned products can also serve as a relatively rapid method for secondary screening. Other methods of analyzing products have been reviewed.[21]

Acknowledgments

We thank J. S. Dasen, W. Wu, and J. Bermingham for invaluable discussions and assistance during the development of these protocols.

Section II

Gene Expression Systems for Functional Analysis

[5] Yeast Three-Hybrid System to Detect and Analyze Interactions between RNA and Protein

By BEILIN ZHANG, BRIAN KRAEMER, DHRUBA SENGUPTA, STANLEY FIELDS, and MARVIN WICKENS

Introduction

RNAs seek protein partners and proteins seek RNAs. We try to shepherd an RNA to a polypeptide partner with whom it can interact meaningfully or to guide a protein to an RNA partner with whom physical association will be of real consequence. Many partners emerge, but the number of meaningless interactions can be extraordinarily high. The challenge is to rapidly identify the rare few that are worthy of commitment and then to hone in on one.

The reason we persist in such searches is that RNA–protein interactions are pivotal in a wide variety of fundamental cellular processes. These include translation, mRNA processing, chromosome replication, and key decisions during early development. Additionally, all RNA viruses, including HIV, hepatitis, and picornaviruses, exploit RNA–protein interactions as a means of regulating their infectivity and replication. Facile means to detect and analyze such interactions would be useful in both basic and applied settings.

Finding a partner in an RNA–protein interaction can take many routes. Consider the common case in which an RNA sequence of interest has been identified and its physiological protein partner now is sought. Among the biochemical methods that can be used to find that polypeptide are conventional purifications based on its enzymatic activity or based solely on its ability to bind to the correct RNA sequence. A variety of standard molecular methods, including gel-shift analysis and UV cross-linking, can guide purifications based on RNA-binding activity. RNA affinity chromatography provides a powerful and direct means to an RNA partner and has been applied successfully in several biological contexts.

Genetic methods provide an alternative approach and immediately yield clones encoding candidate proteins. Several such methods have been described. For example, binding of a protein to its cognate site in an mRNA can repress that mRNA's translation *in vivo,* making selections and screens possible in both yeast and bacteria.[1–3] A system has also been described

[1] R. Stripecke, C. C. Oliveira, J. E. McCarthy, and M. W. Hentze, *Mol. Cell. Biol.* **14,** 5898 (1994).

0076-6879/99 $30.00

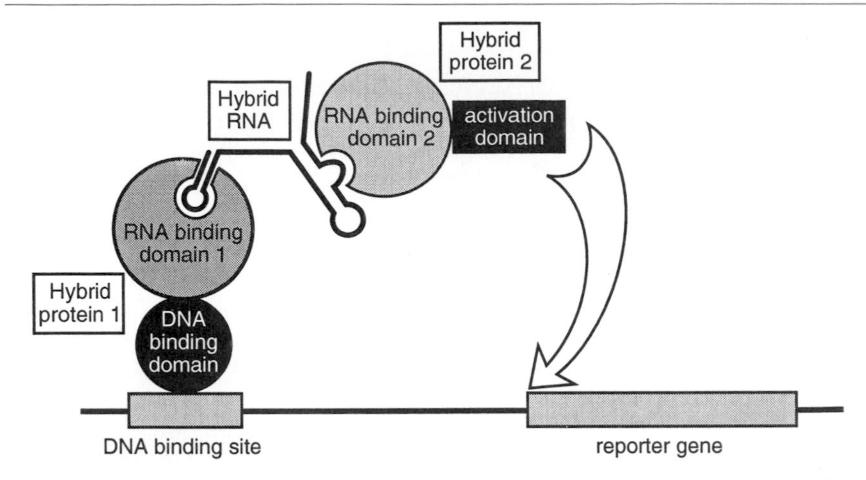

FIG. 1. General scheme of the three-hybrid system. Adapted from SenGupta *et al.*[5]

that exploits the antitermination properties of bacteriophage N protein; when the appropriate portion of N protein is bound to an RNA via an RNA–protein interaction, it causes antitermination, leading to expression of a reporter gene.[4]

This article describes the use of a different genetic approach to detect RNA protein–interactions.[5] In this method, termed the three-hybrid system, the binding of a hybrid RNA to each of two hybrid proteins activates transcription of a reporter gene *in vivo.* The method can be used to detect and analyze RNA–protein interaction when both partners are known or to find a mate when only one is known.[6–8]

Principles of Method

The general strategy of the three-hybrid system is diagrammed in Fig. 1. DNA-binding sites are placed upstream of a reporter gene in the yeast

---

[2] M. P. MacWilliams, D. W. Celander, and J. F. Gardner, *Nucleic Acids Res.* **21,** 5754 (1993).
[3] C. Jain and J. G. Belasco, *Cell* **87,** 115 (1996).
[4] K. Harada, S. S. Martin, and A. D. Frankel, *Nature* **380,** 175 (1996).
[5] D. J. SenGupta, B. Zhang, B. Kraemer, P. Pochart, S. Fields, and M. Wickens, *Proc. Natl. Acad. Sci. U.S.A.* **93,** 8496 (1996).
[6] F. Martin, A. Schaller, S. Eglite, D. Schumperli, and B. Muller, *EMBO J.* **16,** 769 (1997).
[7] Z. F. Wang, M. L. Whitfield, T. C. I. Ingledue, Z. Dominski, and W. F. Marzluff, *Genes Dev.* **10,** 3028 (1996).
[8] B. Zhang, M. Gallegos, A. Puoti, E. Durkin, S. Fields, J. Kimble, and M. P. Wickens, *Nature* **390,** 477 (1997).

chromosome. A first hybrid protein consists of a DNA-binding domain linked to an RNA-binding domain. The RNA-binding domain interacts with its RNA-binding site in a bifunctional ("hybrid") RNA molecule. The other part of the RNA molecule interacts with a second hybrid protein that consists of another RNA-binding domain linked to a transcription activation domain. When this tripartite complex forms at a promoter, even transiently, the reporter gene is turned on. This expression can be detected by phenotype or simple biochemical assays.

The specific molecules most commonly used for three-hybrid screens at the time of writing are depicted in Fig. 2. The DNA-binding site consists of a 17 nucleotide recognition site for the *Escherichia coli* LexA protein and is present in multiple copies upstream of both *HIS3* and *LacZ* genes. Hybrid protein 1 consists of LexA fused to bacteriophage MS2 coat protein, a small polypeptide that binds as a dimer to a short stem–loop sequence in its RNA genome. The hybrid RNA (depicted in more detail in Fig. 5) consists of two MS2 coat protein-binding sites linked to the RNA sequence of interest, X. Hybrid protein 2 consists of the transcription activation domain of the yeast Gal4 transcription factor linked to an RNA-binding protein, Y.

Although the components depicted in Fig. 2 are most commonly used, other RNAs and proteins can be substituted. For example, the histone mRNA 3′ stem loop and the protein to which it binds (SLBP) can replace the MS2 components (B. Zhang and M. Wickens, in press).

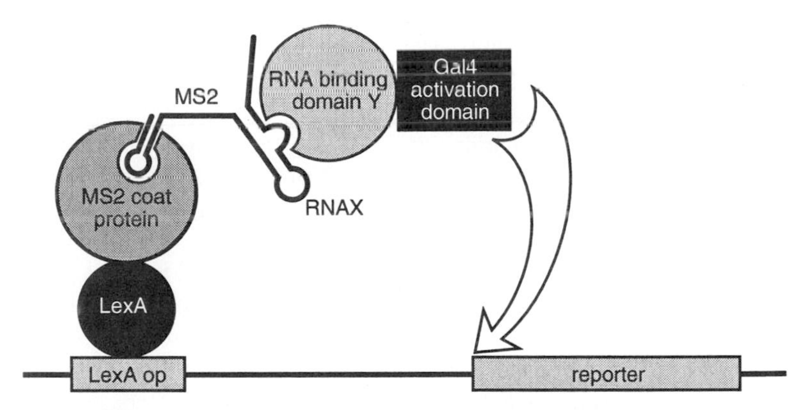

FIG. 2. One array of specific components used commonly in the three-hybrid system. Both *lacZ* and *HIS3* are present in strain L40coat under the control of lexA operators and so can be used as reporters. For simplicity, the following features are not indicated. Multiple LexA operators are present (four in the *HIS3* promoters and eight in the *lacZ* promoter), and LexA protein binds as a dimer. The hybrid RNA contains two MS2-binding sites, and MS2 coat protein binds as a dimer to a single site.

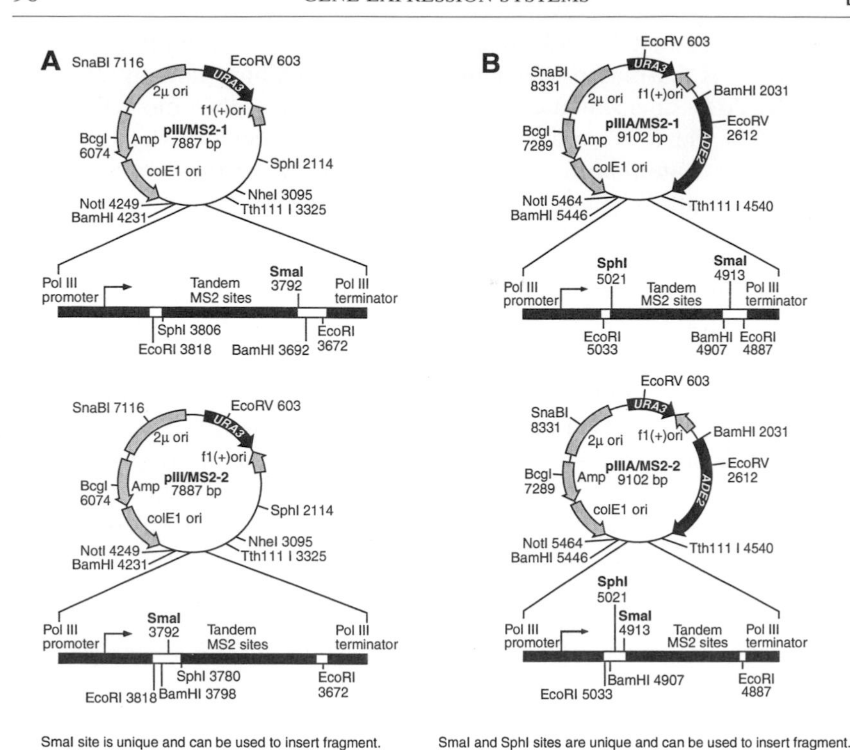

FIG. 3. Plasmid vectors used to express hybrid RNAs in the three-hybrid system. The restriction sites that are indicated are either unique or can be used to verify the presence of various elements on the plasmids. Sites suitable for insertion of sequences of interest are in bold face.

The three-hybrid approach has many of the same strengths and limitations of the two-hybrid system used to detect protein–protein interactions. By introducing libraries of RNA or protein, cognate partners can be identified. As in two-hybrid screens, the challenge then becomes to identify those molecules whose interaction is biologically relevant. For this purpose, mutations in the known RNA or protein component are very useful. The system also makes it possible to identify those regions of an RNA or protein that are required for a known interaction and to test combinations of RNA and protein to determine whether they interact *in vivo.*

Perspective

The three-hybrid system is still in an early stage of development. Many known combinations of RNA and protein partners have been tested, with a very high frequency of success. However, the number of library screens reported to date is small: three successful screens of cDNA libraries have been reported,[6–8] and we have completed a single RNA library screen.

The main thrust of this article is the screening of RNA and protein libraries. However, the system has a number of other potential applications (see Prospects). We offer the protocols here as a starting point for investigators wanting to use the system as is or to develop it further.

Plasmid Vectors

Several plasmids have been constructed to express hybrid RNA sequences and are depicted in Fig. 3. An activation domain plasmid useful in screening RNA libraries, pACTII/CAN, is also shown in Fig. 4. Each of these is a multicopy plasmid containing origins and selectable markers for propagation in yeast and bacteria.

pIII/MS2-1 and pIII/MS2-2

pIII/MS2-1 and pIII/MS2-2 (Fig. 3A) are yeast shuttle vectors derived from PIIIEx426RPR, which was developed by Good and Engelke.[9] Sequences to be analyzed are inserted at the *Sma*I/*Xma*I site. pIII/MS2-1 and pIII/MS2-2 differ only in the relative position of the *Sma*I/*Xma*I site and the MS2-binding sites. Both plasmids carry a *URA3* marker and produce hybrid RNAs from the yeast RNase P RNA (*RPR1*) promoter, an RNA polymerase III promoter. The *RPR1* promoter was chosen for two reasons. First, it is efficient, directing the synthesis of up to several thousand molecules per cell. Second, the transcripts produced from this promoter presumably do not enter the pre-mRNA processing pathway and may not leave the nucleus. An alternative method using an RNA polymerase II promoter to generate the hybrid RNA has also been described.[10]

pIIIA/MS2-1 and pIII/MS2-2

pIIIA/MS2-1 and pIIIA/MS2-2 (Fig. 3B) are similar to pIII/MS2-1 and pIII/MS2-2, but carry the yeast *ADE2* gene (in addition to *URA3*). The *ADE2* gene is exploited in screening for RNA-binding proteins (see later). The two plasmids differ from one another only in the relative position of the *Sma*I/*Xma*I site and the MS2-binding sites.

[9] P. D. Good and D. R. Engelke, *Gene* **151,** 209 (1994).
[10] U. Putz, P. Skehel, and D. Kuhl, *Nucleic Acids Res.* **24,** 4838 (1996).

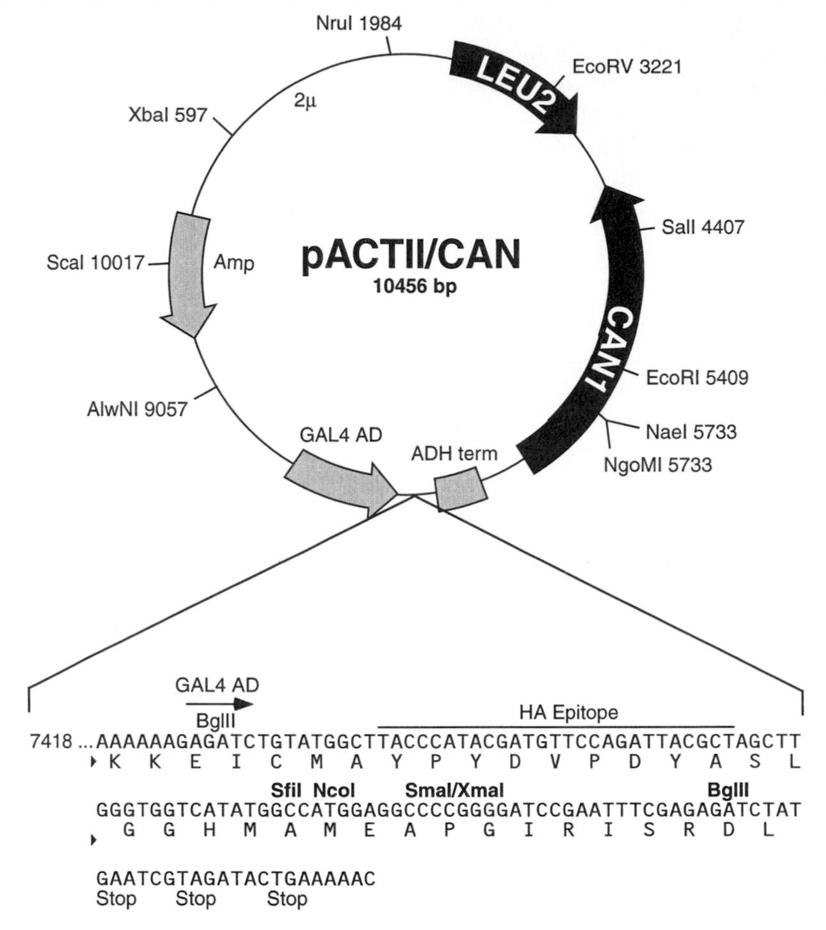

SfiI, NcoI, SmaI/XmaI sites are unique and can be used to insert fragment. BglII also can be used, but will result in loss of the HA epitope.

FIG. 4. Vector used to express the protein "bait" in an RNA library screen. Restriction sites are either unique or can be used to verify the presence of various elements on the plasmid. Sites suitable for insertion of the "bait" protein are in bold face.

pACTII/CAN

The vector pACTII-CAN (Fig. 4) has been generated using the pACTII backbone,[11] by inserting the yeast *CAN1* gene at the unique *Sal*I site of pACTII. The protein sequence to be tested is inserted at the *Sfi*I, *Nco*I,

[11] C. Bai and S. J. Elledge, *in* "The Yeast Two-Hybrid System" (P. L. Bartel and S. Fields, eds.), p. 298. Oxford Univ. Press, London, 1997.

or *Sma*I/*Xma*I sites. The presence of the *CAN1* gene on the plasmid, which encodes an arginine permease, causes cells to die in media containing the arginine analog canavanine. Thus, canavanine selection leads to the loss of this plasmid in a yeast strain that carries a chromosomal *can1* allele, which confers resistance. In this fashion, canavanine can be used to select for loss of the activation domain plasmid, whereas 5-FOA can be used to select for loss of the hybrid RNA plasmid.

Hybrid RNAs

General Considerations

The hybrid RNA and the tripartite nature of the system present unique considerations for three-hybrid vs two-hybrid screens. Among these are issues concerning the length of the RNA that can be analyzed, the ability of cDNA/activation domain proteins to activate transcription independent of binding to the RNA, and of RNAs to activate transcription independent of the cDNA/activation domain protein. These issues are discussed in the following relevant sections.

RNA sequences to be tested are inserted into the unique *Sma*I/*Xma*I site of any of the four RNA vectors depicted in Fig. 3. The hybrid RNA molecule that is transcribed *in vivo* from one of these plasmids consists of the sequence of interest, X, linked to two MS2 coat protein-binding sites and RNase P RNA 5′ leader and 3′ trailer sequences (Fig. 5). The MS2-binding sites are present in two copies because binding to coat protein is cooperative. Each site contains a single nucleotide change that enhances binding.[12]

In our experience, most RNA inserts of suitable size produce RNAs of comparable and high abundance. RNA abundance may be important in determining the level of signal produced from the reporter gene: transferring the hybrid RNA gene from normal high-copy vectors to low-copy vectors reduces levels of *LacZ* expression substantially.

The relative order of the RNA sequence of interest and the MS2 sites can affect signal strength. In the few cases that have been tested systematically, both orientations yield activation and are specific. However, in the IRE/IRP interaction, placing the IRE upstream of the MS2 sites results in two- to threefold more transcription than the alternative arrangement. Although RNA folding programs can be used to determine whether one arrangement is more likely to succeed, the accuracy of their predictions *in*

[12] P. T. Lowary and O. C. Uhlenbeck, *Nucleic Acids Res.* **15,** 10483 (1987).

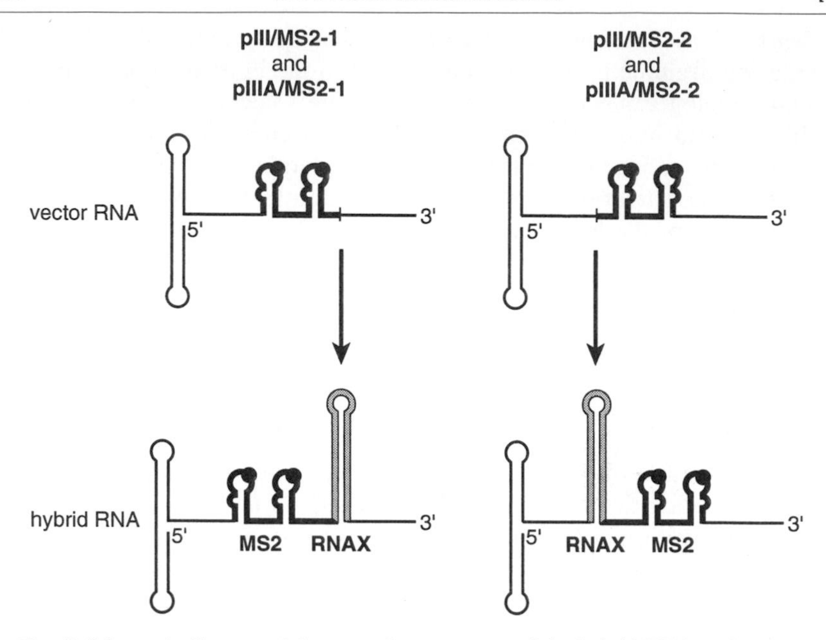

FIG. 5. Schematic diagram of the secondary structure of the hybrid RNA molecules. Thin lines, RNaseP RNA 5′ leader and 3′ trailer sequences; bold lines, tandem MS2 recognition sites, including a point mutation that increases affinity (black dot); gray lines, inserted RNA sequence, depicted as a step loop.

vivo is problematic. In most gases tested, we have placed the RNA sequence of interest upstream of the MS2 sites.

Limitations

RNA sequences to be analyzed are restricted in two respects at present. First, runs of four or more U's in succession can terminate transcription by RNA polymerase III. Second, typically, RNA inserts of lengths less than 150 nucleotides yield higher signals: longer inserts commonly reduce the level of activation of the reporter (see General Considerations). In principle, both of these limitations might be overcome by using a different polymerase, such as that of a bacteriophage.

Runs of U's. Four or more U's in succession can function as RNA polymerase III terminators and can prevent production of the desired hybrid RNA. The efficiency of termination at oligouridine tracts is context dependent, however, so it may be worth testing whether a suspect RNA sequence will function in the system. Northern blotting should always be performed to make certain that the hybrid RNAs are expressed at high levels. For long runs of U's, it may be necessary to eliminate the terminator by mutagenesis.

Size of RNA Element. The size of the RNA insert appears to be an important determinant of three-hybrid activity. In reconstruction experiments using known RNA protein partners, RNA sequences that are 30–100 nucleotides in length (e.g., TAR, IRE) typically yield substantial and specific reporter activation. We have investigated the effects of additional RNA sequences flanking a known protein-binding site. The addition of heterologous sequences to the IRE caused a reduction in the IRE-IRP1 three-hybrid signal. The effect was proportional to the size of the insertion, with the addition of 150–200 nucleotides commonly leading to almost complete abolition of the IRE-IRP1 three-hybrid signal. Similarly, we have tested the effect of an additional natural RNA sequence on the ability to detect the interaction of the yeast Snp1 protein, a homolog of the mammalian U1-70K protein, with the loop I of yeast U1 RNA. Insertion of even the RNA sequences that normally flank loop I, but are not part of the Snp1-binding site, decreased the three-hybrid signal due to Snp1–U1 interaction. Taken together, these experiments suggest that a minimal binding site is preferable and that the optimal length of the RNA insert is less than 150 nucleotides.

Yeast Strains

The yeast reporter strain L40coat is derived from L40-ura3 (a gift of T. Triolo and R. Sternglanz, Stony Brook, NY). The genotype of the strain is *MATa, ura3-52, leu2-3,112, his3Δ200, trp1Δ1, ade2, LYS2::(lexA op)-HIS3, ura3::(lexA-op)-lacZ, LexA-MS2 coat (TRP1)*. The strain is auxotrophic for uracil, leucine, adenine, and histidine. Each of these markers is exploited in the three-hybrid system. Both the *HIS3* and the *lacZ* genes have been placed under the control of *lexA* operators, and hence are reporters in the three-hybrid system. A gene encoding the LexA–MS2 coat protein fusion has been integrated into the chromosome.

Strain R40coat is identical to L40coat, but of opposite mating type.

A canavanine-resistant derivative of L40coat, L40coat-can, carries a *can1* allele. This strain can be rendered canavanine sensitive by transformation with a plasmid such as pACTII/CAN, and it becomes resistant again on loss of the plasmid when plated on media containing canavanine.

Testing Known or Suspected Interactors

The protocol to test a known or suspected RNA–protein interaction is straightforward. The RNA sequence of interest and the gene encoding the

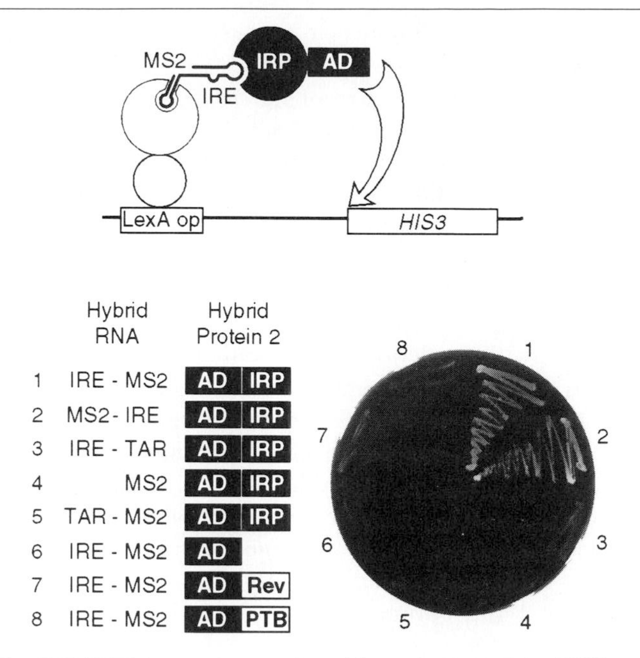

FIG. 6. The IRE/IRP1 interaction monitored by activation of the *HIS3* gene. Plasmids encoding the indicated hybrid RNAs and activation domain fusions were transformed into strain L40coat. After selecting transformants for the presence of plasmids, colonies were restreaked onto media selecting for expression of *HIS3*. Only the two combinations that should lead to IRE/IRP1 interactions (lines 1 and 2) yield growth.

RNA-binding protein are cloned into an RNA plasmid and an activation domain vector, respectively. These are then introduced into yeast by transformation, and the level of expression of a reporter gene is determined. Figure 6 depicts such an analysis of the IRE–IRP1 interaction, using selection for *HIS3* to monitor the interaction.[5,13] Results shown demonstrate that each segment of the system is required for activation of the reporter, which is monitored by growth on selective media.

This approach can be useful in testing candidate interactors obtained by other means. For example, an interaction between mouse telomerase RNA and a newly cloned telomerase protein was confirmed in this fashion (TP1).[14] Similarly, binding of yeast ribosomal protein S9 with *CRY2* pre-

[13] B. Zhang, B. Kraemer, D. SenGupta, S. Fields, and M. Wickens, *in* "The Yeast Two-Hybrid System" (P. L. Bartel and S. Fields, eds.), p. 298. Oxford Univ. Press, London, 1997.

[14] L. Harrington, T. McPhail, V. Mar, W. Zhou, R. Oulton, M. B. Bass, I. Arruda, and M. O. Robinson, *Science* **275,** 973 (1997).

mRNA, deduced by genetic experiments, has been demonstrated using the system (S. Fewell and J. Woolford, personal communication, in press).

In principle, the protocol also provides a facile way to delineate the portions of the RNA or protein that are required for an interaction. Regions required for RNA-binding of FBF1 to its RNA target were inferred through such an analysis.[8] In such experiments, it is important to confirm that protein or RNA is present in cases in which the interaction no longer appears to function. This confirmation can be accomplished by either Northern or Western blotting. In principle, it should be possible to identify single amino acids and nucleotides that affect the interaction.

Finding Protein Partner for Known RNA Sequence

The three-hybrid system can be used to identify a protein partner of a known RNA sequence. A library containing many cDNAs is introduced into a yeast strain that carries a plasmid that encodes the RNA sequence of interest as a hybrid RNA. From such screens emerge cDNAs that are capable of activating the reporter, some of which require the hybrid RNA to do so. As in other such methods, secondary and tertiary screens to eliminate several categories of false positives are crucial.

If the RNA–protein interaction is particularly strong, as in the case of SLBP and its stem–loop target in histone mRNA, then the initial selection can demand high levels of expression of the reporter gene.[5,6] A stringent selection eliminates weak activators. However, a less stringent selection is often preferable because the "strength" of the interaction being sought is not known. In turn, this reduced stringency leads to a higher background and an increased need for subsequent screens.

The following protocol has been used to clone novel RNA-binding proteins.

Step 1. Introduce the RNA Plasmid and cDNA Library

The host strain, L40coat, is normally transformed with the RNA plasmid first. Cells containing the RNA plasmid are then transformed with a cDNA library fused with a transcription activation domain. (Hybrid RNAs are rarely toxic to the host cell, unlike some hybrid protein baits used in two-hybrid screens; as a result, there is no need to cotransform both plasmids.) The transformation mix is plated out on media lacking both leucine (selecting for the cDNA plasmid) and histidine (selecting for *HIS3* gene expression). Maintenance of the RNA plasmid (i.e., selection for *ADE2* or *URA3*) is not demanded. This permits cells that can activate *HIS3* without the RNA to lose the RNA plasmid, permitting the colony color screen below (see "step 2").

Protein Libraries. Activation domain libraries prepared in *LEU2* vectors can be used with the hybrid RNA plasmids and yeast strains described earlier. Many such activation domain libraries have been prepared for use with the two-hybrid system. These may be obtained from individual laboratories and several commercial sources.

3-Aminotriazole in the Initial Selection. We typically add 3-aminotriazole (3AT), a competitive inhibitor of the *HIS3* gene product, to select for stronger interactions. Some RNAs activate reporter genes weakly on their own, and many proteins appear to activate the reporter slightly, independent of the hybrid RNA; both situations yield "false positives." To eliminate weak activation by the RNA "bait," 3AT should be titrated using a strain carrying only the RNA plasmid prior to undertaking the initial transformation. To diminish the number of false protein positives ("RNA-independent" positives; see later), concentrations of 3AT in the range of 2 to 5 m*M* are a good starting point; they offer a reasonable balance between suppressing the background and permiting "real" positives to grow. *In vitro* data on the affinity of the RNA–protein interaction may be valuable on determining the concentration of the 3AT that should be used. In one of the screens that yielded SLBP, whose interaction with its RNA target is particularly robust *in vitro*, 25 m*M* 3AT was included, reducing the background greatly.[6]

Step 2. Eliminate RNA-Independent False Positives by Colony Color

Two classes of positives are obtained from the initial transformation. One class of transformants requires the hybrid RNA to activate *HIS3;* these are termed "RNA dependent." A second class of positives activates *HIS3* with or without the hybrid RNA; these are termed "RNA independent." RNA-independent positives can carry proteins that may bind to the promoter regions of the reporter genes or proteins that interact directly with the LexA-MS2 coat protein fusion. The RNA-independent class of transformants can be very abundant and can account for more than 95% of the total number of colonies.

To facilitate eliminating RNA-independent false positives, we exploit the *ADE2* gene on the RNA plasmid. The host strain is an *ade2* mutant. When the level of adenine in the medium becomes low, cells attempt to synthesize adenine *de novo* and accumulate a red purine metabolite due to lack of the *ADE2*-encoded enzyme. This accumulation renders the cell pink or red in color. In contrast, cells carrying the wild-type *ADE2* gene are white.

In the initial transformation, selection is imposed only for activation of *HIS3*, not for maintenance of the RNA plasmid. For RNA-dependent positives, selection for *HIS3* indirectly selects for the RNA plasmid, which carries the *ADE2* gene; thus these transformants are white and must remain so if they continue to grow in the absence of exogenous histidine. However, RNA-independent positives do not require the RNA plasmid to activate the *HIS3* gene and so can lose that plasmid, which they do with a frequency of a few percent per generation. These false positives therefore yield pink colonies or white colonies with pink sectors.

The initial transformation plates are usually incubated at 30° for a week. This duration allows positives to accumulate *HIS3* gene product and grow, and also provides enough time for the color to develop. If the pink color is not strong after a week, incubation at 4° overnight sometimes helps. Pink or pink-sectored colonies are discarded, and the uniformly white colonies are picked for further analysis. We usually pick all the white colonies (typically a few large, and many small, colonies) and patch onto media selecting again for both the cDNA plasmid and the RNA plasmid. Most of the small white colonies turn out to be RNA independent and fail to grow. White colonies that are able to grow on the selective media are subject to further analysis.

The identification of RNA-dependent positives by colony color is not perfect; many, but not all, RNA-independent positives can be identified and discarded. A majority of the white colonies may still be RNA independent. It is important to rigorously eliminate the remaining RNA-independent activators from among the white colonies (steps 4 and 5) before recovering plasmids in *E. coli.*

Step 3. Assay β-Galactosidase Activity

To corroborate that the white colonies contain cDNAs that activate through the three-hybrid system, the level of expression of the *lacZ* gene is monitored. In strain L40coat, the *lacZ* gene is integrated into the chromosome and placed under the control of LexA-binding sites.

β-Galactosidase can be assayed by measuring the conversion of a lactose analog to a chromogenic or luminescent product. This assay can be performed using colonies permeabilized on either a filter or a cell lysate. The filter assay yields qualitative results, whereas the liquid assay is more quantitative.

Qualitative (Filter) Assays[15]

1. Restreak colonies from step 2 onto the appropriate selective media (SD-Leu-Ura) and grow overnight.

[15] L. Breeden and K. Nasmyth, *Cold Spring Harb. Symp. Quant. Biol.* **50,** 643 (1985).

2. Replica colonies onto plate-sized nitrocellulose filters or filter papers (Whatman, Clifton, NJ, 3MM).
3. Immerse filter in liquid nitrogen for 20 sec.
4. Allow filter to thaw on bench top (approximately 2 min).
5. Prepare petri dish-sized circles of 3MM Whatman paper, place in petri dish, and saturate with Z buffer (60 m*M* Na_2HPO_4, 40 m*M* NaH_2PO_4, 10 m*M* KCl, 1 m*M* $MgSO_4$, 50 m*M* 2-mercaptoethanol, pH 7.0), supplemented with 300 μg/ml X-Gal. The X-Gal should be added fresh. Remove excess buffer.
6. Overlay filter onto Whatman paper and seal dish with Parafilm.
7. Incubate for 30 min to overnight at 30°. Examine the filters regularly.

A strong interaction (such as that between IRE and IRP) should turn blue within 30 min. With protracted incubation, weak interactions eventually yield a blue color. For this reason, it is important to examine the filters periodically to determine how long it takes for the color to develop.

Quantitative (Liquid) Assays. The specific activity of β-galactosidase can be determined in yeast cell lysates using any of a variety of substrates. Colorimetric assays using ONPG[16] or CPRG[17] are common; CPRG is more sensitive, but also more expensive. Alternatively, luminescent substrates provide high sensitivity yet are relatively inexpensive. The following protocol uses a luminescent substrate, Galacton-Plus (Tropix, Bedford, MA). The assay requires an instrument to detect luminescence. The protocol has been designed for a Monolight 2010 luminometer (Analytic Luminescent Laboratories, San Diego, CA). Certain details of the assay, such as sample volumes, will vary with the instrument used.

1. Inoculate 5-ml cultures of selective media in triplicate for each interaction to be tested. Grow overnight.
2. Inoculate fresh selective media to an OD of 0.1
3. Grow to midlog phase (OD ~0.8).
4. Pellet ~1.0 OD units worth of cells for each culture.
5. Resuspend pellet in 100 μl of lysis buffer (100 m*M* potassium phosphate, pH 7.8, 0.2% Triton X-100).
6. Lyse by freeze-thaw. This lysis requires three sequential cycles of freezing in liquid nitrogen for 10 sec followed by incubation at 37° for 90 sec.
7. Vortex each tube briefly.
8. Pellet in microfuge at 12,000*g*.
9. Collect supernatant for luminometer assays. If necessary, samples may be frozen at −70°.

[16] J. H. Miller, *in* Cold Spring Harbor Laboratory, 1972.
[17] K. Iwabuchi, B. Li, P. Bartel, and S. Fields, *Oncogene* **8,** 1693 (1993).

10. Add 10 μl of lysate to luminometer tube. For strong interactions, it may be necessary to dilute the lysate to keep the assay in the linear range.

11. Dilute Galacton reagent 1:100 in reaction buffer (100 m*M* sodium phosphate, 1 m*M* magnesium chloride, pH 8.0). Add 100 μl to each luminometer tube.

12. Incubate at 25° for 60 min.

13. Measure luminescence as directed by luminometer manufacturer. Use light emission accelerator II from Tropix.

14. Measure protein concentration in lysates by Bradford or equivalent assay.

15. Normalize light emission by protein concentration, yielding a specific activity.

Step 4. Cure RNA Plasmid and Test Again

Most but not all of the RNA-independent false positives are eliminated by the colony color screen in step 2. To ensure that positives are genuinely RNA dependent, the RNA plasmid is removed by counterselection against *URA3*. Expression of the reporter gene is then monitored. Candidates that fail to activate the reporter genes are analyzed further.

URA3 Counterselection. To select for cells that have lost the RNA plasmid, cells are plated on media containing 5-fluoroorotic acid (5-FOA). 5-FOA is converted by the *URA3* gene product to 5-fluorouracil, which is toxic. Cells lacking the *URA3* gene product can grow in the presence of 5-FOA if uracil is provided, whereas cells containing the *URA3* gene product cannot.

1. Replica plate the positives from step 2 to SD-Leu plates. Let the cells grow for a day, allowing cells to lose the plasmid.

2. Replica plate onto SD-Leu + 0.1% 5-FOA plates. Incubate at 30° for a few days.

3. Cells that grow can be streaked on SD-Ura plate to confirm the loss of the RNA plasmid. A single pass through 5-FOA counterselection is usually sufficient.

4. Assay β-galactosidase activity.

The *ADE2* marker on the RNA plasmid can be useful in monitoring the loss of the RNA plasmid. Cells that lose the plasmid will turn pink in a few days. If the number of positives is small, then the use of 5-FOA, which is quite expensive, can be avoided. Cells can simply be grown in rich media overnight, then spread onto SD-Leu plates. After a few days, some of the colonies become pink or show pink sectors. Uniformly pink colonies, which have lost the RNA plasmid, can be reassayed for β-galactosidase activity.

Step 5. Determine Binding Specificity Using Mutant and Control RNAs

To test RNA-binding specificity, reintroduce plasmids encoding various hybrid RNAs into the strains that have been cured of their original RNA plasmid. If the number of positives is small, the various RNA plasmids can be introduced by transformation. Otherwise, mating is used to introduce the plasmids. Strain R40coat can be used for this purpose. A sample protocol for the mating assay follows.

1. Grow lawns of separate R40coat transformants carrying a specific hybrid RNA plasmid (mutant vs wild type, for example) on SD-Ura plates.
2. Replica plate the grid of Ura^- colonies from the 5-FOA plate to a YPD plate.
3. Replica plate the lawn from each R40coat strain to the same YPD plate.
4. Incubate the plates overnight at 30° to allow mating.
5. Replica plate to SD-Leu-Ura plate to select for diploids.
6. Assay β-galactosidase activity.

What RNAs Should Be Used in Specificity Tests? Ab initio, positives that have survived step 4 carry proteins that bind preferentially to the hybrid RNA relative to cellular RNAs. Although these positives recognize some features on the hybrid RNA, these need not be the ones that are recognized by the biologically relevant factor(s). Therefore, the ideal controls are subtle (e.g., point) mutations that affect the biological functions or interaction in a sequence-specific manner. In some cases, an antisense bait can be used to control for secondary structure or base composition.

What Fraction of All RNA-Dependent Positives Is "Correct?" The fraction of positives that are physiologically relevant is unpredictable at the outset, as it is a function of the abundance of the protein (in a random library screen), strength of the interaction, and level of 3AT used in the initial transformation, among other parameters. The following guidelines are based on a published screen.

We screened a *Caenorhabditis elegans* cDNA activation domain library using a portion of the *fem-3* 3′UTR as bait, leading to the identification of a regulatory protein, FBF.[8] In this screen, 5 m*M* 3AT was used in the initial transformation, selecting for reasonably strong interactions. In total, 5,000,000 transformants were analyzed, yielding approximately 5000 His^+ Leu^+ colonies. Of these, 100 were white. Sixty of these activated the *lacZ* gene. Twelve of the 60 proved to be genuinely RNA dependent, and 3 of the 12 displayed the appropriate binding specificity. Each of these was FBF, the genuine regulator. Thus, RNA-dependent positives were 0.2% of the total number of colonies obtained in the initial transformation. Importantly,

25% of all RNA-dependent positives were FBF. In screens that identified the protein (SLBP) that binds to the 3′ end of histone mRNAs,[6,7] "correct" positives represented 70 to 100% of the total His^+ $LacZ^+$ transformants.

What if No Subtle RNA Mutations Are Available? Additional screens must be devised to identify those positives that are correct. Clearly, functional tests are ideal, but in many organisms and systems these are time-consuming and labor-intensive. The sequence of the cDNAs may be directly informative by comparison to known RNA-binding proteins or by comparison to the predicted molecular weight of the expected protein (based on, for example, UV cross-linking). Because each case is idiosyncratic, no general discussion will be offered here. However, we caution that such secondary screens are critical.

Step 6. Identify Positive cDNAs

cDNA plasmids that display the predicted RNA-binding specificity are recovered from the yeast cells and introduced into *E. coli* by transformation. The yeast cells can contain multiple cDNA plasmids, only one of which encodes the protein that binds to the RNA. Thus, plasmids should be isolated from multiple *E. coli* transformants and reintroduced into yeast to ensure that the correct plasmid has been obtained. A sample protocol follows. All steps are performed at room temperature.

1. Use a toothpick to transfer a yeast colony to 50 μl lysis solution (2% Triton X-100, 1% SDS, 100 m*M* NaCl, 10 m*M* Tris–Cl, pH 8.0, 1 m*M* EDTA).
2. Add 50 μl phenol/chloroform and ca. 0.1 g acid-washed glass beads.
3. Vortex at high speed for 2 min.
4. Spin at high speed in a microcentrifuge for 5 min.
5. Transfer the supernatant to a clean tube and ethanol precipitate the DNA. Wash with 70% ethanol and dry the pellet.
6. Resuspend the pellet in 10 μl H_2O. Use 1 μl to transform electrocompetent *E. coli*.

For convenience, to determine how many plasmids are present in each yeast transformant, we often perform PCR reactions using lysed yeast colonies. To do so, we use primers that flank the inserts. The following is a protocol for yeast colony PCR.[18]

1. Touch the yeast colony with a sterile disposable pipette tip.
2. Rinse the tip in 10 μl incubation buffer [1.2 *M* sorbitol, 100 m*M* sodium phosphate, pH 7.4, and 2.5 mg/ml zymolyase (ICN Biomedicals,

[18] M. Ling, F. Merante, and B. H. Robinson, *Nucleic Acids Res.* **23,** 4924 (1995).

Inc., Costa Mesa, CA)] by pipetting up and down several times. Incubate at 37° for 5 min.

3. Remove 1 μl for a 20-μl PCR reaction. If desired, the PCR product can be purified by chromatography or purification from a gel and sequenced directly.

Step 7. Functional Tests or Additional Screens

Almost invariably, additional steps will be needed to identify those positives that are biologically meaningful. As stated earlier (see step 5), those screens are idiosyncratic, depending on the interaction and organisms studied. In a screen with an RNA "bait," it is not surprising that one might identify irrelevant RNA-binding proteins, as well as the legitimate interactor.

Finding RNA Partner for Known RNA-Binding Protein

The three-hybrid system can be used to identify a natural RNA ligand for a known RNA-binding protein by screening an RNA library with a protein/activation domain fusion as "bait." Although only one such screen has been performed to date,[19] the objective is sufficiently general that we include discussion of our experience here.

A strain that carries a plasmid expressing an activation domain fusion with a known RNA-binding protein is transformed with a library of plasmids expressing hybrid RNAs, each composed of the MS2 coat protein-binding sites fused to an RNA element. In our initial experiments, the library consisted of fragments of yeast genomic DNA that are transcribed along with the coat protein-binding sites. We demonstrated that we could identify a fragment of the U1 RNA that binds to the yeast Snp1 protein in this type of search.[19] This hybrid RNA library should be useful in assigning RNA ligands to RNA-binding proteins of *Saccharomyces cerevisiae,* particularly those that have been classified as RNA binding solely on the basis of primary sequence data.

A three-hybrid search to identify RNAs follows much the same logic as that described earlier to identify proteins. One particular class of false positives that must be eliminated consists of RNAs that are able to activate expression of the reporter genes on their own, without interacting with the protein fused to the activation domain (i.e., "protein-independent" positives).[19] A step must be included to classify RNAs as protein dependent, and thus worthy of additional analysis, or protein independent, and thus of no further interest in this context. Additionally, the protein-dependent

[19] D. SenGupta, M. Wickens, and S. Fields, *RNA,* in press.

class of RNAs must be tested with the activation domain vector and other RNA-binding protein fusions to identify those RNAs that are specific to the protein of interest.

Step 1. Introduce Activation Domain Plasmid and Hybrid RNA Library

1. Transform L40coat with the plasmid expressing the RNA-binding protein as an activation domain hybrid, based on either the pACTII or the pACTII-CAN vector.
2. Transform cells from a single colony carrying this plasmid with the RNA library, selecting on media lacking tryptophan, leucine, uracil, and histidine and containing 0.5 m*M* 3AT. Without any 3AT, most of the transformants will grow on a plate lacking histidine.

Construction of RNA Library. The yeast RNA library used in our experiments was constructed as follows.

Chromosomal DNA from *S. cerevisiae* was partially digested with the following four enzymes, listed with their recognition sequences: *Mse*I (TTAA), *Tsp*509I (AATT), *Alu*I (AGCT), and *Rsa*I (GTAC). The digests were pooled and fragments in the size range of 50 to 150 nucleotides were purified from a preparative agarose gel. The ends of the digested DNA were filled in with the Klenow fragment of DNA polymerase I where required. The plasmid pIII/MS2-2 was digested with *Sma*I, treated with calf intestine phosphatase, and ligated to the blunt-ended genomic DNA fragments. The ligations were used to transform electrocompetent HB101 *E. coli.* DNA fragments cloned at the *Sma*I site of the pIII/MS2-2 are expressed such that the RNA sequence corresponding to the yeast genomic fragment is positioned 5′ to the MS2 coat protein-binding sites within the hybrid RNA. More than 1.5 million *E. coli* transformants were obtained, pooled, and used to prepare plasmid DNA for the RNA library.

Step 2. Screen for Activation of Second Reporter Gene

To ensure that activation of *HIS3* is not spurious, the level of expression of the *LacZ* gene is monitored.

1. Patch individual colonies that grew in the library transformation onto plates lacking leucine, tryptophan, and uracil.
2. Carry out filter β-galactosidase assays as described earlier.

Step 3. Eliminate Protein-Independent False Positives

The identification of protein-independent false positives requires two successive steps. First, the activation domain plasmid must be removed from the strain. Then the level of LacZ expression must again be determined. Colonies that are LacZ$^+$ are cured of the activation domain plasmid by one of the following methods. If the plasmid is derived from pACTII, the plasmid is cured by growing a transformant overnight in YPD media and then plating for single colonies on SD-Ura to select for the RNA plasmid. These colonies are then replica plated onto an SD-Leu plate to determine which of the colonies lack the *LEU2* marker on the activation domain plasmid. If the activation domain plasmid is derived from pACTII/CAN, the plasmid can be cured by patching colonies onto a canavanine plate. For complete curing of the plasmid, a second patching onto a canavanine plate is generally required. Canavanine is prepared as a stock solution of 20 mg/ml, which is filter sterilized. The selective plates are SD-Arg with 60 μg/ml canavanine.

Assay cells are cured of the activation domain plasmid for β-galactosidase activity by filter assays, using the protocols described earlier. Colonies that have lost activity on loss of the activation domain plasmid contain possible RNA ligands of interest.

How Common Are Protein-Independent Activators? Protein-independent activators include hybrid RNAs that activate transcription when bound to a promoter. In our experiments to date, the frequency of such "activating RNAs" in a genomic library is rather high. For example, 92% of all His$^+$ LacZ$^+$ positives obtained after selection in 0.5 m*M* 3AT were protein independent.

Step 4. Determine Binding Specificity

As in cDNA library screening, specificity tests are necessary in secondary screens to narrow down the number of candidates. Of course, the more subtle the mutation used, the better.

Transform cells from step 3 that contain an RNA plasmid and appear to be protein dependent for three-hybrid activity with control activation domain plasmids such as pACTII and an IRP1-activation domain fusion in pACTII. An RNA ligand that is specific for the RNA-binding protein used in the library screen should not produce a three-hybrid signal with these control plasmids.

Step 5. Sequence RNAs of Interest

Determine the sequence of the RNA ligand and test the binding by alternative methods such as *in vitro* binding.

What Fraction of Protein-Dependent Activators Is "Correct?" In our experience with yeast Snp1 protein as "bait," we screened 2.5×10^6 transformants and obtained 13 that were protein dependent. Of these, the strongest by far was the appropriate segment of U1snRNA. Some of the other positives have weak sequence similarity to the relevant region of U1 and are now being analyzed further.

Prospects

The main focus of this article has been the use of the three-hybrid system to identify partners in an RNA–protein interaction. In principle, with minimal modifications, the system may be modified to detect factors that enhance or prevent an RNA–protein interaction, to identify RNA ligands that enhance or prevent a protein–protein interaction, and to detect an RNA–RNA interaction. Such applications have not yet been reported. As stated earlier, we have written this article in part to facilitate the development of the system by others, as well as to facilitate its use as is. In that spirit, we have included screening of both protein and RNA libraries, although relatively few such experiments have yet been published.

One asset of the three-hybrid method, shared with other genetic approaches to finding protein partners, is that a DNA clone of the interacting protein is obtained. With the proliferation of sequence data bases and genomic information, a small bit of sequence information may be enough to help determine whether that protein is a legitimate partner or shed light on its function. In some organisms, obtaining the clone immediately enables direct tests of its function *in vivo*. In others, alternative screening steps will be needed to identify meaningful mates from among the assembled suitors.

Acknowledgments

We are grateful to the Media Laboratory of the Biochemistry Department of the University of Wisconsin for their help with figures. We also appreciate the helpful comments and suggestions of members of both the Wickens and the Fields laboratories and communications of results from several laboratories prior to publication. Work in our laboratories is supported by grants from the NIH and NSF. S.F. is an investigator of the Howard Hughes Medical Institute.

[6] Determining Subunit Dissociation Constants in Natural and Recombinant Proteins

By LOIS R. MANNING, ANTOINE DUMOULIN, W. TERRY JENKINS, ROBERT M. WINSLOW, and JAMES M. MANNING

Introduction

Dissociation of oligomeric proteins usually leads to changes in their functional properties; for example, with hemoglobin the cooperativity of

0076-6879/99 $30.00

oxygen binding to the tetramer is lost when it dissociates into dimers.[1] Subunit dissociation in proteins depends on the strength of the interfaces between the subunits, which varies considerably for different proteins. The extent to which such dissociations can be measured, i.e., to tetramers, dimers, or monomers, depends on the resolution of the separation procedure and the sensitivity of the detection system. In the procedure described in this article, the tetramer to dimer dissociation for hemoglobin is measured. Mutant hemoglobins, either natural variants or recombinant proteins with amino acid substitutions in the $\alpha_1\beta_2$ subunit interface where tetramer–dimer dissociation occurs, are especially prone to changes in their oxygen-binding affinity. With the advent of recombinant expression systems that produce hemoglobin mutations at any position in either α or β chains, the availability of a rapid and simple method to estimate tetramer dissociation constants on small amounts of protein would be advantageous. The procedure described here, i.e., small zone gel filtration on Superose-12, fulfills this need. When coupled to the Pharmacia (Piscataway, NJ) FPLC (fast protein liquid chromatography) software to measure the peak characteristics accurately, data are treated mathematically to obtain additional information. When applied both to natural and to recombinant hemoglobins containing single or multiple substitutions, this procedure has yielded new insights concerning hemoglobin function. Nonheme proteins could be analyzed by this method by monitoring the effluent at 280 nm with the online detection system of the instrument.

Hemoglobins

Hemoglobin A (HbA, sometimes referred to as HbA_o) is purified as described previously.[2] Hemoglobin F (HbF, sometimes called HbF_o), which was kindly donated by Dr. Robert Bookchin, is isolated from postpartum umbilical cord blood; after purification by FPLC on a Mono S (Pharmacia) column, it is >95% pure as ascertained by isoelectric focusing; no acetylated HbF_1 is detected.[3] Amino acid analysis gives the correct composition for $\alpha_2\gamma_2$, including 8 Ile, an amino acid absent in HbA. Analysis by mass spectrometry shows the correct molecular weight (α chain 15,126; γ chain 15,995, kindly performed at Rockefeller University, New York, by Dr. Urooj Mirza and Dr. Brian Chait). HbF_1, purified from the same source, elutes before HbF on a Pharmacia Mono S column and shows a single

[1] M. F. Perutz, *Quart. Rev. Biophys.* **22,** 139 (1989).

[2] L. R. Manning, W. T. Jenkins, J. R. Hess, K. Vandegriff, R. M. Winslow, and J. M. Manning, *Protein Sci.* **5,** 775 (1996).

[3] A. Dumoulin, L. R. Manning, W. T. Jenkins, R. M. Winslow, and J. M. Manning, *J. Biol. Chem.* **272,** 31326 (1997).

band on isoelectric focusing. It has the correct mass (α chain 15,126 calc.; 15,126 $\pm$ 3 found; γ-Ac chain 16,037 calc.; 16,042 $\pm$ 7 found), thus confirming the presence of a single acetyl group per γ chain. Natural HbS is purified from the red cells of sickle cell anemia patients.[4,5] The recombinant hemoglobins, HbS and the double-sickle mutant L88A(β), are expressed and purified as described previously.[6] In the L88A(β) double mutant, part of the receptor region for the mutation site Val-6(β) on an adjacent tetramer is replaced by Ala. Two recombinant hemoglobins with substitutions in the $\alpha_1\beta_2$ interface, Asp-99(β) $\rightarrow$ Lys[D99K(β)] and Asn-102(β) $\rightarrow$ Ala [N102A(β)], are also expressed and purified as described.[7,8] The naturally occurring hemoglobins Yakima [Asp-99(β) $\rightarrow$ His] and Rothschild [Trp-37(β) $\rightarrow$ Arg] were generously provided by Dr. Ruth Benesch. Each gives a single band by isoelectric focusing, and the molecular weight of the latter is confirmed by mass spectrometry. The preparation and characterization of the cross-linked tetrameric DBBF-HbA with a fumaryl group covalently linked between the two Lys-99(α) side chains in the central dyad axis[9] and DIBS-HbA, which is cross-linked between the N-terminals of the two α-chains,[10] have been described. Bovine Hb is purified from whole bovine blood (Pel Freeze Biological, Rogers, AR) using the same procedure as for HbA.[10]

Measurement of Subunit Dissociation

Superose-12 is a gel filtration support that separates mainly on the basis of molecular weight, but the shape of the protein is also a factor. However, it does possess some residual ionic properties so that the charge on a given protein and the extent to which it interacts with buffer salts could affect its elution time. Therefore, it is necessary to calibrate the column with nondissociating tetrameric, dimers, or monomeric standards corresponding to the protein under study. Two sites of standards are employed to determine the elution position of undissociated hemoglobin tetramers and di-

[4] J. J. Martin de Llano, W. Jones, K. Schneider, B. T. Chait, G. Rodgers, L. J. Benjamin, B. Weksler, and J. M. Manning, *J. Biol. Chem.* **268,** 27004 (1993).

[5] J. J. Martin de Llano, O. Schneewind, G. Stetler, and J. M. Manning, *Proc. Natl. Acad. Sci. U.S.A.* **90,** 918 (1993).

[6] J. J. Martin de Llano and J. M. Manning, *Protein Sci.* **3,** 1206 (1994).

[7] H. Yanase, S. Cahill, J. J. Martin de Llano, L. R. Manning, K. Schneider, B. T. Chait, K. M. Vandegriff, R. M. Winslow, and J. M. Manning, *Protein Sci.* **3,** 1213 (1994).

[8] H. Yanase, L. R. Manning, K. Vandegriff, R. M. Winslow, and J. M. Manning, *Protein Sci.* **4,** 21 (1995).

[9] R. Chatterjee, E. V. Welty, R. Y. Walder, S. L. Pruitt, P. H. Rogers, A. Arnone, and J. A. Walder, *J. Biol. Chem.* **261,** 9929 (1986).

[10] L. R. Manning, S. Morgan, R. C. Beavis, B. T. Chait, J. M. Manning, J. R. Hess, M. Cross, D. L. Currell, M. A. Marini, and R. M. Winslow, *Proc. Natl. Acad. Sci. U.S.A.* **88,** 3329 (1991).

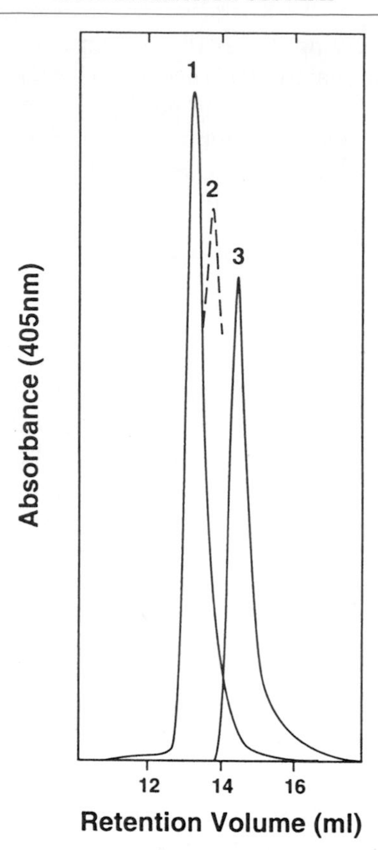

FIG. 1. Elution profile of dissociable hemoglobins. Peak 1 is the elution position of cross-linked tetrameric Hb, DBBF-Hb, or DIBS-Hb; peak 2 is the variable position of dissociable hemoglobin tetramers dependent on the hemoglobin concentration; and peak 3 is the position of dimeric hemoglobin determined with Hb Rothschild. Calculation of hemoglobin concentration is described in the text and by Manning *et al.*[2]

mers. Purified and well-characterized cross-linked hemoglobin A tetramers, DBBF-Hb[9] or DIBS-Hb,[10] are used to determine the elution position of undissociated tetrameric Hb; several concentrations ranging from 8.0 nM to 1.8 μM elute at the same peak positions, 13.28 $\pm$ 0.04 ml (three determinations) (Fig. 1, peak 1). The peak elution position of the $\alpha\beta$ dimer is determined with the natural dimeric mutant Hb Rothschild [Trp-37(β) $\rightarrow$ Arg], which is completely dissociated to dimers in the oxy confirmation[11,12] or

[11] G. Gacon, O. Belkhodja, H. Wajcman, and D. Labie, *FEBS Lett.* **82,** 243 (1977).
[12] V. S. Sharma, G. L. Newton, H. M. Ranney, F. Ahmed, J. W. Harris, and E. H. Danish, *J. Chem. Phys.* **51,** 1856 (1980).

with the D99K(β) recombinant dimeric Hb.[7] Concentrations ranging from 4.3 to 34.5 μM elute at the same positions, 14.28 ± 0.05 ml (three determinations) (Fig. 1, peak 3).

Dissociable hemoglobin in either the O_2 or the CO form is applied to a Superose-12 HR 10/30 column on a Pharmacia FPLC system and eluted with 150 mM Tris–acetate buffer, pH 7.5, at a flow rate of 0.4 ml/min.[2] For HbA, 1–2 mg is used for a complete analysis (see later). Different concentrations of hemoglobin, each in 100 μl, are applied, and the absorbance of the eluent is measured at 405 nm with the Pharmacia on-line mercury lamp detection system with a 5-mm flow cell. Alternatively, the detector could be set at 280 nm for proteins that do not contain chromophores. Dissociable hemoglobins elute as a single peak whose position varies between tetrameric and dimeric hemoglobins (Fig. 1, peak 2) when the concentration is in the range of the K_d value. With higher concentrations, the peak position of dissociable hemoglobins is closer to that of the tetrameric Hb (peak 1), whereas at lower Hb concentrations, it is closer to the dimer position (peak 3). All elution profiles have comparable shapes representing a mixture of dimeric and tetrameric hemoglobin in rapid equilibrium. The percentage of tetramer present can be estimated readily from the elution position of the peak (V) relative to the elution volumes for the dimer (V_d) and tetramer (V_t) [Eq. (3), later]. To obtain recorded peaks of comparable size, both the voltage setting on the recorder and the absorbance setting on the monitor are varied depending on the Hb concentration applied. Peak positions are recorded accurately by the FPLC software; these are the values used in the calculations rather than the absolute absorbance values, which are used to determine high recovery of sample. At the same low concentration, fetal HbF behaves more like the tetrameric species than adult hemoglobin A, which is dimeric at this concentration[3] (Fig. 2), as described in detail later. This difference is a good example of the high resolving capacity of the Superose-12 column.

Calculation of Concentration, Analysis Time, and Care of Column

Elution volumes are analyzed in terms of the average peak concentrations as follows: the total peak volume is estimated to be equivalent to the volume at half-peak height. The average peak concentration is thus (sample volume) × (sample concentration)/(peak volume). The use of these average peak concentrations yields dissociation constants consistent with those obtained by other methods (see later). There is, however, some uncertainty about the correct effective concentration to use in the calculations as the concentration varies throughout the peak. If the effective concentration is different than the average peak concentration, this will affect the K_d values.

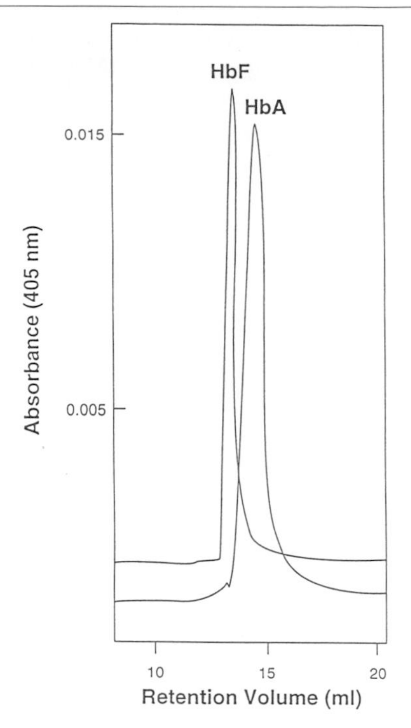

FIG. 2. Elution positions of HbA and HbF. Each hemoglobin was applied at a concentration of 0.1 μM as described in the text and by Dumoulin *et al.*[3]. HbF is tetrameric and HbA is predominantly dimeric as calculated from their K_d values in Figs. 4A and 4B. The detector was set at 405 nm.

Because this possible error affects all the samples to the same extent, ratios of K_d values for different hemoglobins may be more precise than individual K_d values. At hemoglobin concentrations higher than K_d values, peak width is constant (Table I). There is five- to sevenfold overall dilution during the separation (Table I, Fig. 6), most of which appears to occur prior to the sample entering the column (see later). Much of the small degree of dilution is due to mixing both before and after passage through the column and *not* to dilution during gel filtration on Superose-12. Because the spectrum of the eluted hemoglobin is identical to that of the original sample, it is concluded that there is no dissociation of heme from the globin or oxidation to methemoglobin (metHb). The absorbance of the dimer is assumed to be the same as an equivalent weight of the tetramer. A column analysis takes about 1 hr. For each K_d determination, 8–10 such analyses are required, including the standards to determine tetramer and dimer peak elution times. The same Superose-12 column has been used for 2–3 years in our laboratory with practically no change in calibration retention vol-

TABLE I
EFFECT OF SAMPLE VOLUME AND HEMOGLOBIN CONCENTRATION ON ELUTION PROFILES[a]

[Hb] loaded (μM)	Volume loaded (μl)	Peak position (ml)	Peak width at half-height (μl)
A. Constant concentration/varying volume			
20.0	100	13.19	510
20.0	50	13.26	510
20.0	25	13.24	520
B. Varying concentration/varying volume			
10.0	100	13.26	510
20.0	50	13.19	520
40.0	25	13.23	510
		Average: 13.23 ± 0.04	513 ± 7

[a] Either a constant HbA concentration was loaded in varying volumes (A) or varying HbA concentrations were loaded in varying volumes (B) to evaluate the effect on peak position and peak width. In B, the concentration and volume were varied as necessary so that the same total amount of Hb was loaded each time.

umes; when not in use for periods longer than 1 day, it is stored in aqueous 20% ethanol. Periodic cleaning is performed as recommended by Pharmacia.

The buffer used routinely for the study of tetramer–dimer dissociations of the hemoglobins tested is made by adjusting the pH of 150 mM Tris base to pH 7.5 with glacial acetic acid. For lower pH values, the appropriate amount of glacial acetic acid is added. Amino acid analysis after acid hydrolysis of the eluted protein gives the correct amino acid composition for hemoglobin. The concentrations of hemoglobins, which are determined by amino acid analysis of hydrolyzed samples, agree within 3% with the concentrations determined by their visible spectra. For K_d analyses, accurate serial dilutions are made with the Tris–acetate buffers.

Elution Profiles and Curve Analysis

The elution times are highly reproducible for different amounts of the tetrameric and dimeric species (described later), as well as for the same concentration of each dissociable Hb. Within experimental error, the peak heights are found to be related directly to the amount of Hb injected. Thus, a plot of log of detector response versus the log of the Hb concentration injected gives a straight line with slope near unity (Fig. 3). In the concentration range studied, the peaks are somewhat skewed on the trailing edge. There are two possible origins of this elution peak asymmetry. First, it

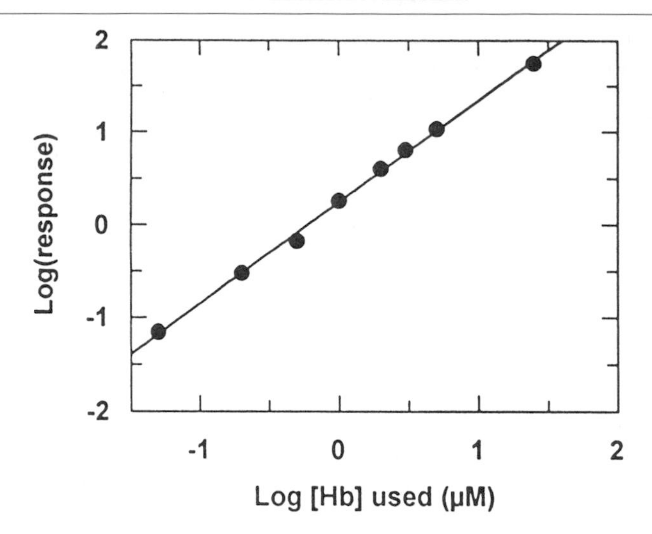

FIG. 3. Detector response as a function of applied hemoglobin concentration. Experimental procedures are described in the text and by Manning *et al.*[2]

could be of instrumental origin due mainly to mixing before and after passage through the column and possibly also due to the electrical damping of the detection system. The second possibility, in the case of equilibrating mixtures of tetramer and dimer species, could be due to retardation at the lower protein concentrations on the leading and trailing edges of the peak due to the relatively greater amount of the dimeric species present at low protein concentrations. We conclude that the observed asymmetry is largely of instrumental origin. The peak width at half-height is somewhat larger near the K_d value for each hemoglobin, as discussed later (Fig. 6).

Dissociation constants are determined by nonlinear regression analyses of the plots of elution volume vs the average peak hemoglobin concentrations. The validity of such analyses is indicated by the fact that, when all the parameters are allowed to vary, the estimated elution volumes for dimers and tetramers are identical to those observed experimentally with calibration standards. Data analysis for proteins for which undissociable standards are not available can be performed in this manner. As shown later, the percentages of tetramer for dissociable hemoglobins can also be calculated from the peak elution volumes of dimers (V_d) and tetramers (V_t) determined with the calibration standards (Fig. 1). The percentages can be transformed to yield linear log/log plots as shown in the figure insets. Such plots have unit slopes and pass through the abscissa ($Y = 1$) at a volume equivalent to log K_d.

Evaluation of K_d Values

The percentage of tetramer present is calculated from the elution curves as follows: Assuming that the elution volume varies with log (molecular weight) and the molecular weight of the dimeric species is M, it follows that

$$\log(2)/(V_d - V_t) = \log(\text{mol. wt./M})/(V_d - V) \tag{1}$$

where V_d, V_t, and V are the elution times for the dimeric, tetrameric, and a mixture of species, respectively. For a mixture of dimers and tetramers, the effective molecular weight is the "weight average" molecular weight, which is given by

$$\begin{aligned} MW_t &= \Sigma(\text{Ni.Mi}^2)/\Sigma(\text{Ni.Mi}) \\ &= M[4(\%T) + 2(100 - \%T)]/[2(\%T) + 2(100 - \%T)] \\ &= M(1 + \%T/100) \end{aligned} \tag{2}$$

where M is the molecular weight of the dimeric species and $\%T$ is the percentage of the tetramer present. Combining Eqs. (1) and (2) gives both

$$\%T = 100(M_W/M - 1) = 100(2^{(V_d - V)(V_d - V_t - 1)}) \tag{3}$$

and

$$V = V_d - (V_d - V_t)[\log(1 + \%T/100)/\log 2] \tag{4}$$

The tetramer dissociation constant K_d is estimated as follows: If the maximum amount of hemoglobin tetramer is [H] and the concentrations of dimeric and tetrameric species are [D] and [T], respectively, so that $\%T = 100[T]/[H]$, it follows that

$$\begin{aligned} K_d &= [D]^2/[T] = 4([H] - [T])^2/[T] \\ &= [(100 - \%T)^2[H]/(25\%T)] \\ &= 0.04(100 - \%T)^2[H]/\%T \end{aligned} \tag{5}$$

Hence

$$\log(K_d) = \log[H] - \log\left(\frac{(\%T)}{0.04(100 - \%T)^2}\right) \tag{6}$$

Thus, a plot of $\log[(\%T/0.04[100 - \%T)^2]$ with respect to log[H] yields a straight line of slope 1. When $[(\%T/0.04(100 - T)^2] = 1$, $K_d = [H]$. The quadraic equation [Eq. (6)] can also be solved in terms of $(\%T)$:

$$\%T = (8[H] + K_d) - (K_d^2 + 16\,K_d[H])^{1/2}/(0.08[H]) \tag{7}$$

From Eq. (7), one can conclude that when $[H] = K_d$, then $\%T = 61\%$ and hence from Eq. (4) that $(V_d - V)/(V_d - V_t) = 69\%$. These values can be used to obtain reliable estimates for K_d for nonlinear regression analyses. Because the elution volume varies empirically with the log of the effective molecular weight, it should be noted that one does not get 50% tetramer when $(V_d - V) = (V_d - V_t)/2$, but rather, from Eq. (3), the elution peak position appears midway between the dimer and tetramer peak positions, when $\%T = 100(\sqrt{2} - 1) = 41.4\%$.

Equations (4) and (7) are used to evaluate K_d and the errors in K_d by nonlinear regression analyses of the values of the elution volumes (V) with respect to the hemoglobin concentrations ([H]). These nonlinear regression analyses are performed with the GraFit PC software program. It should be noted that one cannot use simple linear regression with data plotted as in the figure insets because the variances are not comparable and, furthermore, the analysis dictates that data should lie on a line of unit slope [Eq. (6)].

Potential Systematic Errors

Because the values calculated for K_d depend on the values for the elution volumes of the dimeric (V_d) and tetrameric (V_t) species, we routinely reevaluate V_d and V_t with standard tetrameric and dimeric hemoglobins with each set of results. However, if a wide range of hemoglobin concentrations, [H], is used ($4K_d < [H] < K_d/4$), it is possible to evaluate not only K_d but also V_d and V_t by curve fitting experimental elution volume data for the unknown. This latter approach is used if the protein either interacts to some small degree with the Superose support or calibrating standards for that particular protein are unavailable. When we analyze data in this way, we find very good agreement with the values of V_d and V_t determined with standards. For example, data in Fig. 4C yield values of $K_d = 0.35 \pm 0.11$ μM, $V_d = 14.31 \pm 0.04$ ml, and $V_t = 13.35 \pm 0.05$ ml, compared to the standard dimeric and tetrameric hemoglobins, which give $V_d = 14.3$ ml and $V_t = 13.3$ ml, respectively. However, unless there is a large number of data points, the calculated errors for K_d are much greater if three parameters are allowed to vary than when two of the parameters (V_d and V_t) are determined independently from the standard dimeric Hb Rothschild and tetrameric cross-linked Hb. Furthermore, it is not always possible to use a high enough Hb concentration to convert most of it to the tetrameric form [e.g., the recombinant Hb N102A(β)][2]. The analysis just given pertains to the dissociation of tetramers only to dimers. The good agreement between calculated V_d and V_t values and those obtained with authentic standards provides evidence that the assumption is valid under the experimental conditions described.

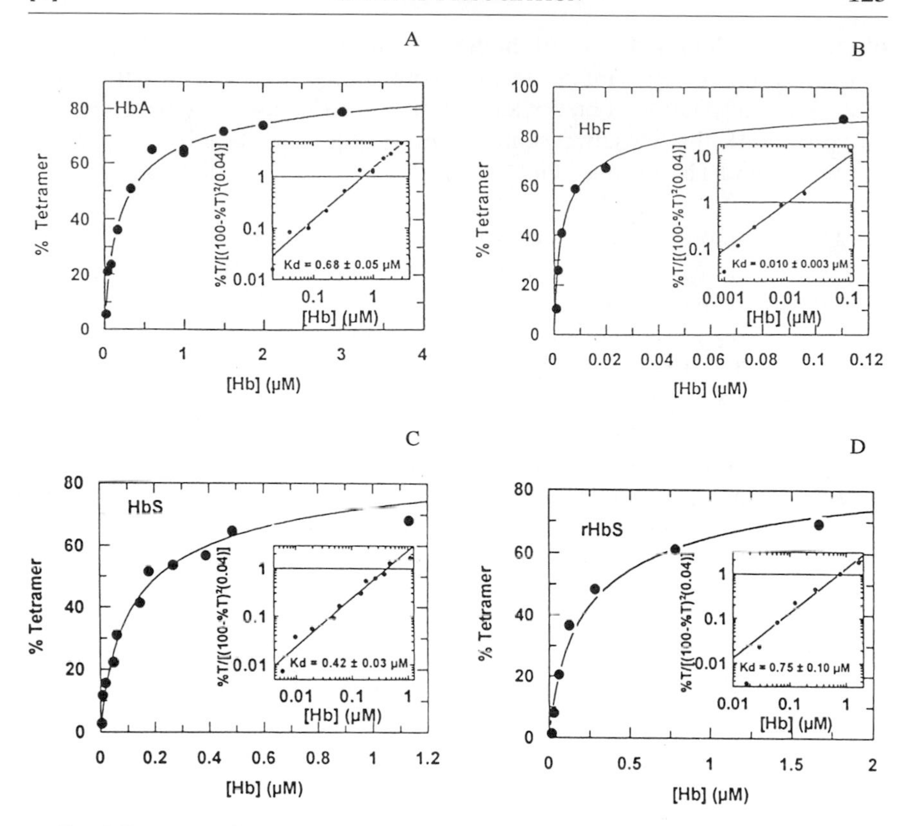

FIG. 4. Tetramer–dimer K_d values of liganded HbA (A), HbF (B), HbS (C), and recombinant HbS (D). Procedures used for these calculations are described in detail in the text and by Manning *et al.*[2]

Hemoglobin A and Hemoglobin S

Adult hemoglobin A and sickle hemoglobin were chosen to evaluate the reliability of the K_d values obtained by this procedure because results for them have been reported by Williams and Kim[13] by a completely independent method, i.e., ultracentrifugal analysis, and also by Ackers and colleagues[14,15] using large-zone gel filtration. For these two hemoglobins, about 200 μl of a 100 μM solution (as tetramer), or 1–2 mg, is used to

[13] R. C. Williams, Jr. and H. Kim, *Biochemistry* **15,** 2207 (1976).

[14] G. K. Ackers, *Proteins* **1,** 2 (1975).

[15] G. J. Turner, F. Galacteros, M. L. Doyle, B. Hedlund, D. W. Pettigrew, B. W. Turner, F. R. Smith, W. Moo-Penn, D. L. Rucknagel, and G. K. Ackers, *Protein Struct. Funct. Genet.* **14,** 333 (1992).

obtain a K_d. Hemoglobins with higher or lower K_d values require proportionally more or less sample, respectively. A plot of the percentage Hb tetramer as a function of hemoglobin concentration gives a hyperbolic-like curve (Fig. 4), that closely resembles the profile described by Andrews,[16] by Ackers and Thompson,[17] and by Chiancone *et al.*[18] for hemoglobin. The figure inset shows the linear transformation of these data.

The tetramer–dimer dissociation constant of liganded HbA is shown in Fig. 4A. Its K_d value, 0.68 μM, obtained by mathematical analysis of data (inset), is in good agreement with values obtained by several different procedures, e.g., Turner *et al.*[15] reported a value of 1.1 μM and Fronticelli *et al.*[19] reported a value of 0.2 μM using the large-zone exclusion method on a Sephadex support, and Benesch and Kwong[20] found 0.7 μM using a method dependent on heme dissociation from dimers. From data of Williams and Kim,[13] who used the highly precise ultracentrifugal method, a K_d value of 0.3 μM for HbA at pH 7.0 can be calculated. The range of published values for the K_d of HbA is consistent with the validity of the high-resolution method described here and the mathematical analysis of data.

For HbS, the K_d is determined to be 0.42 μM (Fig. 4C). Because K_d values for hemoglobin are increased in the presence of chloride and at lower pH values,[3,21] the value we found for HbS is in general agreement with the results reported using ultracentrifugal analysis by Williams and Kim[13] at pH 7.0.

Hemoglobin F

To measure the dissociation constant of liganded HbF, much lower concentrations than those used for HbA are required, as suggested by the results in Fig. 2. At pH 7.5, a K_d value of 0.01 μM is calculated (Fig. 4B), indicating that the HbF tetramer is about 70 times less dissociated to dimers than HbA. The results shown in Fig. 2 indicate that at the same concentrations (0.1 μM), HbF elutes close to the position of undissociable cross-linked tetrameric Hb, whereas HbA elutes close to the dimeric Hb Rothschild,[3] indicating that tetrameric fetal hemoglobin (HbF) does not dissociate to dimers as readily as adult hemoglobin A. The different dissociations for these hemoglobins explain an earlier report that there is no detectable

[16] P. Andrews, *Biochem. J.* **91,** 222 (1964).
[17] G. K. Ackers and T. E. Thompson, *Biochemistry.* **53,** 342 (1965).
[18] E. Chiancone, L. M. Gilbert, G. A. Gilbert, and G. L. Kellett, *J. Biol. Chem.* **243,** 1212 (1968).
[19] C. Fronticelli, M. Gattoni, A. L. Lu, S. W. Brinigar, J. L. G. Bucci, and E. Chiacone, *Biophys. Chem.* **51,** 53 (1994).
[20] R. E. Benesch and S. Kwong, *J. Biol. Chem.* **270,** 13785 (1995).
[21] A. H. Chu and G. K. Ackers, *J. Biol. Chem.* **256,** 1199 (1981).

equilibration between HbF ($\alpha_2\gamma_2$) and HbH (β_4) to form $\alpha_2\beta_2$ under conditions where β_4 equilibrated readily with other hemoglobins containing $\alpha\beta$ subunits.[22] Similarly, Hb Bart's (γ_4) is found to be tightly associated and not subject to equilibrium with other hemoglobins. Frier and Perutz[23] have solved the structure of deoxy HbF and located the five amino acid differences between the $\alpha\beta$ and the $\alpha\gamma$ subunit contact sites in the tetramer. However, it is not currently known which are responsible for the significantly decreased dissociation of HbF compared to HbA reported earlier or whether additional factors are involved. To our knowledge, the structure of oxyHbF has not been reported. We have expressed a penta-substituted recombinant hemoglobin (HbA/F) that contains the HbF interface amino acids and the remaining sequence of HbA.[3] Its tetramer–dimer dissociation value is 0.14 μM, which is between that of HbF and HbA. The reasons for this behavior are currently under study.

Hemoglobin F_1

The presence of an acetyl group on the N-terminal of each of its γ chains has a profound effect on the tetramer–dimer dissociation of liganded HbF_1; it has a K_d of 0.33 μM.[3] Hence, this specific acetylation nearly completely reverses the 70-fold decreased tetramer–dimer dissociation of HbF (Fig. 4B).

Recombinant Sickle Hemoglobins

Recombinant sickle hemoglobin, expressed, purified, and characterized as described previously,[4–6] has a K_d value of 0.75 μM (Fig. 4D); this value is very similar to that of natural HbS (Fig. 4C). These results are consistent with our earlier conclusion that hemoglobin expressed in the yeast system is identical to the corresponding natural hemoglobin purified from erythrocytes[4–6] and forms the basis for studies on the other recombinant hemoglobins described in this article.

The aggregation of sickle Hb in the deoxy conformation is initiated by the interaction of the external Val-6(β) substitution with a hydrophobic site comprising Leu-88(β) on an adjacent tetramer.[4,5] In order to ascertain the strength of this primary site of tetramer aggregation, we have reported that its substitution by Ala results in a significant increase (30%) in the concentration needed for gelation, i.e., aggregation is impeded.[6] Although it is the deoxyHbS tetramers that aggregate in the erythrocyte, any increased

[22] E. R. Huehns, G. H. Beaven, and B. L. Stevens, *Biochem. J.* **92,** 440 (1964).

[23] J. A. Frier and M. F. Perutz, *J. Mol. Biol.* **112,** 97 (1977).

dissociation of the oxyHb would affect the overall equilibrium to give a decreased concentration of deoxyHbS tetramers and hence a reduced extent of aggregation. Because it is conceivable that the amino acid replacement of Ala for Leu-88(β) affects its structure in the oxy conformation to give an increased tetramer–dimer dissociation, it is important to determine the K_d of this recombinant HbS L88A(β) double mutant. The value is found to be 0.56 μM, which is of the same order of magnitude as the values for natural and recombinant HbS described earlier. These results show that this amino acid replacement in the double mutant has little effect on the dissociation behavior of the $\alpha_1\beta_2$ interface in the oxy conformation of hemoglobin. Thus, the amino acid substitution in the recombinant Hb affects the aggregation process directly at the level of the tetrameric interaction.

Recombinant Hemoglobins with Substitutions in $\alpha_1\beta_2$ Subunit Interface

Two recombinant hemoglobins D99K(β) and N102A(β), with substitutions in the $\alpha_1\beta_2$ interface at Asp-99(β) and Asn-102(β), respectively, were expressed in order to study their roles in the oxy $\rightarrow$ deoxy transition.[7,8] Using the method described earlier, the oxygenated form of the N102A(β) mutant is about 400 times more dissociated than HbA, like the natural Hb Kansas with a Thr at position 102 of the β chain.[24]

For the D99K(β) recombinant Hb,[7] there is no concentration dependence of the peak elution position as a function of hemoglobin concentrations ranging from 0.05 to 4.2 μM. At each concentration, the peak elution is in the position of the $\alpha\beta$ dimer, with no indication of any movement into the position of the tetramer.

The natural hemoglobin mutant, Hb Yakima, was chosen for study because its substitution is at the same site as that for the D99K(β) recombinant Hb, but with a His rather than a Lys.[7] Its K_d value, 0.10 μM, indicates that it is less dissociated than HbA, in agreement with the value of 0.06 μM reported for this natural Hb mutant by Benesch and Kwong[20] and 0.09 μM found by Turner *et al.*[15] Thus, Hb Yakima displays dissociation properties that bear little similarity to those of the D99K(β), a mutant at the same site. Even though both substitutions contain side chains that bear some positive charge, it is apparently their relative size that causes the extreme disparity in their dissociation behavior.

[24] J. Bonaventura and A. Riggs, *J. Biol. Chem.* **243,** 980 (1968).

Bovine Hemoglobin

Bovine hemoglobin is of interest because it is very stable to the conditions of alkali denaturation. In order to determine its tetramer–dimer dissociation, we used the standard calibrating proteins described earlier for HbA. A K_d value of 0.62 μM was found (Fig. 5), showing that bovine Hb dissociates to a similar extent as HbA. Hence its marked stability toward alkali is not due to a highly stable tetrameric structure.

Discussion

Because small zone gel filtration studies on Sephadex supports have been criticized because of the possibility of sample dilution on the column and peak broadening leading to erroneous K_d values,[14] we have tested the extent to which this occurs on the Superose-12 support used in our studies. Two types of analysis were performed (Table I): one using Hb concentrations within the K_d range and another using Hb concentrations higher than the K_d. The values for peak widths and positions were determined by the FPLC software packages; reproducibility was 1% or better. Data in Table

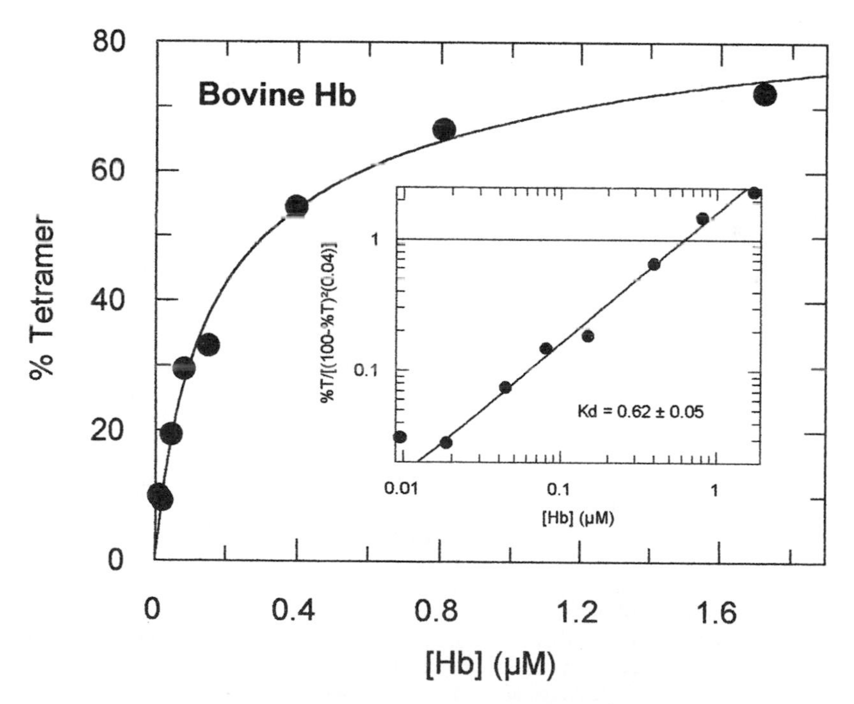

FIG. 5. Tetramer–dimer K_d value of bovine Hb.

I show that neither peak position nor peak width is influenced by sample load or volume consistent with the absence of dilution during the process of gel filtration on Superose-12. Erroneous K_d values due to changes in peak positions and widths reported for Sephadex support as a function of the sample volume and concentration[14] are thus not encountered to a significant extent on the Superose supports under these conditions.

There is a detectable increase in peak width for liganded HbF, HbF_1 and HbA within the K_d concentration range (Fig. 6) and it occurs maximally at concentrations close to their K_d values determined from the peak position (Fig. 6, arrows). Minimum peak widths occur when dimers or tetramers are the predominant species. Bell-shaped curves are found between these two extrema, i.e., at concentrations where maximal equilibration between

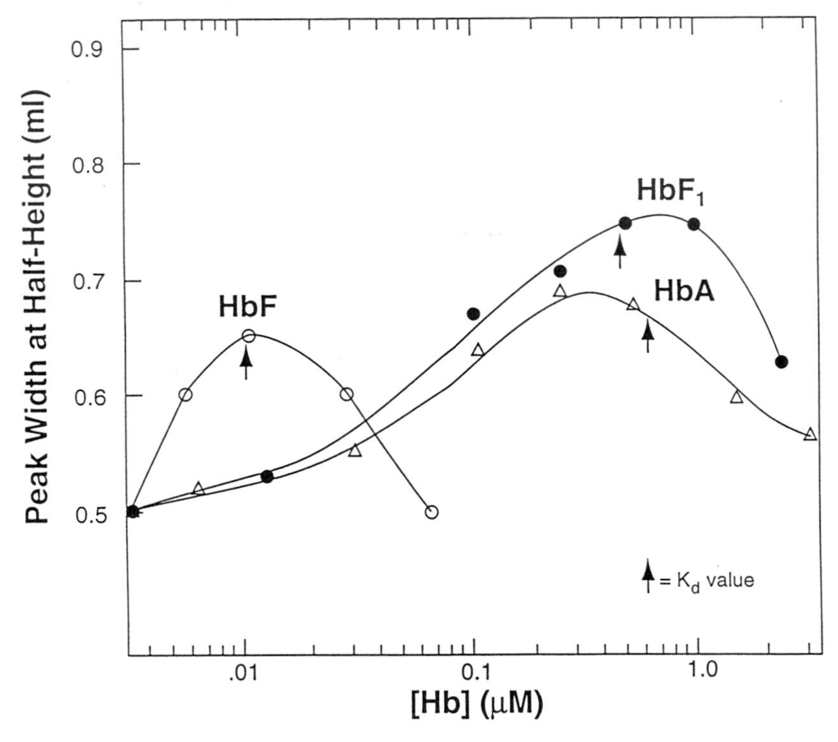

FIG. 6. Relationships of peak broadening and tetramer–dimer dissociation constants. The peak widths and positions at pH 7.5 were determined by the FPLC software package and were reproducible to within 7 μl for the width at half-height and 0.04 ml for the peak position (see Table I for precision of measurements) as described by Dumoulin *et al.*[3] K_d values as calculated from the peak positions are shown by the arrows and are described in detail in the text.

dimers and tetramers occurs (the K_d value). Possibilities for this behavior involve a simple mechanism with compensating trailing of dimers and tetramers leading to increased peak widths in a symmetrical fashion to give a bell-shaped curve but not affecting peak position. Alternatively, a more complex dynamic mechanism involving the dissociation–association process itself could be involved. We are evaluating these possibilities for this behavior, which has not been observed by other methods for measuring protein dissociation but which can be analyzed accurately by the procedure described in this article.

Comments

Two aspects of determining subunit dissociation constants by FPLC on Superose-12 should be emphasized. First, the question of whether the gel-filtration procedure itself influences the equilibrium between tetramers and dimers has essentially been answered by Uversky[25] in a careful analysis of equilibrating native, molten globule and denatured states of proteins. He has shown in convincing fashion for many such protein systems that Superose-12 is inert and does not affect these equilibria. This conclusion is in agreement with ours regarding K_d values of hemoglobins calculated by this method compared with published values obtained by other methods as described in this article. With respect to proteins for which standard elution positions of equilibrating species are not known, curve fitting has been used as described in the section "Evaluation of K_d Values" to obtain accurate subunit dissociation constants.

[25] V. N. Uversky, *Biochemistry* **32,** 13288 (1993).

[7] Ectopic Expression Systems in *Drosophila*

By PIER PAOLO D'AVINO and CARL S. THUMMEL

Introduction

The development of higher organisms is dependent on the appropriate spatial and temporal patterns of gene expression. In the fruit fly, *Drosophila melanogaster,* misexpression of a single gene is often sufficient to disrupt normal development, generating dominant phenotypes that may provide important clues regarding gene function. In the past, the only means of ectopically expressing a gene in *Drosophila* was through the generation of

0076-6879/99 $30.00

mutants, usually dominant, that caused misexpression and/or overexpression of a specific gene product. However, a variety of new tools for ectopic expression have been developed over the past few years, thanks to extensive efforts by the *Drosophila* research community. With these techniques in hand, it is now possible to manipulate gene expression both spatially and temporally, as well as to combine different systems in order to direct the expression of any gene in a predetermined stage- or tissue-specific manner. The use of these ectopic expression systems, combined with classical genetic studies in *Drosophila,* has led to a detailed understanding of a variety of key developmental pathways, including the establishment of cell fate, tissue patterning, and intercellular signaling mechanisms. These systems have also provided an invaluable means of obtaining markers for specific tissues, cells, or cell lineages, as well as a means of targeting cell ablation.[1,2]

Drosophila ectopic expression systems utilize a promoter that drives either constitutive or regulated expression of the gene of interest. These constructs are prepared in a *P*-element vector and are introduced into the fly genome by germ line transformation.[3] On establishing a transgenic line, ectopic expression of the gene of interest can be analyzed in either a wild-type or a mutant genetic background. This article provides a description of the different ectopic expression systems that have been developed in *Drosophila,* listing the strengths and limitations of each technique. Further information regarding the vectors described in this article can be accessed through FlyBase at http://flybase.bio.indiana.edu:82/transposons/.

Constitutive and Uniform Expression

Promoters from five *Drosophila* genes have been used in an effort to achieve constitutive and uniform expression during development: *tubulin, actin5C, armadillo, polyubiquitin,* and *EF-1α.* Initially, the promoters from two cytoskeleton protein genes, *actin5C* and *tubulin,* were used for ectopic expression.[4,5] Both promoters, however, have two important limitations: they direct relatively low levels of transcription and their expression is not uniform, especially during embryonic development.[4–8] Nevertheless, the *actin5C* promoter has been widely used and has been shown to direct

[1] J. W. Sentry, M. M. Yang, and K. Kaiser, *BioEssays* **15,** 491 (1993).
[2] D. E. Buenzow and R. Holmgren, *Dev. Biol.* **170,** 338 (1995).
[3] G. M. Rubin and A. C. Spradling, *Science* **218,** 348 (1982).
[4] S. Bialojan, D. Falkenburg, and R. Renkawitz-Pohl, *EMBO J.* **3,** 2543 (1984).
[5] E. A. Fyrberg, J. W. Mahaffey, B. J. Bond, and N. Davidson, *Cell* **33,** 115 (1983).
[6] J. E. Natzle and B. J. McCarthy, *Dev. Biol.* **104,** 187 (1984).
[7] T. C. Burn, J. O. Vigoreaux, and S. L. Tobin, *Dev. Biol.* **131,** 345 (1989).
[8] J. O. Vigoreaux and S. L. Tobin, *Genes Dev.* **1,** 1161 (1987).

uniform expression in imaginal discs.[9] Two versions of a *P*-element vector containing *actin5C* sequences, pCaSpeR-act, are available.[10] This vector contains the *actin5C* proximal promoter and first exon as well as its 3′ polyadenylation signals. The two versions of this vector differ in the unique cloning sites, *Eco*RI or *Bam*HI, that are present between the promoter and 3′ sequences.

The promoter of the segment polarity gene *armadillo* has been used in an effort to overcome the limitations of the *actin5C* and *tubulin* promoters. High levels of *armadillo* expression have been observed during embryogenesis and pupal development.[11] Similarly, transcription from the *armadillo* promoter is uniform during early embryogenesis and in imaginal discs.[12] However, its level is not constant throughout development. For example, when the *armadillo* promoter was used to drive expression of an *Escherichia coli lacZ* reporter gene, β-galactosidase staining was not detected in the larval midgut.[12] This observation suggests that the usefulness of the *armadillo* promoter for uniform ectopic expression may be restricted to embryonic development and imaginal discs.

Most recently, the *polyubiquitin* promoter has been employed for uniform ectopic expression.[13] This promoter appears to direct higher levels of transcription than the other promoters just mentioned, in many tissues and throughout development. However, although a *polyubiquitin-lacZ* reporter gene is expressed uniformly in early embryos and imaginal discs, its expression pattern has not yet been characterized at other developmental stages.[14] A useful *P*-element–*polyubiquitin* vector, Pwum2, has been described.[15] This vector contains the *polyubiquitin* promoter fused to a small fragment encoding a *myc* epitope tag. Two restriction sites, *Kpn*I and *Not*I, are located downstream from the *myc*-coding region and upstream from an *hsp70* 3′-untranslated region and polyadenylation signals (described later). The *myc* tag provides a useful means of detecting the encoded protein in transformed animals by immunolocalization of the *myc* epitope.

Finally, the peptide synthesis elongation factor *EF-1α* promoter has been used for ectopic expression in *Drosophila*.[16] This promoter was re-

[9] G. Struhl and K. Basler, *Cell* **72,** 527 (1993).
[10] C. S. Thummel, A. M. Boulet, and H. D. Lipshitz, *Gene* **74,** 445 (1988).
[11] B. Riggleman, E. Wieschaus, and P. Schedl, *Genes Dev.* **3,** 96 (1989).
[12] J.-P. Vincent, C. H. Girdham, and P. H. O'Farrell, *Dev. Biol.* **164,** 328 (1994).
[13] H. S. Lee, J. A. Simon, and J. T. Lis, *Mol. Cell. Biol.* **8,** 4727 (1988).
[14] J.-P. Vincent and C. Girdham, *in* "Methods in Molecular Biology" (R. Tuan, ed.), p. 385. Humana Press, Totowa, NJ, 1997.
[15] M. M. S. Heck, A. Pereira, P. Pesavento, Y. Yannoni, A. C. Spradling, and L. S. B. Goldstein, *J. Cell Biol.* **123,** 665 (1993).
[16] R. Ackermann and C. Brack, *Nucleic Acids Res.* **24,** 2452 (1996).

ported to direct high levels of uniform expression in embryos and adults, although no supporting data were presented.

In conclusion, no promoter provides completely uniform and constitutive expression throughout development. This is not surprising given that every promoter is subject to some level of spatial and temporal regulation. The choice of a promoter for widespread expression should, therefore, be based on the particular needs of each experiment. For example, the *armadillo, polyubiquitin,* or *EF-1α* promoters should be useful for achieving uniform gene expression during embryogenesis or in imaginal discs.

Regulated Expression

One problem associated with constitutive ectopic gene expression is that this unregulated expression may lead to lethality, preventing the isolation of a transformed line. In addition, uniform expression may lead to undesirable phenotypes in tissues that do not normally express the gene of interest. These problems can be overcome by methods that allow stage- and/or tissue-regulated ectopic expression. Some of these methods are relatively simple whereas others are more sophisticated and require combining one or more different techniques.

The simplest method for achieving controlled expression of a foreign gene is to use regulatory sequences from a well-characterized promoter.[17–20] Such regulatory elements will drive gene expression in a well-defined tissue and/or stage-specific manner. Use of the *glass* multimer reporter (GMR) to drive ectopic gene expression in the developing *Drosophila* eye provides an ideal example of this system.[20] The GMR vector contains a pentamer of binding sites for the GLASS transcriptional activator, derived from the *Drosophila Rh1* promoter. GLASS expression is restricted to the developing eye, the larval photoreceptor organs, and a few cells in the larval brain, providing tight spatial regulation on ectopic gene expression.

The merit of using characterized regulatory sequences is that only one transgenic line has to be established in order to precisely drive gene expression in both time and space. This method is, however, limited by the availability of well-studied and cloned promoters. One means of overcoming this limitation is the "enhancer piracy" approach developed by Noll *et al.*[21] This method is very similar to the "enhancer trap" technique in which the *lacZ* gene, fused to a minimal promoter, is inserted randomly at multiple

[17] S. M. Parkhurst and D. Ish-Horowicz, *Development* **111,** 1121 (1991).
[18] S. M. Parkhurst, D. Bopp, and D. Ish-Horowicz, *Cell* **63,** 1179 (1990).
[19] C. S. Zucker, D. Mismer, R. Hardy, and G. M. Rubin, *Cell* **53,** 475 (1988).
[20] B. A. Hay, T. Wolff, and G. M. Rubin, *Development* **120,** 2121 (1994).
[21] R. Noll, M. A. Sturtevant, R. R. Gollapudi, and E. Bier, *Development* **120,** 2329 (1994).

locations into the *Drosophila* genome. In over half of these transformed lines, the *lacZ* gene falls under the influence of one or more flanking genomic enhancers that activate its transcription in a spatially and temporally regulated pattern.[22–24] In their strategy, Noll and colleagues substituted the *rhomboid* (*rho*) gene for the *lacZ* reporter gene present in enhancer trap vectors. Upon introducing this modified construct into multiple sites in the *Drosophila* genome, *rho* expression came under the influence of flanking genomic enhancer elements. This led to novel dominant phenotypes that could provide insights into *rho* developmental functions, which may also be useful for the development of genetic screens for suppressors or enhancers.[21]

Temporally Regulated Expression: Use of Heat Shock Promoters

Since its first applications, the use of heat shock promoters has been one of the most important and powerful tools for directing temporally regulated expression in *Drosophila.*[25–27] The level of ectopic gene expression can also be modulated easily by altering the temperature and/or the duration of heat treatment. Indeed, this is one of the unique advantages of working with *Drosophila,* insofar as similar temporal regulation of ectopic gene expression is not currently available in other higher organisms.

Although heat shock ectopic expression experiments have been essential for elucidating many developmental mechanisms, there are at least three potential drawbacks in the use of this system. First, heat shock promoters have a low, but sometimes significant, basal level of transcription under non-heat shock conditions. This can be a problem when even small amounts of a gene product are toxic or lead to a premature mutant phenotype. Second, heat shock alone, at some developmental stages, can phenocopy certain mutations. This is because endogenous cellular gene expression is disrupted upon heat treatment, and the reduced levels of some key regulatory gene products lead to the generation of recognizable mutant phenotypes.[28] For example, a high temperature heat shock of 4-hr *Drosophila*

[22] C. J. O'Kane and W. J. Gehring, *Proc. Natl. Acad. Sci. U.S.A.* **84,** 9123 (1987).

[23] E. Bier, H. Vaessin, S. Shepherd, K. Lee, K. McCall, S. Barbel, L. Ackerman, R. Caretto, T. Uemura, E. Grell, L. Y. Jan, and Y. N. Jan, *Genes Dev.* **3,** 1273 (1989).

[24] H. J. Bellen, C. O'Kane, C. Wilson, U. Grossniklaus, R. K. Pearson, and W. J. Gehring, *Genes Dev.* **3,** 1288 (1989).

[25] J. T. Lis, J. A. Simon, and C. A. Sutton, *Cell* **35,** 403 (1983).

[26] G. Struhl, *Nature* **318,** 677 (1985).

[27] S. Schneuwly, R. Klemenz, and W. J. Gehring, *Nature* **325,** 816 (1987).

[28] N. S. Petersen and H. K. Mitchell, *in* "Comprehensive Insect Physiology, Biochemistry and Pharmacology" (G. A. Kerkut and L. I. Gilbert, eds.), p. 347. Pergamon Press, New York, 1982.

embryos can reproduce the four-winged phenotype associated with *Bithorax Complex* mutations.[29] Finally, heat shock promoters are active in virtually all cell types and thus may not be useful for the analysis of tissue-specific functions.

There are a variety of ways in which the investigator can circumvent these problems. By isolating a number of transformed lines, it is often possible to select one or two transformants that provide a low level of background expression along with a relatively high level of heat-induced expression. Furthermore, it is possible to maintain transformed flies at low temperature (18–21°), reducing the background level of heat shock promoter activity. A non-heat-shocked control should also be included in each experiment. By maintaining one set of animals at 25°, while the second is treated at 35–37° for the desired period of time, the investigator can identify effects that are due solely to heat-induced expression.

Mutant phenocopies can be avoided by using relatively mild heat shocks (≤37°) as well as avoiding developmental stages when the animal is most sensitive to heat shock.[28] Heat treatment of a nontransformed control stock, in parallel with the transformant of interest, can provide an effective control to identify phenotypes that are due solely to heat shock.

Techniques have also been developed that allow restricted heat treatment of specific target cells. Monsma *et al.*[30] reported the use of a heated needle to induce local heat shock induction at virtually any developmental stage. The temperature and diameter of the needle can be varied in order to determine the size of the affected region. More recently, Halfon *et al.*[31] described the use of a finely regulated laser microbeam burst to induce heat shock treatments at the single cell level. These techniques, however, require that the cells of interest be easily accessible, and thus are primarily useful during embryogenesis in *Drosophila.*

In addition to ectopic expression experiments, heat-induced expression can be used to examine later phenotypes of some mutants that die early during development. For example, heat-induced expression of *spaghetti squash* (*sqh*), which encodes the nonmuscle myosin II regulatory light chain, can effectively rescue the early lethality of *sqh* mutants, facilitating the analysis of phenotypes later in development.[32] This method requires that the gene of interest be widely expressed, reflecting the pattern of heat-

[29] H. Gloor, *Rev. Suissue Zool.* **54,** 637 (1947).

[30] S. A. Monsma, R. Ard, J. T. Lis, and M. F. Wolfner, *J. Exp. Zool.* **247,** 279 (1988).

[31] M. S. Halfon, H. Kose, A. Chiba, and H. Keshishian, *Proc. Natl. Acad. Sci. U.S.A.* **94,** 6255 (1997).

[32] K. A. Edwards and D. P. Kiehart, *Development* **122,** 1499 (1996).

induced transcription. In this way, the investigator can obtain the equivalent of a conditional lethal mutant when such mutants are not directly available.

Heat Shock Vectors and Protocols

Several *P*-element vectors are currently available for heat-inducible ectopic expression. These vectors have been constructed using *hsp26, hsp70,* and *hsp82* promoters and 5′-untranslated leaders, although the vectors that contain *hsp70* sequences are used most frequently. The presence of 5′ leader sequences from *hsp* genes is crucial because these sequences are necessary for efficient translation at high temperatures.[33] Most of these vectors contain a polylinker for inserting the gene of interest, followed by a 3′-untranslated sequence derived from the *hsp70* or *Actin5C Drosophila* genes or SV40. The *hsp70* 3′ trailer directs rapid mRNA degradation under non-heat shock conditions and is therefore recommended only when rapid turnover of the gene product is required. Two vectors that carry the *hsp70* promoter have been widely used, and their structures are presented in Fig. 1.

Many protocols have been published for heat-induced ectopic expression, and a good protocol for embryo treatment can be found in Brand *et al.*[34] This article describes an efficient protocol for heat treatment of late larvae and pupae. This protocol has been successful for the analysis of genetic regulatory hierarchies triggered by the steroid hormone ecdysone at the onset of metamorphosis.[35–37]

The inherent asynchrony of *Drosophila* larval development makes it difficult to stage larvae accurately. One simple and efficient method for staging third instar larvae is to grow them on food containing 0.05–0.1% bromphenol blue.[38] Late third instar larvae staged by this method can be used to select newly formed prepupae at 15-min intervals. These synchronized prepupae can then be allowed to develop for the appropriate period of time in order to obtain closely staged animals throughout pupal development.

[33] T. J. McGarry and S. Lindquist, *Cell* **42,** 903 (1985).

[34] A. H. Brand, A. S. Manoukian, and N. Perrimon, *in "Drosophila melanogaster*: Practical Uses in Cell and Molecular Biology" (L. S. B. Goldstein and E. A. Fyrberg, eds.), p. 635. Academic Press, New York, 1994.

[35] G. T. Lam, C. Jiang, and C. S. Thummel, *Development* **124,** 1757 (1997).

[36] C. T. Woodard, E. H. Baehrecke, and C. S. Thummel, *Cell* **79,** 607 (1994).

[37] J. C. Fletcher, P. P. D'Avino, and C. S. Thummel, *Proc. Natl. Acad. Sci. U.S.A.* **94,** 4582 (1997).

[38] A. J. Andres and C. S. Thummel, *in "Drosophila melanogaster*: Practical Uses in Cell and Molecular Biology" (L. Goldstein and E. Fyrberg, eds.), p. 565. Academic Press, New York, 1994.

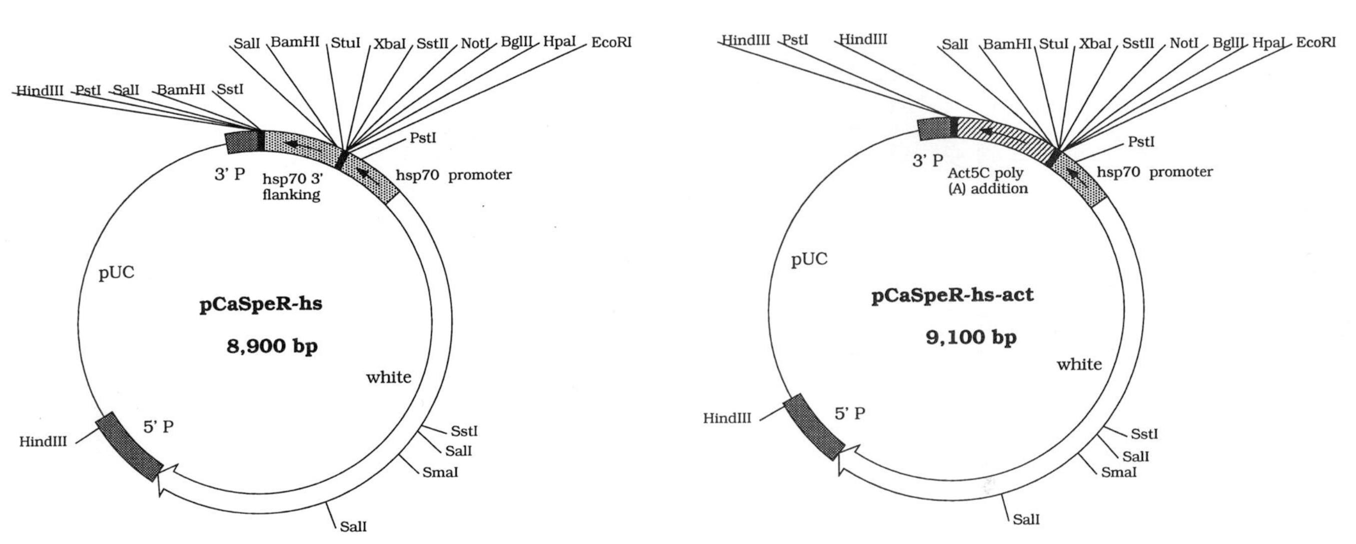

FIG. 1. Maps of *P*-element vectors for heat-inducible ectopic expression. pCaSpeR-hs has been described previously.[50] pCaSpeR-hs-act is a derivative that was constructed by C. Woodard and CST. An 800-bp *Sal*I–*Pst*I fragment from pCaSpeR-act, carrying the *actin5C* 3′ trailer and polyadenylation signals, was used to replace the *hsp70* 3′ region present in pCaSpeR-hs. This vector should increase the stability of ectopically expressed genes under non-heat shock conditions. The *white* gene provides a selectable marker to identify transformants. The following unique restriction sites are available for inserting foreign genes into either of these vectors: *Eco*RI, *Hpa*I, *Bgl*II, *Not*I, *Sst*II, *Xba*I, and *Stu*I. GenBank accession numbers: pCaSpeR-hs, U59056; pCaSpeR-hs-act, U60735. These vectors are available on request.

1. Collect third instar larvae, prepupae or pupae, staged as described earlier, with a wet paintbrush and wash briefly in water at room temperature (no higher than 25°). As described earlier, it is best to use three stocks for each experiment: nontransformed control animals, transformants maintained at 25°, and transformants subjected to heat shock.
2. Transfer 10–20 animals to a 1.5-ml microcentrifuge tube. A second microcentrifuge tube, with small holes at the bottom, can be inserted into the first tube in order to prevent the larvae from escaping.
3. Transfer the tube to a 35–37° water bath for 30–60 min. The temperature and time may vary depending on the experiment, but $\leq$35° is recommended if incubations longer than 30 min are required.
4. Transfer the vials to room temperature and maintain at 25° for about 2 hr in order to allow the animals to recover from heat treatment. This period of time is usually sufficient to allow maximal accumulation of the protein of interest. It is best to try different heat shock temperatures, times of heat treatment, and recovery times in order to identify the appropriate conditions for optimal ectopic expression. This can be assessed easily by Northern or Western blot analysis.

Spatially Regulated Ectopic Expression: GAL4 System

The yeast transcriptional activator *GAL4* provides the most powerful method currently available for directing cell type or tissue-specific ectopic expression.[39] This method, including vectors, protocols, uses, strengths and limitations, has been reviewed previously.[34] We will, therefore, limit our description to the basic concepts of this technique.

GAL4 can activate transcription of any target gene by binding to a GAL4 UAS (upstream activation sequence) positioned near a minimal promoter.[40] In this system, two different *P*-element transformant lines are established, one of which expresses GAL4 in a particular spatial and temporal pattern and the other of which carries a target gene under the control of multiple UAS elements. Expression of GAL4 has no detectable effects on *Drosophila* development, and the target gene is not expressed in the absence of GAL4. Only on crossing these two lines will the target gene be activated, in a temporal and spatial pattern that reflects that of GAL4 (Fig. 2).

Two approaches have been used to generate lines that express the GAL4 activator. First, GAL4 can be expressed using either a defined *Drosophila* promoter or the *hsp70* promoter, as described earlier. Second,

[39] A. H. Brand and N. Perrimon, *Development* **118,** 401 (1993).

[40] J. A. Fischer, E. Giniger, T. Manaitis, and M. Ptashne, *Nature* **332,** 853 (1988).

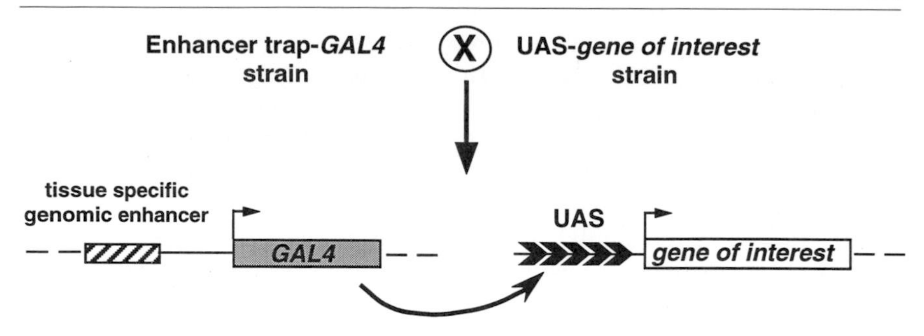

FIG. 2. Use of the GAL4 system to drive tissue-specific ectopic expression. A fly strain carrying *GAL4* under the control of a tissue-specific genomic enhancer is crossed to a strain containing the gene of interest under the control of UAS sequences. In the progeny, the gene of interest will be expressed in the same pattern as its activator, GAL4.[39]

collections of GAL4-expressing lines have been established using the "enhancer trap" technique, in which the GAL4 expression pattern is determined by flanking genomic enhancer elements. These transformant lines have been characterized by crossing them with flies that carry a UAS–*lacZ* reporter gene and analyzing the pattern of β-galactosidase expression at different stages of development. In this way, a large number of lines have been established that provide a wide range of different GAL4 expression patterns.[39,41,42] It should be noted that the level of ectopic expression can also be modulated by varying the number of UAS sequences. Finally, there are three possible drawbacks of the GAL4 technique. First, GAL4-mediated expression cannot be detected in the female germ line and before gastrulation in embryos.[34] Second, the level of GAL4-mediated transcription is often variable from cell to cell within the same expression domain.[34] Finally, the effects of GAL4-mediated expression at later stages of development could be due to earlier expression of the GAL4 activator. Some attention thus needs to be paid to the temporal and spatial patterns of GAL4 expression up to the stage selected for phenotypic analysis.

The GAL4 system has been developed into a method of screening for phenotypes caused by misexpression of endogenous genes.[43] In this method, a minimal promoter under the control of UAS elements was inserted into a *P*-element vector such that transcription from this promoter was directed into flanking genomic sequences. This element was introduced into multiple sites in the *Drosophila* genome and activated with tissue-specific GAL4

[41] K. Gustafson and G. L. Boulianne, *Genome* **39,** 174 (1996).

[42] L. Manseau, A. Baradaran, D. Brower, A. Budhu, F. Elefant, H. Phan, A. V. Philp, M. Yang, D. Glover, K. Kaiser, K. Palter, and S. Selleck, *Dev. Dyn.* **209,** 310 (1997).

[43] P. Rorth, *Proc. Natl. Acad. Sci. U.S.A.* **93,** 12418 (1996).

expression. Under these conditions, the UAS-regulated promoter is induced and a flanking endogenous gene will be expressed ectopically in a pattern that reflects that of the GAL4 activator. The resultant dominant phenotypes provide a novel means of screening for new regulatory genes.

Combined Techniques

Many combinations of one or more techniques described earlier have been established in order to overcome some of the limitations of each approach. Some of these techniques are very elegant and provide a means of ectopically expressing a given gene in clones of a single cell at a desired developmental stage.

For example, one method combines the temporal regulation provided by heat shock with the spatial regulation provided by an activator such as GAL4.[20,39] In these methods, the gene of interest is under the control of responsive elements of a specific transcriptional activator (yeast GAL4 or *Drosophila* GLASS have been used), which, in turn, is under the control of an *hsp70* heat shock promoter. The transcriptional activator and the target gene are, however, separated in two different transgenic lines. In one line, the target gene remains inactive because of the absence of the activator protein, whereas in the other line the activator is present but there is no target gene to induce. The gene of interest will only be expressed after the two strains are crossed and the progeny are subjected to heat treatment (Fig. 3). This method is useful for expressing highly toxic genes in a spatially and temporally regulated manner.

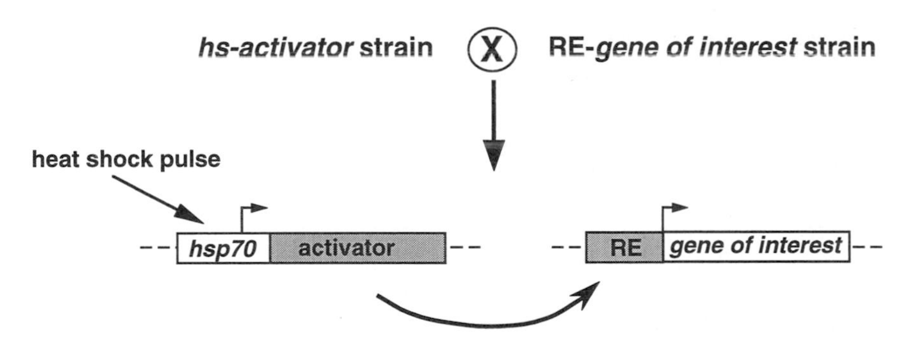

FIG. 3. Combining the heat shock technique with the use of characterized promoters. In this system, one fly strain carries a transgene in which a transcriptional activator is under the control of the *hsp70* heat shock promoter. Another transgenic fly strain carries the gene of interest under the control of response elements (RE) for the activator. After crossing these two lines, the progeny will carry both transgenes and thus the gene of interest can be expressed by heat induction of the transcriptional activator.

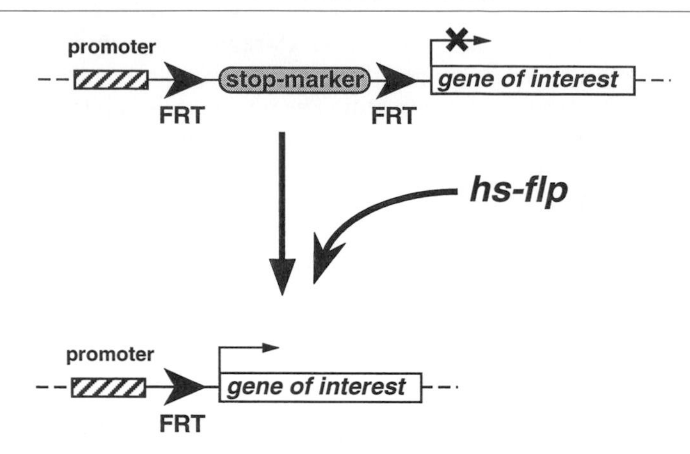

FIG. 4. The *FLP*-out technique. The gene of interest is separated from either a regulated or a constitutive promoter by a DNA fragment called a *FLP*-out cassette. This cassette contains one or more transcriptional stop signals (stop) and a genetic marker flanked by two directly repeated FRT sites. The *FLP*-out cassette thus prevents expression of the gene of interest. After heat-induced expression of FLP recombinase, the *FLP*-out cassette is excised randomly in the organism, juxtaposing the promoter and the gene of interest. In this way, clones of cells are generated that express the gene of interest under the control of a known promoter. These clones can be recognized easily by loss of the genetic marker.[9]

An elegant technique to direct patches of ectopic gene expression within a single tissue has been reported by Struhl and Basler.[9] This method is based on the use of the FLP recombinase from the 2-μm plasmid of *Saccharomyces cerevisiae.* FLP can efficiently catalyze site-specific recombination between two FLP recombination targets (FRTs) in flies.[44] When two FRTs are arranged as direct repeats, FLP-mediated recombination leads to the excision of the intervening DNA, leaving only one FRT and joining the sequences on either side.

Figure 4 shows how FLP-mediated recombination can be used in order to generate inducible ectopic expression. A promoter is separated from a given gene by a so-called "*FLP*-out cassette." This cassette contains a transcriptional stop signal along with a visible marker gene that is flanked by two direct FRT repeats. Transcription from the promoter is thus terminated upstream from the gene of interest, maintaining the target gene in a repressed state. The presence of the stop sequence can be followed by observing the phenotype of the marker gene (in this case, a *yellow* bristle marker). Upon ubiquitous activation of *FLP* expression by a heat pulse, the *FLP*-out cassette is excised randomly in the organism, establishing cell clones

[44] K. G. Golic and S. Lindquist, *Cell* **59,** 499 (1989).

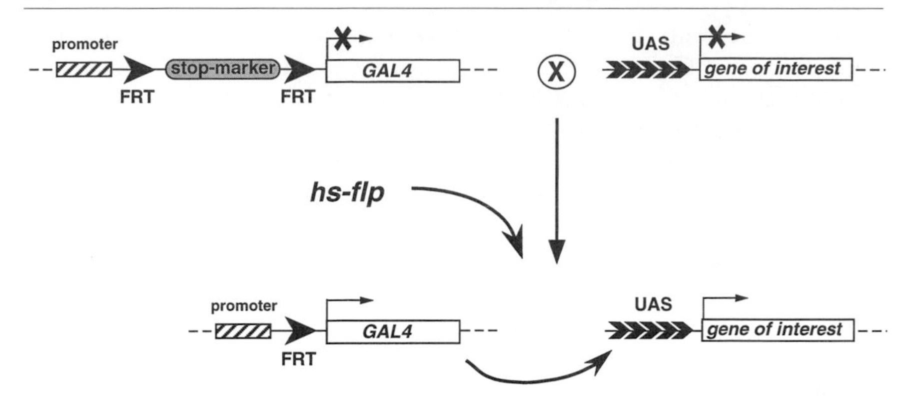

FIG. 5. Combining the GAL4 system with the *FLP*-out technique. In one transgenic line, a *FLP* out cassette divides a promoter from the *GAL4* gene, whereas in another line the gene of interest is under the control of UAS sequences. The two lines are crossed and, in the progeny, heat-induced FLP recombinase expression generates cell clones that express GAL4 under the control of the promoter. GAL4, in turn, activates the gene of interest through its binding to UAS sequences.[46,47]

that express the gene of interest (see Fig. 4). Such clones can be recognized easily by the loss of the marker gene. Either a constitutive or a regulated promoter can be used with this technique, depending on the purpose of the experiment. Because this method directs patches of ectopic gene expression within a tissue, it is extremely helpful for analyzing the function of genes involved in establishing positional information during development.

A modification of the *FLP*-out method has been described that allows the analysis of gene expression at very early stages of development.[45] In this system, called IT for immediate and targeted, *FLP* expression is under the control of the β_2-tubulin promoter. Because this promoter functions only in maturing spermatocytes, the target gene will only be active in progeny derived from sperm in which the *FLP*-out cassette has been excised. Furthermore, the use of an early promoter, such as that from the *hunchback* gene, restricts ectopic expression to early embryonic stages. The IT system can be used to direct the expression of a gene at earlier times than are possible with either the *hsp70* or GAL4 systems.

The *FLP*-out method has also been combined with the GAL4 system in order to facilitate the ectopic expression of target genes.[46,47] In this method, outlined in Fig. 5, the *FLP*-out cassette divides a constitutive or regulated promoter from the *GAL4* gene, and different genes of interest

[45] E. A. Wimmer, S. M. Cohen, H. Jackle, and C. Desplan, *Development* **124,** 1509 (1997).
[46] F. Pignoni and S. L. Zipursky, *Development* **124,** 271 (1997).
[47] J. F. de Celis and S. Bray, *Development* **124,** 3241 (1997).

are under the control of UAS sequences. After FLP-mediated excision of the cassette, the *GAL4* gene is transcribed, which, in turn, activates expression of the gene of interest. Although this is formally identical to the *FLP*-out system depicted in Fig. 4, it provides more flexibility in that a variety of different UAS-target gene transgenes can be activated using different crosses. de Celis and Bray have devised a modification of this system in which they introduced a *GAL4–lacZ* dicistronic gene to mark cell clones not only by the loss of the marker, but also by the expression of *lacZ*.

Conclusions

The last few years have seen the development of elegant systems to ectopically express a gene of interest at virtually any time and place during development. These systems draw together the precise temporal regulation of the *hsp70* promoter with the spatial regulation provided by the GAL4 system. In addition, judicious use of the FRT/FLP recombinase system provides an elegant means of further restricting the patterns of ectopic gene expression. It seems likely that future combinations of these systems, along with the CRE/loxP recombination system,[48] will provide additional ectopic expression systems over the next few years.

Despite the power of these techniques, however, it is important to bear in mind that ectopic expression provides only a partial understanding of gene function. These methods provide insights into what functions a particular gene can provide, not necessarily what the gene does in its normal developmental context. Furthermore, the high levels of ectopic expression associated with some systems may allow a factor to feed into a developmental pathway that it does not normally regulate. Despite these considerations, however, ectopic gene expression provides an essential and invaluable tool to study gene function. By combining gain-of-function studies with the analysis of loss-of-function mutant phenotypes, the investigator has an unparalleled opportunity to understand gene function during development.[49]

Acknowledgments

P.P.D. is a Research Associate and C.S.T. an Associate Investigator with the Howard Hughes Medical Institute. We thank J. Broadus and S. Sakonju for critical comments on the manuscript.

[48] M. L. Siegal and D. L. Hartl, *Genetics* **144,** 715 (1996).
[49] Y. N. Jan and L. Y. Jan, *Cell* **75,** 827 (1993).
[50] C. S. Thummel and V. Pirrotta, *Dros. Inf. Serv.* **71,** 150 (1992).

Section III

Yeast Expression Systems

[8] Copper Ion Inducible and Repressible Promoter Systems in Yeast

By Simon Labbé and Dennis J. Thiele

Introduction

The bakers' yeast *Saccharomyces cerevisiae* is an excellent eukaryotic model system to understand the molecular mechanisms underlying protein structure and function in cell growth, regulation, and metabolism. Indeed, yeast cells have been used successfully for isolating and defining the function of cDNAs from many organisms involved in a variety of homeostatic processes, particularly those associated with human diseases.[1] The controlled expression of both yeast and heterologous proteins in *S. cerevisiae* has relied on the use of plasmids harboring a yeast promoter, polylinker, and transcriptional terminator. Two classes of promoters are generally used. The first class is considered to be expressed constitutively and includes promoters such as *CYC1, ADH1, TEF2,* and *GPD.*[2] Although these promoters are not inducible, their relative strengths are different, with the *CYC1* promoter being the weakest and the *GPD* promoter the strongest. The second class includes regulatable promoters such as those derived from *MET25* and *GAL1-10* genes.[3] Although the regulation by these promoters is tight, the concentration of the corepressor, methionine, required to trigger the down-regulation of gene expression from the *MET25* promoter is considerable, demanding as much as 1 m*M*. Furthermore, to observe the full repression of protein expression under the control of the *MET25* promoter, 12 to 24 hr is required.[4] Although the exact time course of repression is dependent on the half-life of the expressed protein, this interval of time is important. In the case of the *GAL1* promoter, the complete repression of the target gene expression typically requires 12 to 24 hr and involves an often inconvenient shift of the carbon source between glucose and galactose. Carbon source shifts are well established to have pleiotropic effects on yeast gene expression and are therefore undesireable in many experiments. For both the *MET25* and *GAL1* promoters, the maximal induction of protein expression is observed after 4 hr.[4] Alternative *S. cerevisiae* expres-

[1] D. Botstein, S. A. Chervitz, and J. M. Cherry, *Science* **277,** 1259 (1997).

[2] D. Mumberg, R. Müller, and M. Funk, *Gene* **156,** 119 (1995).

[3] D. Mumberg, R. Müller, and M. Funk, *Nucleic Acids Res.* **22,** 5767 (1994).

[4] V. Rönicke, W. Graulich, D. Mumberg, R. Müller, and M. Funk, *Methods Enzymol.* **283,** 313 (1997).

0076-6879/99 $30.00

sion vectors with a tetracycline-regulatable promoter system have been developed.[5] Although this tetracycline operator (*tetO*)-driven expression system controls the target gene expression tightly, the concentration of tetracycline or its derivatives, which cause the repression of gene expression, is problematic to maintain at the appropriate level.[6] Furthermore, the maximal induction of protein expression observed by using the tetracycline-regulatable promoter system requires a long period of time,[5] due to the fact that the effector (e.g., antibiotic) is persistent and needs to be removed completely to allow the derepression.

In *S. cerevisiae,* the presence of the trace element copper (Cu) is tightly regulated by at least two distinct biochemical pathways. One involves a high-affinity copper transport system and the other a detoxification system.[7] In the latter system, when copper concentrations reach a level that is above the range needed for optimal cell growth, copper ions (micromolar range or above) are sensed by the metalloregulatory transcription factor Ace1. In the presence of Cu(I), Ace1 specifically binds a recognition DNA sequence HTHXXGCTC (H = A, C, or T; X = any residue), denoted MRE (metal regulatory element), activating the expression of the *CUP1* and *CRS5* metallothionein genes and the Cu,Zn-superoxide dismutase gene (*SOD1*).[7] Among these three genes, *CUP1* is the most potent for copper induction. The *CUP1* promoter contains four MREs and exhibits a rapid induction with the maximal level of transcript detected 30 min after exposure to copper.[8] It has been shown that the *CUP1* promoter can simultaneously tightly regulate the expression of the transcriptional repressor Rox1 and the N-end-recognizing (E3) protein UBR1 in a conditional system used to demonstrate that TBP-associated factors (TAFs) are not generally required for transcriptional activation in yeast.[9] In this conditional system, copper via Ace1 induces the biosynthesis of the transcriptional repressor Rox1, which can subsequently shut down the TAFs mRNA synthesis. In parallel, the presence of copper also induces the biosynthesis of the UBR1 protein, which can then initiate the degradation of TAF proteins through the ubiquitin-dependent pathway. Based on the tightness of the *CUP1*-dependent regulation, we used the *CUP1* promoter to create a series of versatile vectors to allow its utilization to control the expression of heterologous genes in *S. cerevisiae*.

[5] E. Gari, L. Piedrafita, M. Aldea, and E. Herrero, *Yeast* **13,** 837 (1997).

[6] M. Gossen, S. Freundlieb, G. Bender, G. Müller, W. Hillen, and H. Bujard, *Science* **268,** 1766 (1995).

[7] K. A. Koch, M. M. O. Peña, and D. J. Thiele, *Chem. Biol.* **4,** 549 (1997).

[8] M. M. O. Peña, K. A. Koch, and D. J. Thiele, *Mol. Cell. Biol.* **18,** 2514 (1998).

[9] Z. Moqtaderi, Y. Bai, D. Poon, P. A. Weil, and K. Struhl, *Nature* **383,** 188 (1996).

Previous studies[10] have identified an exquisitely sensitive and selective metal responsive system that regulates genes that play a critical role in high-affinity copper transport in yeast. Each of these genes, such as *FRE1*,[11] *CTR1*,[12] and *CTR3*,[13] is transcriptionally regulated as a function of copper availability: repressed when cells are grown in the presence of copper ions and highly derepressed during copper starvation. This gene regulation is dependent of the presence of two strictly conserved repeated copper-responsive *cis*-acting element, TTTGCTC, denoted CuREs, and is controlled by a copper-sensing transcription factor, Mac1. To illustrate the exquisite copper responsiveness, we fused 1116 bp of the 5′-noncoding region and the first 10 codons of the *CTR3* gene to the *S. cerevisiae URA3* gene. In the presence of 20 μM $CuSO_4$, yeast cells harboring the fusion gene on a low-copy number plasmid or integrated into their genomes were not capable to grow on drop-out media without uracil (Fig. 1). Conversely, under copper-limiting conditions using the same media, the cells were growing as well as the control cells, indicating that the *CTR3* promoter was derepressed, thereby allowing the expression of the *URA3*-encoded enzyme required for uracil biosynthesis. Therefore, based on this observation, we used the *CTR3* and *CTR1* promoters, specifically the upstream regions encompassing the CuREs, to develop a series of vectors in which the expression of heterologous genes can be tightly modulated as a function of copper availability.

This article describes 24 expression vectors based on the p4XXprom series developed by Mumberg *et al.*[2] The vectors that we have constructed, denoted pCu4XXprom, can be utilized to tightly and rapidly up- and down-regulate gene expression depending on cellular copper status. To create these copper-regulated expression plasmids, we use two different types of copper-responsive promoter systems. First, we use the metallothionein *CUP1* promoter,[14] which is inducible by the presence of copper ions. Under copper-limiting conditions the basal level of expression of the *CUP1* promoter is very low. This is an important difference between the yeast *CUP1* promoter and the mammalian mouse *MT-I* promoter, which has been described to possess a high basal level of expression, thereby making its utilization disadvantageous when a tight regulation is required.[15] The sec-

[10] S. Labbé, Z. Zhu, and D. J. Thiele, *J. Biol. Chem.* **272,** 15951 (1997).

[11] A. Dancis, D. G. Roman, G. J. Anderson, A. G. Hinnebusch, and R. D. Klausner, *Proc. Natl. Acad. Sci. U.S.A.* **89,** 3869 (1992).

[12] A. Dancis, D. S. Yuan, D. Haile, C. Askwith, D. Eide, C. Moehle, J. Kaplan, and R. D. Klausner, *Cell* **76,** 393 (1994).

[13] S. A. B. Knight, S. Labbé, L. F. Kwon, D. J. Kosman, and D. J. Thiele, *Genes Dev.* **10,** 1917 (1996).

[14] D. J. Thiele and D. H. Hamer, *Mol. Cell. Biol.* **6,** 1158 (1986).

[15] R. J. Kaufman, *Methods Enzymol.* **185,** 487 (1990).

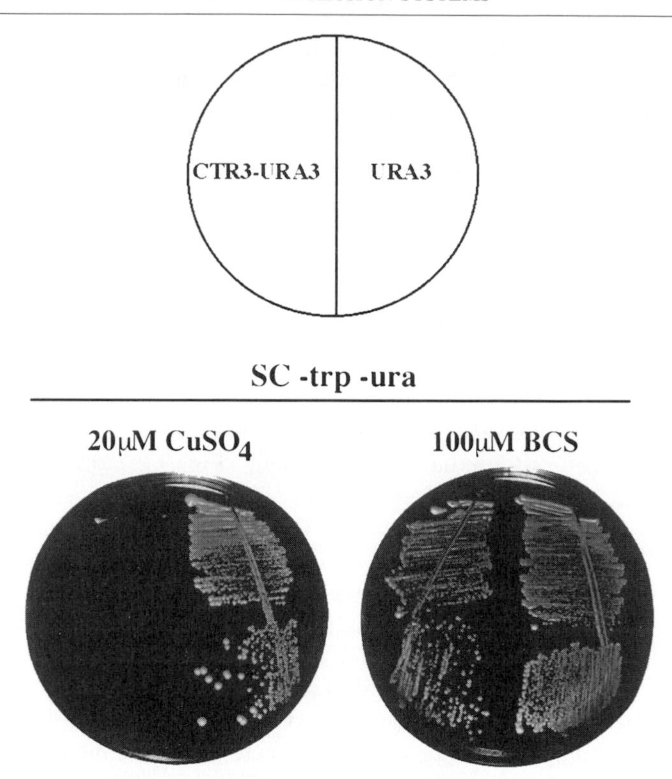

FIG. 1. Uracil auxotrophy in response to copper. Strain SLY1 transformed with either pPSG*CTR3-URA3* (denoted CTR3-URA3) or pPSG A (plasmid alone, denoted URA3) was evaluated for uracil auxotrophy. The medium consisted of SC supplemented with all necessary auxotroph nutrients except tryptophan (for plasmid selection) and uracil and either 20 μM $CuSO_4$ (copper replete) or 100 μM BCS (copper starvation). The right plate shows absence of growth when cells are in the presence of copper since the *CTR3* promoter is repressed, thereby blocking uracil biosynthesis. The left plate shows growth when cells are in the presence of BCS since the *CTR3* promoter is induced, thereby allowing uracil biosynthesis. The plates were incubated at 30° for 4 days prior to photography.

ond type of promoter that we use is regulated by copper in the opposite direction from the *CUP1* promoter. The copper transporter promoters, *CTR3* and *CTR1,* are repressed by the presence of copper and highly activated under copper-starvation conditions. Furthermore, copper repression occurs with just a trace amount of copper ions. Indeed, nanomolar copper concentrations are sufficient to down-regulate the *CTR3* and *CTR1* promoters.[10] Moreover, the time course of copper-starvation activation and copper repression of a reporter gene under the control of this type of promoter is rapid, with 91% of the maximal level of transcription detected

10 min after treatment with the copper chelator bathocuproine disulfonate (BCS) and virtually no detectable transcript after a 60-min exposure to copper, respectively.[10] Based on the exquisite sensitivity and selectivity of both types of copper-responsive promoter systems, we used the metallothionein and the copper transporter promoters for developing a series of expression vectors in which the levels of expression of heterologous proteins can be tightly positively or negatively regulated after a short interval of time depending on the cellular copper status. The copper ion concentrations required to regulate the vectors described herein are sufficiently low such that no inhibition of cell growth is observed during the incubation times.

Copper-Regulated Expression Vectors

A schematic map of the plasmids is illustrated in Fig. 2. Furthermore, Fig. 2 shows the regions of the three copper-responsive promoters used to create these regulatable expression vectors. To avoid the elimination of unique cloning sites, we fused to the minimal *CYC1* promoter short regions of the *CTR3* and *CTR1* promoters that encompass the CuREs. These fusion promoters are able to down- and up-regulate gene expression in the presence of copper or BCS, respectively, as well as the endogenous *CTR3* and *CTR1* promoters.[10,16] The transcriptional terminator used in these vectors was isolated from the *CYC1* gene.[3] Because the *HIS3, TRP1, LEU2,* and *URA3* markers for selection and maintenance in yeast share few restriction sites with the multiple cloning sites (MCS), we identified restriction sites for each set of vectors that are unique in the MCS, thereby available for cloning the gene of interest (Table I).

Regulation of β-Galactosidase Levels Expressed from *CUP1, CTR3,* and *CTR1* Copper-Responsiveness Promoters

To assess the efficacy of the copper-regulatable promoter systems that we developed, we inserted the *Escherichia coli* gene encoding β-galactosidase into the *Hin*dIII site of pCu416*CUP1,* pCu416*CTR3-CYC1,* and pCu416*CTR1-CYC1.* We first analyzed the steady-state of β-galactosidase mRNA and protein levels when expressed from the *CUP1* promoter. When yeast cells were grown in minimal medium, which contains 16 n*M* copper, a very low basal level of mRNA and protein was observed (Fig. 3, left-hand side). In contrast, when cells were treated with 1 and 100 μ*M* of $CuSO_4$, the levels of both mRNA and protein were induced approximately

[16] Y. Yamaguchi-Iwai, M. Serpe, D. Haile, W. Yang, D. J. Kosman, R. D. Klausner, and A. Dancis, *J. Biol. Chem.* **272,** 17711 (1997).

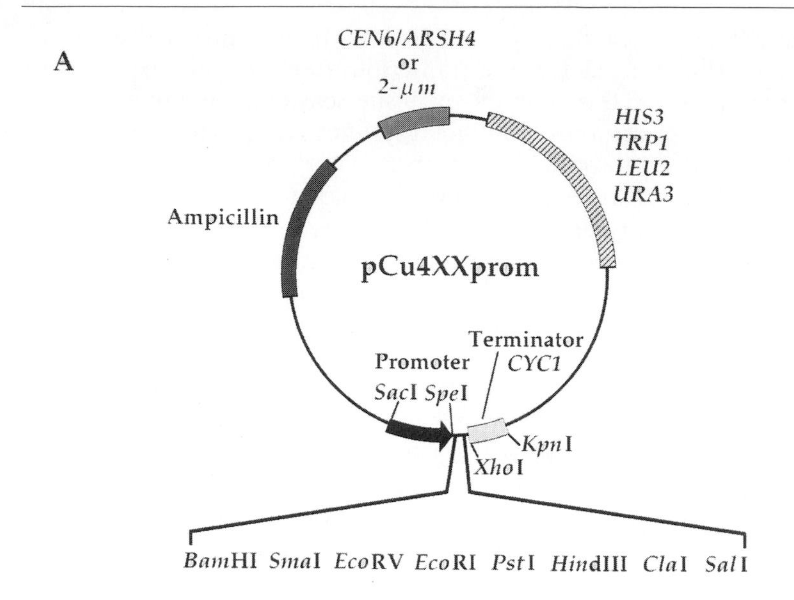

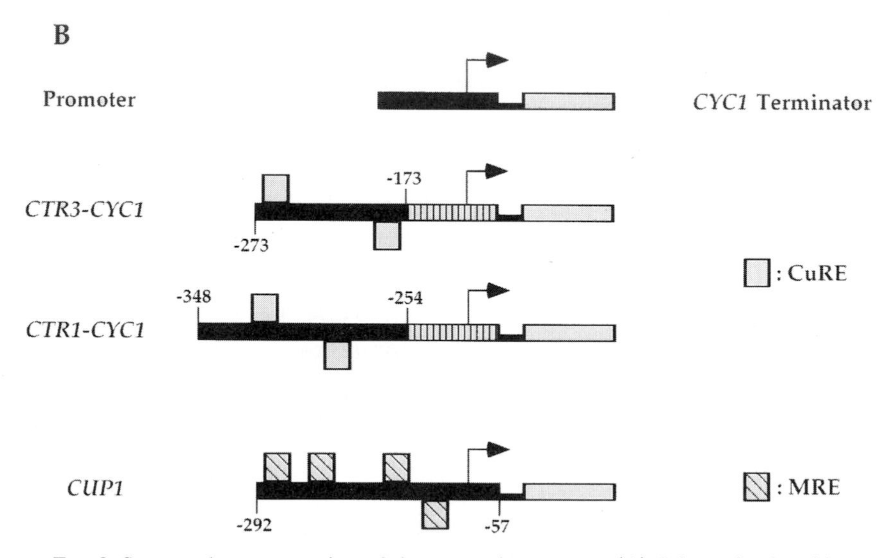

FIG. 2. Structural representation of the expression vectors. (A) Schematic plasmid map of the pCu4XXprom series. The multiple cloning site polylinker (MCS) is shown enlarged and is preceded by *CTR3, CTR1,* or *CUP1* promoter, which is delineated by *Sac*I and *Spe*I sites. The *CYC1* terminator is located immediately downstream of the MCS and is delimited by *Xho*I and *Kpn*I sites. The map also shows the relative positions of the selectable markers and the origins of replication. The nomenclature used is based on the expression vectors described by Mumberg *et al.*[2] For instance, plasmid pCu416*CUP1* harbors the *CUP1* promoter

TABLE I
UNIQUE RESTRICTION SITES IN EXPRESSION VECTORS

Plasmid[a]	Selectable marker	Unique restriction sites
pCu4X3	*HIS3*	*Spe*I, *Bam*HI, *Sma*I, *Eco*RV, *Eco*RI, *Cla*I, *Sal*I, and *Xho*I
pCu4X4	*TRP1*	*Spe*I, *Bam*HI, *Sma*I, *Eco*RI, *Pst*I, *Cla*I, *Sal*I, and *Xho*I
pCu4X5	*LEU2*	*Spe*I, *Bam*HI, *Sma*I, *Pst*I, *Hin*dIII, *Sal*I, and *Xho*I
pCu4X6	*URA3*	*Spe*I, *Bam*HI, *Sma*I, *Eco*RI, *Hin*dIII, *Cla*I, *Sal*I, and *Xho*I

[a] See Fig. 2.

3- and 5-fold, respectively. As would be expected, both mRNA and protein levels of β-galactosidase were found to be very low in the presence of the copper chelator BCS. Although a clear induction of β-galactosidase expression was observed after a 3-hr exposure to copper (Fig. 3), the optimal time incubation observed for having a maximal induction (approximately 20-fold) is 75 min.[8] Analogous experiments were performed with yeast cells expressing β-galactosidase under the control of the *CTR3-CYC1* and *CTR1-CYC1* promoters. When yeast cells were treated with 1 and 100 μM of $CuSO_4$, the levels of both mRNA and protein were repressed approximately 3- and 4-fold, respectively. Conversely, both β-galactosidase mRNA and protein levels were robustly derepressed approximately 10-fold, when cells were treated with BCS (Fig. 3, right-hand side). For simplicity, herein just the expression levels of the β-galactosidase under the control of the *CTR1-CYC1* promoter are shown as the expression levels observed from the *CTR3-CYC1* promoter were nearly identical (data not shown). The steady-state levels of yeast heat shock transcription factor (HSF) were assessed as an internal control and as a marker of the integrity of the protein extract preparations[17] (Fig. 3). Also, the steady-state levels of actin (*ACT1*) were measured simultaneously as an internal control for the mRNA analyses.

[17] Z. Zhu, S. Labbé, M. M. O. Peña, and D.J. Thiele, *J. Biol. Chem.* **273,** 1277 (1998).

---

and is based on the vector pRS416, which has the *URA3* gene and the *CEN6/ARSH4* origin of replication. (B) Schematic representation of the different promoters. For the *CTR3* and *CTR1* promoter regions the striped bars represent the fusion with the minimal *CYC1* promoter. Open boxes indicate the copper-response element, called CuRE. For the *CUP1* promoter region, hatched boxes illustrate the metal regulatory element, called MRE. The nucleotide numbers refer to the position relative to the A of the start codon of the *CTR3,*[13] *CTR1,*[12] and *CUP1,*[20,21] open reading frames.

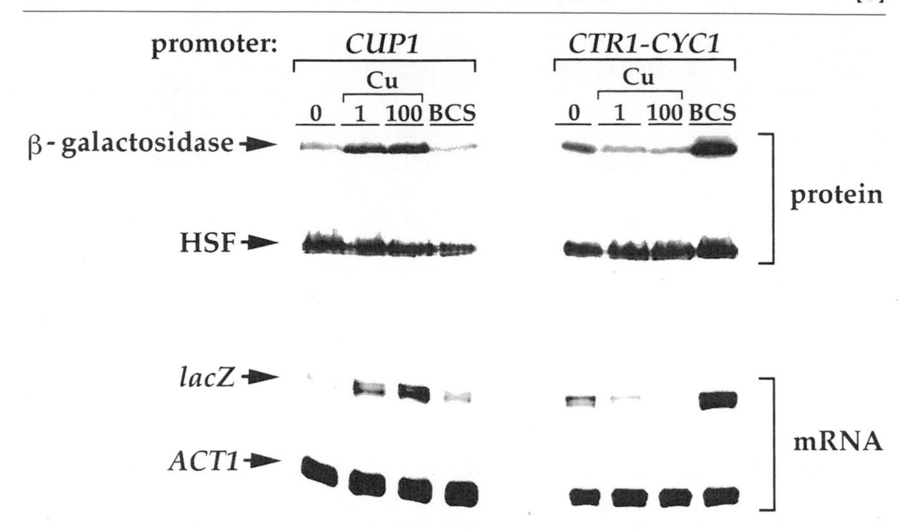

FIG. 3. Strain SLY1 transformed with pCu416*CUP1-lacZ* and pCu416*CTR1-CYC1-lacZ* was grown to early log phase under a very low copper ion condition (16 n*M* copper). $CuSO_4$ (0, 1, and 100 μ*M*) or 100 μ*M* BCS was added, and after a 3-hr incubation at 30°, whole cell protein extracts and total RNA were prepared from culture aliquots. (Top left) β-Galactosidase steady-state levels expressed from the plasmid pCu416*CUP1-lacZ*. (Top right) β-Galactosidase steady-state levels expressed from the plasmid pCu416*CTR1-CYC1-lacZ*. Steady-state levels of heat shock factor (HSF) were examined as an internal control. The positions of the β-galactosidase and HSF are indicated with arrows. Shown at the bottom is an RNase protection analysis of *lacZ* and *ACT1* mRNA steady-state levels. *lacZ* and *ACT1* mRNAs are denoted with arrows.

Experimental Procedures and Vector Constructions

Strain SLY1 (*MAT***a** *gal1 trp1-1 his3 ade8 CUP1 ura3::KAN*^{r}), transformed with the expression plasmids, was grown in a modified minimal medium (SD) that contains 0.67% yeast nitrogen base minus copper and iron (Bio101, Inc., La Jolla, CA), 2% dextrose, 50 m*M* MES buffer, pH 6.1, and 10 μ*M* $NH_4Fe(SO_4)_2$. Copper induction or copper starvation of yeast cells was carried out by adding the indicated amount of $CuSO_4$ or bathocuproine disulfonate (Aldrich, Milwaukee, WI) as described previously.[10] RNA analyses were performed by RNase protection,[10] and the presence of β-galactosidase and HSF proteins was assessed by Western blotting using standard protocols.[18]

The fusion plasmid pPSG*CTR3-URA3* was constructed by the insertion in-frame of the *CTR3* promoter (1116 bp of the 5′-noncoding region) and

[18] J. Sambrook, E. F. Fritsch, and T. Maniatis, "Molecular Cloning: A Laboratory Manual," 2nd Ed. Cold Spring Harbor Laboratory, Cold Spring Harbor, NY, 1989.

the first 10 codons of the *CTR3* gene to the *S. cerevisiae URA3* gene. To accomplish this construct, the *CTR3* sequence from −1116 to +10 was inserted into the *Bam*HI and *Eco*RI sites of pPSG A.[19] The *E. coli lacZ* gene encoding for the β-galactosidase used in this study was isolated from a derivative of YEp*CTR3-lacZ* described previously.[10] To create a new series of yeast copper-responsive expression vectors, plasmids p413-, p414-, p415-, p416-GPD and p423-, p424-, p425-, p426-GPD[2] were expunged of their GPD promoter using the *Sac*I and *Spe*I sites. Then, three copper-responsive promoters derived from the *CTR3,*[13] *CTR1,*[12] and *CUP1*[20,21] genes were inserted successively into each centromeric or 2-μm plasmid series to replace the GPD promoter. The *CTR3* and *CTR1* copper responsive sequences were isolated by polymerase chain reaction (PCR) from p*CTR3-CYC1-lacZ* and p*CTR1-CYC1-lacZ* fusion plasmids described previously.[10] The *CUP1* promoter was also generated by PCR from pYEp*CUP1-393.*[22] The PCR products designed with *Sac*I and *Spe*I recognition sites were sequenced in their entirety to ascertain their correctness. Therefore, a total of 24 vectors, denoted pCu4XXprom, have been developed for cloning and copper-regulated gene expression in yeast.

Acknowledgments

We are thankful to our colleagues, especially members of the Thiele research group, for valuable comments on this work. All expression vectors described in this article have been deposited at the American Type Culture Collection (ATCC, Rockville, MD). This study was supported in part by National Institutes of Health Grant RO1 GM41840 to D.J.T. S.L. is a recipient of a Centennial Fellowship from the Medical Research Council of Canada (MRC). D.J.T. is a Burroughs Wellcome Toxicology Scholar.

[19] P. Silar and D. J. Thiele, *Gene* **104,** 99 (1991).

[20] M. Karin, R. Najarian, A. Haslinger, P. Valenzuela, J. Welch, and S. Fogel, *Proc. Natl. Acad. Sci. U.S.A.* **81,** 337 (1984).

[21] T. R. Butt, E. J. Sternberg, J. A. Gorman, P. Clark, D. Hamer, M. Rosenberg, and S. T. Crooke, *Proc. Natl. Acad. Sci. U.S.A.* **81,** 3332 (1984).

[22] N. Santoro, N. Johansson, and D. J. Thiele, *Mol. Cell. Biol.* **18,** 6340 (1998).

[9] Use of *Pichia pastoris* for Expression of Recombinant Proteins

By STUART A. ROSENFELD*

Introduction

The methylotrophic yeast, *Pichia pastoris,* was initially evaluated as a potential source of single cell protein due to the ability of this yeast to utilize methanol as a sole carbon source.[1] As this system was being investigated, recombinant vector construction, methods for transformation, selectable marker generation, and scale-up fermentation were developed.[2–5] It became evident that *P. pastoris* could serve as a eukaryotic host for heterologous protein expression in that this yeast can provide a suitable environment for folding and secreting foreign proteins as well as carrying out many posttranslational protein modifications characteristic of mammalian cells. Table I compares the general cell composition of various biological systems that have been used for expressing recombinant proteins. It is evident that the maximum protein attainable via *P. pastoris* fermentation is superior to other expression systems. Of course the expression level, activity, and structure of proteins produced by *P. pastoris* can only be evaluated on a case-by-case basis; however, since 1994, at least 100 recombinant proteins (e.g., enzymes, antigens, regulatory proteins, membrane proteins, proteases, antibodies) have been expressed in this organism based on a review of the literature.

Several features of *P. pastoris* make it attractive as a host for expressing recombinant proteins. The characterization and exploitation of the alcohol oxidase (*AOX*) promoter to drive heterologous gene expression was instrumental to the successful implementation of the *P. pastoris* expression sys-

* For any commercial use of the information obtained in this article, or to reproduce or otherwise use this article in whole or in part, permission must be obtained from The DuPont Merck Pharmaceutical Company, the author, and Academic Press.

[1] E. H. Wegner, U.S. Patent 4414329 (1983).

[2] J. M. Cregg, T. S. Vedvick, and W. C. Raschke, *Bio/Technology* **11,** 905 (1993).

[3] J. Cregg and D. R. Higgins, *Can. J. Bot.* (Suppl. 1), S891 (1995).

[4] C. P. Hollenberg and G. Gellissen, *Curr. Opin. Biotechnol.* **8,** 554 (1997).

[5] M. Romanos, *Curr. Opin. Biotechnol.* **6,** 527 (1995).

0076-6879/99 $30.00

TABLE I

COMPARATIVE CELL COMPOSITION: RECOMBINANT PROTEIN EXPRESSION SYSTEMS

Cell type	Dry weight (pg/cell)[a]	Protein[b] (pg/cell)	Maximum cell density (cells/ml)	Maximum protein (g/liter)
E. coli	0.3[c]	0.15	1×10^{11}	15
Pichia	18[d]	9	7×10^{9}	65
Insect	600[e]	300	1×10^{7}	3
Mammalian	300–1500[f]	150–750	1×10^{7}	2

[a] Dry cell weight ≈ 0.3× wet cell weight.
[b] Protein ≈ 0.5× dry cell weight.
[c] F. C. Neidhardt, ed., *in* "*Escherichia coli* and *Salmonella typhimurium,*" p. 4. American Society for Microbiology, Washington, DC, 1987.
[d] Based on *S. cerevisiae* data F. Sherman, *Methods in Enzymol* **194,** 17 (1991).
[e] J. P. Ferrance, A. Goel, and M. Ataai, *Biotechnol. Bioeng.* **42,** 697 (1992).
[f] R. I. Freshney, "Culture of Animal Cells." Wiley-Liss, New York, 1994.

tem.[6,7] This promoter is inducible on methanol addition to the culture, and exceptionally high levels of recombinant proteins (e.g., 12 g/liter for tetanus toxin,[8] 4 g/liter secreted human serum albumin[9]) have been reported. The yeast is capable of growing on defined minimal media and to extremely high cell densities (450 g/liter wet cell weight), and the fermentation process is scaleable to industrial levels of production.

The goal of this article is to emphasize the strategies and methods for optimizing the selection, screening, and scale-up production of recombinant proteins expressed using this system. The basic principles for manipulating *P. pastoris* have been described by Invitrogen Corporation (San Diego, CA) handbooks and manuals and provide an excellent source of practical information to the scientist beginning to use *Pichia*. In addition, several general reviews of this system serve as a useful starting point for understanding the various facets of *P. pastoris* biology and utility.[2–5]

[6] S. B. Ellis, P. F. Brust, P. J. Koutz, A. F. Waters, M. M. Harpold, and T. R. Gingeras, *Mol. Cell Biol.* **5,** 1111 (1985).
[7] J. F. Tschop, P. F. Brust, J. M. Cregg, C. A. Stillman, and T. R. Gingeras, *Nuceic Acids Res.* **15,** 3859 (1987).
[8] J. J. Clare, F. B. Rayment, S. P. Ballantine, K. Sreekrishna, and M. A. Romanos, *Bio/ Technology* **9,** 455 (1991).
[9] K. A. Barr, S. A. Hopkins, and K. Sreekrishna, *Pharm. Eng.* **12,** 48 (1992).

Pichia pastoris Cloning Vectors

Common features of all *P. pastoris* cloning vectors that have been used for expressing recombinant proteins include (a) a marker for selecting plasmid transformants in *Escherichia coli* and *P. pastoris;* (b) a *Pichia* promoter that drives transcription of the recombinant gene insert; and (c) a transcriptional terminator sequence (typically *AOX*1tt). In addition to these features, a signal sequence typically derived from either the pre-pro α mating factor of *Saccharomyces cerevisiae* or the *PHO1* signal sequence of the *P. pastoris* acid phosphatase gene may be incorporated to facilitate secretion of the desired protein to the medium.

Transformation of *P. pastoris* via these vectors occurs through integration of the transforming DNA into the host genome by homologous recombination.[8,10,11] The site of integration can be predetermined by digesting the vector DNA such that its ends are homologous to the corresponding chromosomal site of integration. Based on markers available on a variety of vector DNAs, useful sites of chromosomal integration include the *HIS4* gene, the *AOX1* gene, and the glyceraldehyde-3-phosphate dehydrogenase promoter (pGAP).[12] Plasmid digestion that results in homology to the 5′ end and 3′ regions of the *AOX1* chromosomal locus leads to transplacement (i.e., replacement of the genomic *AOX1* gene with the incomplete plasmid-derived gene). Transplacements may be recognized by a Mut^s phenotype, their slow growth on methanol as a sole carbon source, due to disruption of the chromosomal *AOX1* gene. However, only 5–30% of the selected transformants are true transplacements, the remainder are generated by single site crossover integrations or are a consequence of gene conversion events. Directed single crossover recombination events within the *AOX1, HIS4,* or pGAP may be achieved by digesting the plasmid DNA such that a single cut is introduced within these genes. Resulting transformants have a Mut^+ phenotype, an unimpeded ability to grow on methanol as a sole carbon source. These two modes of vector integration, single crossover integration and transplacement, are illustrated in Figs. 1 and 2, respectively. Recently developed vectors (Invitrogen Corporation) incorporate a zeocin resistance determinant such that digestion with the appropriate restriction endonuclease ensures that virtually all zeocin-resistant transformants are generated by single crossover events at the *AOX1* gene or GAP promoter and all transformants have an integrated copy (or copies) of the desired gene.

[10] J. M. Cregg, K. J. Barringer, and A. Y. Hessler, *Mol. Cell Biol.* **5,** 3376 (1985).

[11] J. M. Cregg, K. R. Madden, K. J. Barringer, G. Thill, and C. A. Stillman, *Mol. Cell Biol.* **9,** 1316 (1989).

[12] H. R. Waterham, M. E. Digan, P. J. Koutz, S. V. Lair, and J. M. Cregg, *Gene* **186,** 37 (1997).

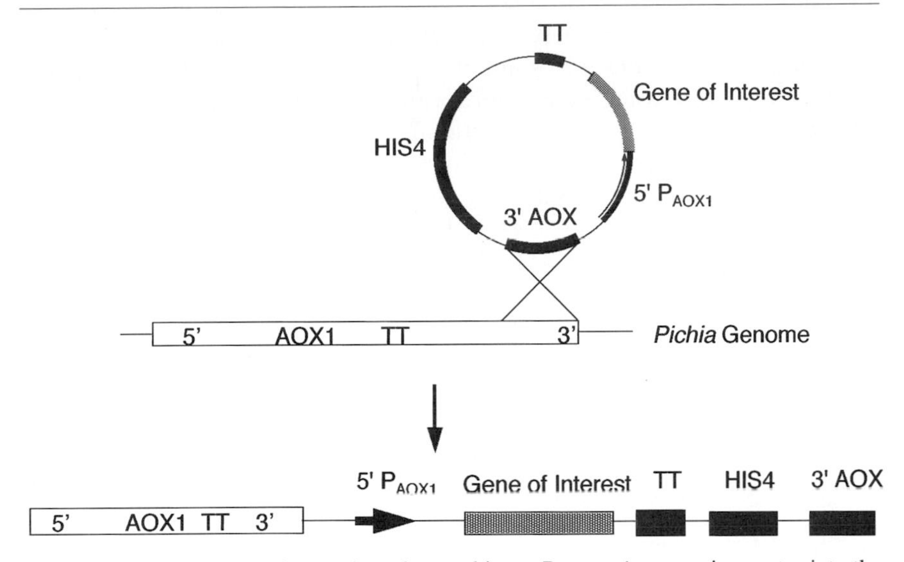

FIG. 1. Single crossover integration of recombinant *P. pastoris* expression vector into the host genome illustrating the integration of an expression vector at the homologous 3′ *AOX1* site on the *P. pastoris* chromosome. If the genomic copy of the *AOX1* gene is a wild type (*AOX1*$^+$), the integrant has a Mut$^+$ phenotype (see text) as a wild-type copy of *AOX1* remains intact.

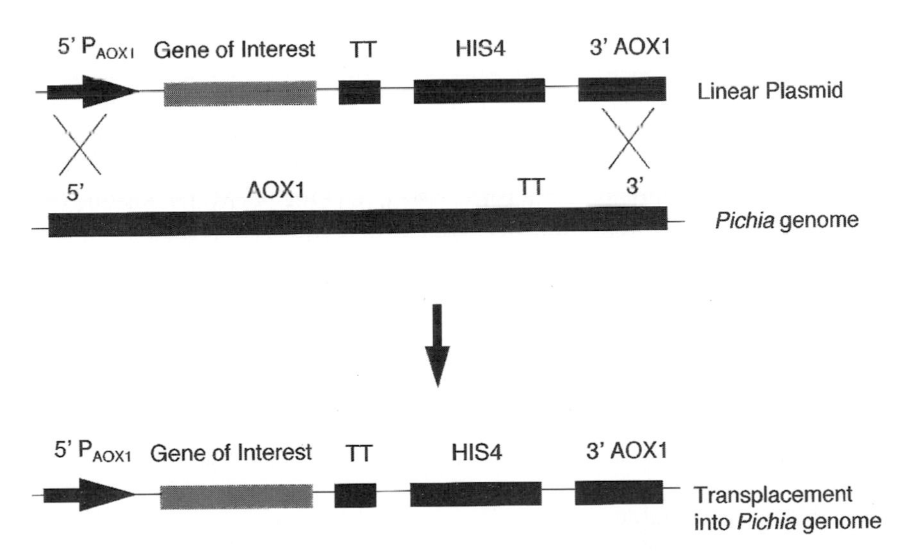

FIG. 2. Transplacement of *AOX1* gene from a *P. pastoris* expression vector into the host genome. In this example, the incoming recombinant vector replaces the chromosomal copy of the *AOX1* gene. The resulting transformant would have a Mut^s phenotype (see text).

Another inherent property of the *P. pastoris* expression system is that multiple gene insertions at a single locus occur with a low but detectable frequency (1–10%), and gene copies ranging from 1 gene copy up to 30 gene copies may be recovered.[8] Screening for such "jackpot" clones has resulted in higher recombinant protein expressing clones due to gene dosage effects. A useful scheme for selecting multiple gene inserts based on integration of a vector pPic9K (Table II) bearing the $Tn903Kan^r$ determinant resistance has been developed by Scorer *et al.*[13]

Table II summarizes some of the pertinent commercially available (Invitrogen Corporation) vectors used for the expression of recombinant proteins in *P. pastoris*. It should be mentioned that most of the vectors contain the *AOX1* promoter, which drives heterologous gene expression on induction with methanol addition to the medium. A constitutive promoter, the glyceraldehyde-3-phosphate dehydrogenase gene promoter (pGAP), has been described.[12] pGapZ vectors (Table II) may assume a larger role in the *P. pastoris* expression system, as heterologous gene expression may be facilitated by growing cells on glucose or glycerol without the need of methanol feeding.

Pichia pastoris Host Strains

GS115 (*his4*) is the most commonly used *P. pastoris* strain for transformation and subsequent protein expression studies. Transformation of this strain with vectors bearing the wild-type *HIS4* and partial *AOX1* gene (Table II) allows one to integrate plasmid-born genes within the chromosomal *HIS4* gene or *AOX1* locus. A HIS^+ phenotype arises from such an integration or transplacement event. Scoring the transformants for their ability to use methanol as a sole carbon source (Mut^+ vs Mut^s) permits one to determine if the integration occurred within the *AOX1* locus.

X-33, a wild-type *P. pastoris* strain, is a useful host for transformations with pPicZ or pGapZ vector constructs (Table II) as transformants are selected on the basis of zeocin resistance. The sites of integration occur at either the *AOX1* locus (using pPicZ) or the glyceraldehyde-3-phosphate dehydrogenase promoter locus (using pGapZ). All such transformants will be Mut^+.

KM71 (*his4, aox1::arg4*) may be used with vectors containing the *HIS4* or zeocin resistance genes as selectable markers where a Mut^s genetic background is desirable.

[13] C. A. Scorer, J. J. Clare, W. R. McCombie, M. A. Romanos, and K. Sreekrishna, *Bio/Technology* **12,** 181 (1994).

TABLE II
COMMERCIALLY AVAILABLE *Pichia* EXPRESSION VECTORS

Vector	Unique cloning site	Selection marker	Application	Reference
pHIL-D2	*Eco*RI	*HIS4*	Gene replacement or insertion at *AOX1* locus; gene insertion at *HIS4* locus	Invitrogen handbook
pPic3.5	*Bam*HI, *Sna*BI, *Eco*RI, *Avr*II, *Not*I	*HIS4*	Multicloning site vector	13
pPic3.5K	*Bam*HI, *Sna*BI, *Eco*RI, *Avr*II, *Not*I	*HIS4* and kan^r	Multicloning site vector, multicopy insert selection with G418	Invitrogen handbook
pHIL-S1	*Xho*I, *Eco*RI, *Sma*I, *Bam*HI	*HIS4*	*Pichia* PHO1 signal sequence	Invitrogen handbook
pPic9	*Xho*I, *Sna*BI, *Eco*RI, *Avr*II, *Not*I	*HIS4*	*S. cerevisiae* pre-pro α-mating factor signal sequence for protein secretion	13
pPic9K	*Sna*BI, *Eco*RI, *Avr*II, *Not*I	*HIS4* and kan^r	*S. cerevisiae* pre-pro α-mating factor signal sequence for protein secretion and multicopy insert selection with G418	13
pPicZ A,B,C	*Sfu*I, *Eco*RI, *Pml*I, *Sfi*I, *Bsm*BI, *Asp*718I, *Kpn*I, *Xho*I, *Sac*II, *Not*I, *Apa*I[a]	Zeocin^r	Gene insertions at *AOX1* locus; myc epitope-$(HIS)_6$-tagged fusion peptides; pPICZα A,B,C, plasmids contain the pre-pro α mating signal sequence for secretion of protein	Invitrogen handbook
pPICZα A,B,C	*Eco*RI, *Pml*I, *Sfi*I, *Bsm*BI, *Asp*718I, *Kpn*I, *Xho*I, *Sac*II, *Not*I, *Xba*I			
pGAPZ A,B,C	*Sfu*I, *Eco*RI, *Pml*I, *Sfi*I, *Bsm*BI, *Asp*718I, *Kpn*I, *Xho*I, *Sac*II, *Not*I, *Apa*I[a]	Zeocin^r	Gene insertions at GAP locus; myc epitope-$(HIS)_6$-tagged fusion peptides; pGAPZα A,B,C plasmids contain the *S. cerevisiae* pre-pro α mating signal sequence for secretion of protein	Invitrogen handbook, 12
pGAPZα A,B,C	*Eco*RI, *Pml*I, *Sfi*I, *Bsm*BI, *Asp*718I, *Kpn*I, *Xho*I, *Sac*II, *Not*I, *Xba*I			

[a] *Apa*I in version A, *Xba*I in version B, and *Sna*BI in version C.

SMD1168 (*his4, pepA*) is a strain that contains a protease deficiency and may be useful for protein expression studies where the desired recombinant protein is susceptible to proteolytic cleavage.

AOX1, AOX2-deficient strains cannot utilize methanol as a carbon source but may be employed for expression studies where induction of the *AOX1* promoter may be facilitated by the addition of methanol as a gratuitous inducer, analogous to the role of isopropylthio-β-galactoside (IPTG) induction of the *lac* promoter in bacterial systems.

The choice of using Mut^+ vs Mut^s strains is largely empirically determined from the expression level of the specific protein of interest. *A priori* there is no way to predict which recombinant strain will be more productive; however, if scale-up fermentations are to be conducted, Mut^+ strains are generally advantageous due to their faster growth rate and less rigorous methanol addition regimens. Transformants generated from pGAPZ vectors circumvent methanol utilization considerations because the expression of recombinant proteins from the GAP promoter is constitutive and does not rely on methanol induction.

Transformation Methodologies

The two most widely used methods for *Pichia* transformation are spheroplast transformation and electroporation.[10,13] Both methods yield about 10^3–10^4 transformants/μg of DNA. The spheroplast method for preparing DNA-competent cells is laborious and more variable than electroporation. However, spheroplasting appears to yield more transformants with multiple gene inserts and may be desirable under certain circumstances. It has been noted that spheroplasting is not recommended when transforming *Pichia* with plasmids containing the zeocin resistance marker (Invitrogen instruction manual). Electroporation of *P. pastoris* has the advantage of speed, simplicity, and reproducibility. Methods for electroporation are described in the Methods section.

Screening *Pichia pastoris* Transformants for Recombinant Protein Expression

As discussed earlier, the diversity of transformants arising from a *P. pastoris* transformation necessitates a means of discriminating transformants that produce the desired recombinant protein at high levels. A variety of methods have been described to select and screen for *P. pastoris* transformants that express high levels of the heterologous protein of interest. Some methods are based on the premise that multicopy integrants are more likely to express higher levels of the recombinant protein compared to single-copy clones. Scorer *et al.*[13] described the use of the *Tn903Kan*^{r} determinant contained on the pPic9K and pPic3K cloning vectors to select for *P. pastoris* integrants that are resistant to increasingly higher levels of G418. Higher levels of G418 resistance correlates with higher gene dosage. The increased gene copy number may result in higher expression of the recombinant protein of interest, although an optimal rather than the highest copy number may elicit the most recombinant protein production. Similarly, rapid DNA dot-blot hybridization techniques have been used to identify transformants containing multicopies of the recombinant gene of interest.[8] Although both these methods have been employed successfully to discern *P. pastoris* transformants that express high levels of recombinant protein, another approach involves using immunological methods to directly identify clones expressing the specific protein of interest.[14–16] Basically, the desired recombinant protein or an easily identifiable epitope (e.g., c-*myc,* Table II) fused to the protein is detected via antibody blotting techniques similar to those used to screen *E. coli* colonies expressing recombinant proteins. Using this methodology, up to 10,000 *P. pastoris* transformants expressing the trimeric CD40L glycoprotein have been screened rapidly to identify the 60 most intensely antibody reactive clones.[16] In addition to the methods described earlier, high throughput screening of *Pichia* clones grown in 96-well plate formats have been described.[17] Sensitive assays or high-performance liquid chromatography (HPLC) analysis may be used to quantitate the level of recombinant protein produced.

Shake Flask Evaluation of Clones Expressing Recombinant Protein

Following the transformation of *P. pastoris* with the appropriate vector encoding the recombinant gene of interest, multiple clones are typically screened for relative recombinant protein expression via shaker flask evaluation. Shaker flask growth conditions can affect the expression/stability of the specific protein under investigation dramatically. Medium (rich vs minimal), pH, and induction methodologies all influence the expression level of a specific recombinant protein. Perhaps most important, high levels of aeration have been shown to improve the performance of shake flask culture growth and protein expression. Because there are no general conditions that are applicable to all proteins to be expressed, a variety of growth conditions are typically employed to determine those that are most suitable for the production of a given recombinant protein. One useful guiding principle when evaluating different growth conditions is to match as closely as possible the growth medium to that which will ultimately be used for larger scale fermentation studies. Another consideration is to use a minimal based medium, where possible, when the intended protein is to be secreted and purified from *Pichia* cultures. Minimal medium does not contain proteins, peptides, and so on that complicate downstream purification efforts.

[14] C. H. M. Kochen and A. W. Thomas, *Anal. Biochem.* **239,** 111 (1996).

[15] J. L. Wung and N. R. J. Gascoigne, *BioTechniques* **21,** 811 (1996).

[16] J. T. McGrew, D. Leiske, B. Dell, R. Klinke, D. Krasts, S.-F. Wee, N. Abbott, R. Armitage, and K. Harrington, *Gene* **187,** 193 (1997).

[17] Y. Laroche, V. Storme, J. DeMeutter, J. Messens, and M. Lauwereys, *Bio/Technology* **12,** 119 (1994).

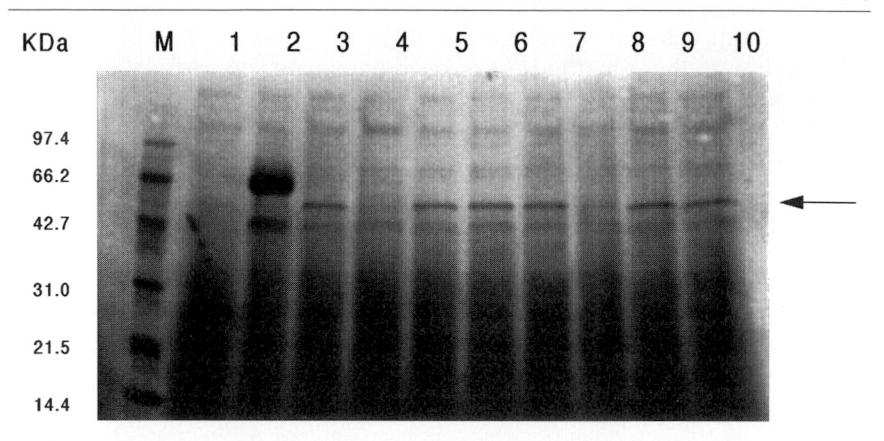

FIG. 3. Coomassie-stained SDS–PAGE of supernatant samples from tube cultures of *P. pastoris* transformed with pPic9 plasmid DNA encoding human fibroblast collagenase (MMP-1). M, molecular weight marker. Lane 1, *Pichia* strain GS115; lane 2, GS115 expressing bovine serum albumin (Invitrogen, Inc.); lanes 3–10, selected His^+ Mut^s transformants. The arrow indicates the position of the MMP-1 protein band. Reproduced from S. A. Rosenfeld, O. H. Ross, M. C. Hillman, J. I. Corman, and R. L. Dowling, *Prot. Expr. Purif.* **7,** 423 (1996).

An example of rapid identification of *P. pastoris* transformants expressing a secreted protein of interest is illustrated in Fig. 3. His^+ Mut^s isolates obtained following transformation with DNA encoding human fibroblast collagenase (MMP-1) were screened by SDS–PAGE analysis of supernatants following methanol induction of test tube cultures.[18] As depicted in Fig. 3, six clones were identified that expressed a secreted protein having the predicted molecular mass for MMP-1 (52 kDa). Depending on the expression levels of the particular secreted protein, this method of analysis is useful for qualitatively evaluating and comparing numerous transformants under a variety of screening media and growth conditions.

Fermentation of Recombinant Protein Expressing *Pichia pastoris* Clones

Perhaps one aspect of the *P. pastoris* expression system that has been shown to have the greatest impact on the successful production of recombinant proteins using this organism is the ability to perform high cell density

[18] S. A. Rosenfeld, O. H. Ross, M. C. Hillman, J. I. Corman, and R. L. Dowling, *Prot. Expr. Purif.* **7,** 423 (1996).

fermentations.[19–22] As discussed earlier (Table I), the high levels of biomass that can be attained during *P. pastoris* fermentations enables one to produce large quantities of recombinant protein at low cost. Careful control of medium, dissolved oxygen, pH, temperature, and induction leads to exceptionally high cell density fermentations (450 g/liter wet cell weight). Although these advantages are also conferred on expression of intracellularly produced proteins, the ability of *P. pastoris* to secrete large amounts of recombinant proteins to the growth medium is arguably its greatest attribute. Secretion of human serum albumin (HSA) at 4 g/liter has been reported,[9] and recovery of secreted proteins from culture supernatants is considerably easier than the recovery of intracellular proteins from cell pastes. Some secreted proteins are degraded or unstable during *Pichia* fermentations and in these cases several approaches have been used to minimize this protein breakdown. The addition of amino acids or peptides (e.g., casamino acids),[23] buffering the medium to a pH value where protein degradation is minimized,[24,25] or using a *pep4* strain[24,26] have all been employed successfully to obviate the proteolysis of various secreted proteins.

Methods

Growth Media

A complete description of media that are used for *Pichia* growth, transformation, and shake flask analysis is detailed in Invitrogen Corporation handbooks. In general, the media formulations vary, depending on the presence or absence of complex protein components, selection marker requirements, pH buffering capabilities, and primary carbon source.

[19] R. S. Siegel and R. A. Brierley, *Biotechnol. Bioeng.* **34,** 403 (1989).

[20] R. A. Brierley, R. S. Siegel, C. M. Bussineau, W. S. Craig, G. C. Holz, G. R. Davis, R. G. Buckholz, G. P. Thill, L. M. Wondrack, M. E. Digan, M. M. Harpold, S. V. Lair, S. B. Ellis, and M. E. Williams, International Patent Application, Publication No. WO 90/03431 (1989)

[21] R. A. Brierley, C. Bussineau, R. Kosson, A. Melton, and R. S. Siegel, *Ann. N.Y. Acad. Sci.* **589,** 350 (1990).

[22] R. S. Siegel, R. G. Buckholz, G. P. Thill, and L. M. Wondrack, International Patent Application, Publication No. WO 90/10697 (1990)

[23] J. J. Clare, M. A. Romanos, F. B. Rayment, J. E. Rowedder, M. A. Smith, M. M. Payne, K. Srekrishna, and C. A. Henwood, *Gene* **105,** 205 (1991).

[24] R. A. Brierley, G. R. Davis, and G. C. Holz, U.S. Patent 5,324,639 (1994).

[25] S. A. Rosenfeld, D. Nadeau, J. Tirado, G. F. Hollis, R. M. Knabb, and S. Jia, *Prot. Expr. Purif.* **8,** 476 (1996).

[26] R. G. Brankamp, K. Sreekrishna, P. L. Smith, D. T. Blankenship, and A. D. Cardin, *Prot. Expr. Purif.* **6,** 813 (1995).

Electroporation of Pichia pastoris

Preface: All solutions and bottles should be sterile. All solutions should be kept ice cold.

1. Grow a 5-ml culture of the appropriate *Pichia* host overnight at 30° in yeast extract peptone dextrose (YEPD) medium in a sterile 50-ml conical tube with good aeration.
2. Inoculate 250 ml of YEPD medium in at least a 1-liter flask for good aeration with 0.25 ml of culture. Incubate at 30° overnight.
3. Culture should be about A_{600} = 1–10. Dilute culture in fresh 250 ml of YEPD to A_{600} = 0.5. Grow to A_{600} = 1.5 at 30°. Generation time is approximately 2–3 hr.
4. Place culture on ice and centrifuge in a sterile 250-ml polycarbonate centrifuge bottle at 5000*g* (about 6000 rpm with a GSA rotor) for 10 min at 4°. If pellet is not tight, respin longer at a slightly higher centrifugal force.
5. Resuspend pellet in 250 ml *ice-cold* sterile Milli-Q water (Millipore Corp., Bedford, MA) by hand swirling or gentle vortexing.
6. Repeat spin. Pellet may be noticeably looser at this step, so use caution when decanting supernatant.
7. Resuspend pellet in 125 ml water as described in step 5. Centrifuge as in step 4.
8. Resuspend pellet in 10 ml ice-cold sterile 1 *M* sorbitol. Transfer to 50-ml sterile round-bottom centrifuge tube.
9. Centrifuge as in step 4 and resuspend pellet in 0.5 ml of ice-cold 1 *M* sorbitol.
10. Mix 80 μl of cells with 10 μg of linearized vector DNA (in about 5 μl TE) in an ice-cold 0.2-em gap electroporation cuvette. Include electroporation of cells alone without DNA addition as a negative control.
11. Incubate cuvette in ice for 5 min.
12. Pulse cells at the following settings on a Bio-Rad (Richmond, CA) gene pulser (charging voltage 1500, capacitance 25, resistance 400). Using these parameters, a pulse length of about 8–9 msec should be obtained.
13. Immediately add 1.0 ml of ice-cold 1 *M* sorbitol to cuvette and mix.
14. Spread 200 μl on each of five selection plates and incubate at 30° until colonies arise (about 3 days).

Additional Screening Protocols

Screening of Pichia pastoris Transformants for High-Level G418 Resistance. As mentioned earlier, multiple plasmid integration events may be selected by incorporating the gene for G418 resistance onto the transforming DNA (see pPIC9K, Table II); higher levels of G418 resistance often correlate with a higher gene dosage or copy number.[13] The increased gene copy number may result in higher expression of the recombinant protein of interest.

1. Electroporate *P. pastoris* with digested recombinant vector of choice as described earlier.
2. Transformants that arise on the selection plate are replica plated onto YPD medium [1% yeast extract (Difco Laboratories, Detroit, MI), 2% peptone, 2% dextrose (w/v), 2% agar] containing increasing amounts of G418 (0.25, 0.5, 1.0, 2.0, and 4.0 mg/ml) and incubated at 30°. A decreasing number of G418-resistant colonies will arise at increasing concentrations of the antibiotic.
3. Pick representative colonies from plates containing each concentration of G418 and streak for single-colony isolation on agar medium containing G418 at the concentration from which the colonies were picked.
4. Grow and induce representative transformants in shake flasks and determine which ones express the recombinant protein to the highest level.

Screening Pichia pastoris Transformants with Antibodies to Secreted Recombinant Protein. This screening method relies on growing transformed *P. pastoris* colonies on blotting membranes under uninduced conditions (e.g., without methanol in the medium) followed by transfer of the membranes to inducing medium (methanol-containing medium). The membranes are then treated with antibodies to the protein of interest followed by an appropriate secondary antibody and detection reagents according to standard immunological methods. In this fashion, thousands of transformed colonies may be assessed rapidly for their secretion of recombinant protein. Variations on this theme may be applied, depending on the vector–host combination employed.

1. Electroporate an appropriate *P. pastoris* host with the digested recombinant vector of choice and plate on selective medium. Incubate at 30° for 3–5 days until transformants arise.
2. Colonies are either replica plated or spotted onto cellulose acetate membranes (Schleicher & Schuell, Keene, NH) that have been placed on minimal glycerol medium agar plates (Invitrogen handbook) or supplemented medium, depending on the host. Make sure there are no air bubbles beneath the membrane and that it is completely wetted. Incubate the plates at 30° for 48 hr.
3. Place the cellulose membrane on top of a prewetted nitrocellulose membrane (Schleicher & Schuell) that is on medium-containing methanol (0.5%, v/v) and incubate for 2–6 days at 30°, during which time the membranes are lifted and placed on fresh plate medium every 24 hr.

4. Block the nitrocellulose membranes according to standard methods[27] and use directly for immunostaining with an appropriate primary antibody, secondary antibody, and detection reagents. Colonies displaying the highest immunoreactivity are retained for shake flask studies.

Fermentation Considerations

As described previously, *P. pastoris* can be grown to extremely high cell density and is capable of producing high levels of recombinant protein both intracellularly and secreted to the growth medium. However, to achieve such high production capabilities, fermentors must be employed to maintain and control growth parameters (e.g., pH and dissolved oxygen) not achievable with shake flask culturing. Therefore, the microbiologist/molecular biologist must become somewhat familiar with fermentation equipment and its operation. A variety of vendors provide fermentors from bench-top to production scale, and investigators inexperienced in fermentation should consult with the vendors directly concerning specific facility and utility requirements.

The following guidelines for fermentation of *P. pastoris* are designed for bench-scale (1–5 liter) fermentors, a scale that is generally applicable to many processes; however, the concepts are applicable to larger volume fermentations if necessary. Also the medium composition and feed considerations are based on the use of the alcohol oxidase promoter driving recombinant protein expression in a Mut^+ strain.

1. Prepare an inoculum (5–10% of the initial fermentation volume) by growing an overnight culture of the desired *P. pastoris* recombinant in either minimal glycerol (MGY) medium or buffered glycerol complex (BMGY) medium (see Invitrogen handbook for media recipes). The inoculum is typically grown in baffled flasks at 30° at 300 rpm until $A_{600} \approx 5$ (16–24 hr).
2. Prepare the fermentation medium by sterilizing the fermentation vessel containing fermentation basal salts medium (FBS) [add per liter: $CaSO_4 \cdot 2H_2O$, 0.93 g; K_2SO_4, 18.2 g; $MgSO_4 \cdot 7H_2O$, 14.9; KOH, 4.13 g; phosphoric acid (85%), 26.7 ml; glycerol, 40.0 g]. Remember that following the later stages of the fermentation (methanol feeding), the fermentation volume will be approximately twice the initial fermentation volume, therefore the initial volume should take this into account when starting the run.

[27] E. Harlow and D. Lane, "Antibodies: A Laboratory Manual." Cold Spring Harbor Laboratories, Cold Spring Harbor, NY, 1988.

3. After sterilization and cooling, set the fermentation temperature to 30°, and set the agitation and aeration to operating conditions. Typically these conditions require keeping the dissolved oxygen above 20% by adjusting the agitation and air/O_2 (> 1 vvm) as needed. Adjust the fermentor medium to pH 5.0 with NH_4OH (undiluted 30% stock), which also serves as the sole nitrogen source during fermentation. Control foaming by adding a foam control agent [≅ 1.0 ml/liter fermentation volume of J673 or polypropylene glycol, Struktol Company of America (Stow, OH)] as required.
4. After autoclaving the fermentor, add 2.0 ml of *Pichia* trace elements solution (PTM-1) [add per liter: $CuSO_4 \cdot H_2O$, 6.0 g; NaI, 0.08 g; $MnSO_4 \cdot H_2O$, 3.0 g; $NaMoO_4 \cdot 2H_2O$, 0.2 g; H_3BO_3, 0.02 g; $CoCl_2 \cdot 6H_2O$, 0.5 g; $ZnCl_2$, 20.0 g; $FeSO_4 \cdot 7H_2O$, 65.0 g; biotin, 0.2 g; H_2SO_4, 5.0 ml] per liter of fermentation basal salts medium. Filter sterilize the PTM-1 solution and store at room temperature.
5. Add the inoculum to the vessel.
6. The batch culture stage continues until the glycerol in the medium is consumed (18–24 hr), as evidenced by a sharp increase in the dissolved oxygen, a dissolved oxygen spike. Cell culture samples are taken and typically a cellular yield of about 100 g/liter wet cells is expected.
7. Following the glycerol batch phase of growth, add an additional 4.0% (w/v) glycerol to the vessel. After the second glycerol batch exhausts (detected by a dissolved oxygen spike about 2.5 hr after the glycerol is added), the cell weight will be ≈200 g/liter.
8. A delay of about 15–30 min between the termination of the glycerol batch and the start of the methanol fed batch is advised to allow the residual glycerol to be consumed completely. A more rapid adaptation to the new carbon source has been observed when this delay is performed.
9. Although *P. pastoris* is methylotrophic, the methanol feed (100% methanol) must be slow to adapt the cells to this carbon source. The initial methanol feed rate is set to 4.5 ml/hr/liter initial fermentation volume. Allow the cells to adapt for 4 hr while maintaining the DO at 25–40%. Adaptation is observed by turning off the methanol feed and detecting a DO spike within a minute.
11. At the end of the initial methanol feed stage, increase the methanol feed rate to 9.0 ml/hr/liter initial fermentation volume. Check the feed rate periodically by shutting the feed pump off and observing dissolved oxygen spiking. Allow this feed to continue overnight.
12. Increase the methanol feed rate to 15 ml/hr/liter initial fermentation volume and maintain this rate throughout the remainder of the

fermentation. Maintain this feed rate until the methanol delivered is about three-fourths of the initial fermentation volume. For example, if the initial fermentation volume was 2 liter, about 1.5 liter of methanol would be added to the vessel, resulting in a final fermentor volume of 3.5 liter. The cell density at the termination of the fermentation approaches 350–450 g/liter. Although the entire fermentation lasts about 3 days, recombinant protein expression will vary depending on the specific protein studied, and each fermentation must be optimized.

Comments

Careful monitoring and control of temperature, pH, and dissolved oxygen is essential for a successful *Pichia* fermentation. The fermentation process requires high levels of oxygen to maintain a DO level between 20 and 40%. The metabolic consequence of this high oxygen consumption rate is significant heat generation. Therefore, facilities must have adequate cooling and oxygen delivery capabilities to fully exploit *P. pastoris* high cell density fermentations. Slowing the methanol feed rates described earlier throttles back on the high oxygen demand and subsequent heat generation and provides a means to bring a fermentation back into control if necessary.

Care has to be exercised when working with Mut^s cultures, as their ability to optimally utilize methanol is impaired; in fact, methanol toxicity may be encountered if the methanol concentration is too high in the medium. Therefore, Mut^s cultures are typically grown such that the methanol concentration never exceeds 0.3%. During the methanol induction stage, the methanol feed rate is initiated at 0.5 ml/liter/hr for the first 2 hr then increased about 10% every 30 min until a final feed rate of about 3.0 ml/liter/hr is attained. The duration of the fermentation is determined empirically based on the specific expression of the desired recombinant protein during the course of the fermentation.

A recently developed *Pichia* expression construct that drives gene expression from the glyceraldehyde-3-phosphate dehydrogenase promoter simplifies the fermentation regimen. Constitutive expression permits a single glucose (or glycerol) feed to be conducted in much the same way as the glycerol/methanol feeding described earlier.[12] A more thorough characterization of this system will ensue as additional investigators utilize it increasingly to express their proteins of interest. Particularly in large-scale fermentation settings, where delivery of large volumes of methanol pose fire and safety issues, alternative expression methods are desirable.

Discussion

This article has presented the basic elements of the *P. pastoris* expression system. While no single expression system will satisfy all research needs, *P. pastoris* has proven to be a useful addition to existing prokaryotic and eukaryotic protein production systems. What does the future hold for this system? *Pichia pastoris* may be used more extensively in the production of uniformly labeled proteins with heavy isotopes of carbon (^{13}C) and nitrogen (^{15}N) for structural and protein crystallography studies.[17,28] In addition, multisubunit protein complexes have been produced using this system, which may facilitate more involved protein chemistry/enzymology and characterization.[16,29] The high biomass and protein productivity attained by *P. pastoris* may be exploited for the production of biocatalysts.[30] *Pichia pastoris* may be useful for the expression of pharmacologically interesting receptors, including the important G-protein-coupled receptors.[31,32] This ability to express diverse proteins in *P. pastoris* suggests that the system may be amenable to high throughput drug-screening efforts in addition to its role as a source of recombinant proteins. As the molecular biology and genetics of *P. pastoris* matures, novel applications of this system to research problems will continue to evolve.

[28] A. B. Mason, R. C. Woodworth, R. W. A. Oliver, B. N. Green, L.-N. Lin, J. F. Brandts, B. M. Tam, A. Maxwell, and R. T. A. MacGillivray, *Prot. Expr. Purif.* **8,** 119 (1996).

[29] A. Kalandadze, M. Galleno, L. Foncerrada, J. L. Strominger, and K. W. Wucherpfennig, *J. Biol. Chem.* **271,** 20156 (1996).

[30] M. S. Payne, K. L. Petrillo, J. E. Gavagan, R. Dicosimo, L. W. Wagner, and D. L. Anton, *Gene* **194,** 179 (1997).

[31] H. M. Weiß, W. Haase, H. Michel, and H. Reiländer, *FEBS Lett.* **377,** 451 (1995).

[32] F. Talmont, S. Sidobre, P. Demange, A. Milon, and L. J. Emorine, *FEBS Lett.* **394,** 268 (1996).

[10] Use of Yeast *sec6* Mutant for Purification of Vesicles Containing Recombinant Membrane Proteins

By LARRY A. COURY, MARK L. ZEIDEL, and JEFFREY L. BRODSKY

Introduction

Although novel open reading frames and homologs of existing genes continue to be identified from the sequencing of whole organism DNA, the functional analysis of the corresponding gene products lags behind the sequence analysis. For this reason, the need to express and characterize

0076-6879/99 $30.00

proteins in heterologous systems is greater than ever. Prokaryotic expression systems offer the advantages of fast growth rates, easy manipulation of the cultured organisms, multiple selectable markers, a large variety of genetic backgrounds to facilitate gene expression, and the potential ability to produce large quantities of protein. However, prokaryotes lack an intracellular secretory pathway that may be essential to process eukaryotic gene products posttranslationally. On the contrary, tissue culture cells do post translationally modify proteins and harbor second messenger signaling pathways that may be essential for protein function, although these advantages may be outweighed by the expense of growing large quantities of cells in culture and by the inability to regulate gene expression tightly.

A heterologous expression system that combines advantages of both prokaryotic and eukaryotic expression systems is yeast. Yeast can glycosylate and phosphorylate proteins that are overexpressed,[1,2] and they contain the identical secretory pathway and many of the second messenger signaling pathways that exist in higher eukaryotic cells, as reviewed.[3-6] Yeast can also be grown inexpensively in large quantities, contain many genetic backgrounds that may increase protein expression, and can be manipulated genetically if a novel background is desired. In addition, yeast have several inducible promoters that allow regulated expression of specific genes.[7-9] Finally, it is possible to exploit the properties of specific yeast mutants, such as mutants in the secretory pathway, that may facilitate the enrichment of heterologous proteins in defined subcellular compartments.

This article describes the use of the yeast mutant, *sec6-4,* that contains a temperature-sensitive mutation in a gene product required for the transport of secretory vesicles from the *trans*-Golgi network to the plasma membrane.[10] When *sec6* yeast are grown at the restrictive temperature of 37°, post-Golgi, plasma membrane-targeted vesicles accumulate and eventually the cells die; however, cells grow normally at the permissive temperature of 25°.[11] This mutant has the particular advantage that gene expression

[1] K. J. Blumer, J. E. Reneke, and J. Thorner, *J. Biol. Chem.* **263,** 10836 (1988).

[2] R. L. Johnson, R. A. Vaughan, M. J. Caterina, P. J. M. Van Haastert, and P. N. Devreotes, *Biochemistry* **30,** 6982 (1991).

[3] P. Catty and A. Goffeau, *Biosci. Rep.* **16,** 75 (1996).

[4] B. Errede, R. M. Cade, B. M. Yashar, Y. Kamada, D. E. Levin, K. Irie, and K. Matsumoto, *Mol. Reprod. Dev.* **42,** 477 (1995).

[5] S. J. Kron and N. A. Gow, *Curr. Opin. Cell Biol.* **7,** 845 (1995).

[6] W. H. Mager and A. J. DeKruijff, *Microbiol. Rev.* **59,** 506 (1995).

[7] D. Lohr, P. Venkov, and J. Zlatanova, *FASEB J.* **9,** 777 (1995).

[8] J. O. Mascorro-Gallardo, A. A. Covarrubias, and R. Gaxiola, *Gene* **172,** 169 (1996).

[9] G. Weitzel and G. C. Li, *Int. J. Hyperther.* **9,** 783 (1993).

[10] D. R. TerBush, T. Maurice, D. Roth, and P. Novick, *EMBO J.* **15,** 6483 (1996).

[11] P. Novick, C. Field, and R. Schekman, *Cell* **21,** 205 (1980).

from an inducible promoter can be initiated concomitant with the arrest of vesicle fusion and membrane protein insertion at the plasma membrane. Thus, gene expression begins at the same time that secretory vesicles become unable to fuse with the plasma membrane, ensuring that the desired gene product accumulates in the membranes of these vesicles. The purification of these vesicles is rapid and simple, thereby facilitating the subsequent characterization of the desired gene product. Finally, because the Sec6 protein is known to be involved only in the fusion of these vesicles with the plasma membrane,[10,11] translocation and processing of proteins in the endoplasmic reticulum and processing in the Golgi are largely unaffected by the *sec6* mutation.

The *sec6-4* mutant has been used by several groups to express and study membrane proteins from yeast and other organisms. The mutant was first utilized to study mutations in the yeast plasma membrane H^+-ATPase.[12] This group manipulated the wild-type, chromosomal copy of the ATPase so that it was under the control of the *GAL1* promoter (a promoter active in the presence of galactose and in the absence of glucose), and mutant forms of the enzyme were introduced on a plasmid under control of a heat shock promoter. Therefore, if the yeast were switched from galactose-containing to glucose-containing medium and switched simultaneously from 25 to 37°, expression of the wild-type gene was repressed and expression of the mutant gene was induced, leading to the accumulation of the mutant ATPase in the secretory vesicles. This permitted the expression and characterization of mutant forms of the ATPase in secretory vesicles lacking the wild-type enzyme[12]; this analysis represented a technical breakthrough because a functional form of the H^+-ATPase is normally essential for cell viability.[13]

The *sec6-4* mutant has also been used to express the three mouse isoforms of the multidrug resistance P-glycoproteins (Mdr1, Mdr2, and Mdr3) using the constitutive promoter for alcohol dehydrogenase.[14,15] Assaying these proteins in mammalian cells is difficult, but their expression and characterization in secretory vesicles provide many advantages: the vesicles are tightly sealed, uniform in size, functionally polarized with the cytoplasmic face directed outward, and maintain a strong electrochemical gradient from the endogenous proton ATPase, which allows for the study of electrogenic transport.

[12] R. K. Nakamoto, R. Rao, and C. W. Slayman, *J. Biol. Chem.* **266,** 7940 (1991).
[13] R. Rao and C. W. Slayman, *Biophys. J.* **62,** 228 (1992).
[14] S. Ruetz and P. Gros, *Cell* **77,** 1071 (1994).
[15] S. Ruetz and P. Gros, *J. Biol. Chem.* **269,** 12277 (1994).

Finally, we[16] and others[17] chose to use the *sec6* expression system to examine the biophysical properties of human water channels, or aquaporins. Whereas Laize *et al.*[17] used a constitutive promoter to express CHIP28 (AQP1) in *sec6* vesicles, we utilized the inducible *GAL1* promoter to maximize protein expression and to avoid any potential toxic effects from constitutive protein expression. Historically, the mammalian aquaporins have been difficult to study because these proteins, with the exception of AQP1, are often present in low abundance and are difficult to purify.[18,19] In addition, the only system that has previously allowed for heterologous protein expression and measurements of the permeabilities of the aquaporins utilized *Xenopus* oocyte injection.[20,21] This system has the disadvantage that it is difficult to control the electrochemical gradients across the membrane because the composition of the cytosol is regulated by the oocyte. In contrast, aquaporin expression in yeast secretory vesicles provides larger quantities of protein and an *in vitro* system that is manipulated easily. The secretory vesicles can be assayed directly and more extensively, and detailed biophysical measurements of water and solute permeabilities can be measured using previously described methods.[22] In sum, the expression of the desired membrane protein in yeast secretory vesicles using the *sec6* mutant provides a unique and advantageous system for gene expression and characterization *in vitro*.

Materials and Methods

A number of variations in the protocol for gene expression and subsequent isolation of yeast secretory vesicles have been reported.[12,14–17,23–25] The procedures vary according to the desired purity of the vesicles, the specifics of the assay procedure, and the necessity to tightly regulate gene expression. The procedures described in this article represent the conditions

[16] L. A. Coury, J. C. Mathai, G. V. R. Prasad, J. L. Brodsky, P. Agre, and M. L. Zeidel, *Am. J. Physiol.* **274** (*Renal Physiol.* **43**), 734 (1998).

[17] V. Laize, G. Rousselet, J. M. Verbavatz, V. Berthonaud, R. Gobin, N. Roudier, L. Abrami, P. Ripoche, and F. Tacnet, *FEBS Lett.* **373,** 269 (1995).

[18] S. Nielsen, B. L. Smith, E. I. Christensen, M. A. Knepper, and P. Agre, *J. Cell Biol.* **120,** 371 (1993).

[19] Y. Maeda, B. L. Smith, P. Agre, and M. A. Knepper, *J. Clin. Invest.* **95,** 422 (1995).

[20] G. M. Preston, T. P. Carroll, W. B. Guggino, and P. Agre, *Science* **256,** 385, (1992).

[21] G. M. Preston, J. S. Jung, W. B. Guggino, and P. Agre, *J. Biol. Chem.* **268,** 17, (1993).

[22] M. B. Lande, J. M. Donovan, and M. L. Zeidel, *J. Gen. Physiol.* **106,** 67 (1995).

[23] C. L. Holcomb, T. Etcheverry, and R. Schekman, *Anal. Biochem.* **166,** 328 (1987).

[24] C. L. Holcomb, W. J. Hansen, T. Etcheverry, and R. Schekman, *J. Cell Biol.* **106,** 641 (1988).

[25] N. C. Walworth and P. J. Novick, *J. Cell Biol.* **105,** 163 (1987).

for gene expression and vesicle isolation employed routinely in our laboratories. Useful variations in these procedures are referenced accordingly.

Plasmid Construction

If the synthesis of a mammalian protein is desired, it is important to use the corresponding cDNA for protein expression in yeast as differences in the splicing machinery between yeast and higher eukaryotes are well documented.[26] Also, it is best to minimize the length of the 5′-untranslated region[27–29] (also see "Discussion"). The literature on heterologous protein expression in yeast is extensive, and presumably a host of promoters are satisfactory. However, we find that expression of a cDNA in the *sec6-4* yeast mutant is best controlled and may be maximized with an inducible promoter. Although many inducible promoters are used in yeast, we chose the *GAL1* promoter, which is one of the most commonly utilized.

The pYES2 expression vector (available from Invitrogen, San Diego, CA) contains the *GAL1* promoter followed by a multiple cloning site and has been used extensively by us for the expression of aquaporins and other heterologous proteins in yeast.[16,30] Other commonly used inducible promoters include the metallothionein *CUP1* promoter,[8] which is tightly controlled by copper; promoters activated in response to heat shock, which are of particular interest for expression in the temperature-sensitive *sec6-4* mutant[9]; and the *PHO5* promoter,[31] which is derepressed at low phosphate concentrations. Excellent reviews of yeast vectors[32] and constitutive and inducible promoters[33] are available.

Growth Conditions

Introduction of the plasmid into yeast cells may be accomplished either by electroporation or LiCl-mediated transformation.[34] In general, electroporation requires less time, although it is often less efficient. Isolation of transformants requires selection of yeast containing the plasmid on media

[26] J. D. Beggs, J. VandenBerg, A. Van Ooyen, and C. Weissman, *Nature* **283,** 835 (1980).
[27] R. Serrano and J.-M. Villalba, *Methods Cell Biol.* **50,** 481 (1995).
[28] A. G. Hinnebusch and S. W. Liebman, "The Molecular and Cellular Biology of the Yeast *Saccharomyces*" (J. R. Broach, J. R. Pringle, and E. W. Jones, eds.), Vol. I, p. 627. Cold Spring Harbor Laboratory Press, Cold Spring Harbor, NY, 1991.
[29] T. F. Donahue and A. M. Cigan, *Methods Enzymol.* **185,** 366 (1990).
[30] E. D. Werner, J. L. Brodsky, and A. A. McCracken, *Proc. Natl. Acad. Sci. U.S.A.* **93,** 13797 (1996).
[31] W. E. Raymond and N. Kleckner, *Mol. Gen. Genet.* **238,** 390 (1993).
[32] J. C. Schneider and L. Guarente, *Methods Enzymol.* **194,** 373 (1991).
[33] M. Schena, D. Picard, and K. R. Yamamoto, *Methods Enzymol.* **194,** 389 (1991).
[34] D. M. Becker and L. Guarente, *Methods Enzymol.* **194,** 182 (1991).

lacking a specific nutrient or containing a specific drug to which only the yeast transformants are resistant. For example, yeast that are *ura3* auxotrophs are able to grow on media lacking uracil when they contain the pYES2 expression vector that contains the wild-type *URA3* gene. Other selectable markers used commonly in yeast include enzymes in the adenine, histidine, leucine, lysine, and tryptophan biosynthetic pathways. The preparation of yeast media lacking specific nutrients may be found elsewhere.[35] We cloned the AQP1 and AQP2 cDNAs into the pYES2 expression vector and selected for transformants on plates with synthetic complete (SC) medium[36] lacking uracil but containing 2% raffinose as the carbon source (SC −Ura raff medium).

Following transformation, single colonies were isolated and grown overnight to saturation in 2 ml of SC −Ura raff medium at 25° with constant shaking. In a subsequent step we switched the yeast to medium containing galactose as the carbon source as the *GAL1* promoter initiates gene expression only when galactose is the predominant carbon source. We use 2% raffinose instead of glucose at this stage of the protocol, however, because if yeast are switched directly from glucose to galactose, a "galactose lag" occurs whereby the yeast deplete preexisting energy stores before derepressing the genes that are required for the utilization of galactose.[7,37] Therefore, it is advantageous to grow the yeast in a nonglucose carbon source, such as raffinose or glycerol, so that the subsequent induction of galactose-regulated gene expression is rapid. It should be noted, however, that the doubling time of yeast grown in raffinose is somewhat slower than for yeast grown in glucose.

The 2-ml starter culture in SC −Ura raff medium is added to a 1-liter culture of the same growth medium and incubated at 25° with constant shaking. When these cultures reach an OD_{600} (optical density at a wavelength of 600 nm) of about 1.0 (usually about 12 hr), the cultures are centrifuged at 4000*g* (5000 rpm in a Sorvall GSA rotor) at 4° for 5 min, resuspended in 4 liters of SC −Ura gal induction medium (containing 2% galactose instead of 2% raffinose as the carbon source), and shifted to 37° for 2–3 hr to induce protein expression in the *sec6* vesicles. Protein yields may be increased, however, by resuspending the cells in rich medium (instead of SC −Ura), which consists of 1% yeast extract, 2% peptone, and 2% galactose (YP-Gal), and by incubating at 25° for 2–3 hr prior to the 37° temperature shift. Although rich medium carries the risk of increased

[35] F. Sherman, *Methods Enzymol.* **194,** 3 (1991).

[36] M. D. Rose, F. Winston, and P. Heiter, "Methods in Yeast Genetics." Cold Spring Harbor Laboratory Press, Cold Spring Harbor, NY, 1990.

[37] M. Johnston, *Microbiol. Rev.* **51,** 458 (1987).

plasmid loss and growth in induction medium at the permissive temperature may allow potentially toxic proteins to reach the cell surface (see Discussion), such problems were not encountered.[16]

Following growth at 37°, the cells are collected by centrifugation at 4000*g* at 4° for 5 min and washed once in ice-cold water. Pellets are resuspended in an absolute minimum volume of water and quick frozen in liquid nitrogen. Cultures may then be stored indefinitely at −70°.

Vesicle Isolation

Although Holcomb *et al.*[23,24] first described a method for obtaining an extremely pure fraction of plasma membrane-targeted secretory vesicles, we and others have found that this level of purification is not necessary for assaying heterologously expressed proteins.[12,14–17,25] In addition, yields are reduced greatly when vesicles are purified to this extent, and because the procedure is lengthy, requiring multiple rounds of centrifugation as well as a 12-hr electrophoresis step, a loss of enzymatic activity may be encountered. Thus, we describe here a modification of a previously described method[12,25] used by Brodsky *et al.*[38] for the isolation of yeast microsomes.

Thawed cultures are resuspended to a final concentration of 50 OD_{600} units/ml (e.g., a 1-liter culture at OD_{600} = 1.0 is resuspended in 20 ml) in 10 m*M* dithiothreitol (DTT) and 100 m*M* Tris–Cl, pH 9.4. The resuspended culture is shaken gently at room temperature for 10 min. This step increases the efficiency of spheroplast lysis at a later step by reducing disulfide bonds in the yeast cell wall. We then collect the cells by centrifugation at 4000*g* at 4° for 5 min and resuspend them in spheroplast buffer to a final concentration of 50 OD_{600} units/ml. Spheroplast buffer consists of 1.4 *M* sorbitol, 50 m*M* K_2HPO_4, pH 7.5, 10 m*M* NaN_3, and 40 m*M* 2-mercaptoethanol. Spheroplasts are generated by digesting the cell wall with lyticase (or zymolyase) for 45 min at 37°. The amount of bacterially expressed, recombinant lyticase needed to form spheroplasts may be determined empirically; after 45 min the OD_{600} of a 10-μl aliquot of the yeast suspension diluted into 1 ml of 0.1% sodium dodecyl sulfate (SDS) should be ~20% of the OD_{600} of the initial dilution measured at 0 min. Alternatively, 5 mg (250 mg/ml stock solution) of lyticase purified from *Arthrobacter luteus* may be used at ~10 units per OD_{600} unit of cells. Both the bacterially expressed, recombinant form and the purified *A. luteus* enzyme are available commercially (Sigma, St. Louis, MO).

[38] J. L. Brodsky, S. Hamamoto, D. Feldheim, and R. Schekman, *J. Cell Biol.* **120,** 95 (1993).

The spheroplasts are then harvested at 3000*g* (5000 rpm, SS-34 rotor) for 5 min at 4°, and the cells are resuspended gently with a pipette or Teflon rod in spheroplast buffer containing 10 m*M* $MgCl_2$ to a final concentration of 50 OD_{600} units/ml. Concanavalin A (Sigma, St. Louis, MO) is then added to a final concentration of 0.78 to 1.25 mg/ml and incubated with rotation or gentle shaking at 4° for 15–30 min. A concanavalin A stock solution (25 mg/ml) is prepared in spheroplast buffer containing 1 m*M* $MnCl_2$ and 1 m*M* $CaCl_2$ and may be frozen in 1-ml aliquots. Concanavalin A cross-links the glycoprotein-rich plasma membrane and increases its density above that of the secretory vesicles, facilitating the subsequent separation of secretory vesicles from plasma membrane.

Lectin-coated spheroplasts are harvested at 3000*g* as described earlier for 5 min at 4° and then resuspended in lysis buffer to a final concentration of 60–70 OD_{600} units/ml. The suspension is homogenized using the loose pestle of a Dounce homogenizer and 30–40 strokes of the pestle at 4° (or on ice). Lysis buffer consists of 0.8 *M* sorbitol, 10 m*M* triethanolamine (TEA), and 1 m*M* EDTA. The pH is adjusted to 7.2 with acetic acid or TEA.

When measurements of the water permeability of the vesicles are desired, 20 m*M* (7.5 mg/ml) 5- or 6-carboxyfluorescein (Molecular Probes, Eugene, OR) is included in the lysis buffer. In addition, we add the following protease inhibitors (final concentrations are indicated): 0.25 m*M* phenylmethylsulfonyl fluoride (PMSF) and, 1 μg/ml pepstatin.

If problems with proteolysis arise, other protease inhibitors that may be used are 1 μg/ml aprotinin, 1 m*M* diisopropyl fluorophosphate, 2 μg/ml chymostatin, and 1 μg/ml leupeptin.

Unlysed cells, cell debris, mitochondria, and nuclei are pelleted at 20,000*g* (11,000 rpm in a Sorvall SS34 rotor) for 10 min at 4°. The supernatant is removed with a pipette and centrifuged at 144,000*g* (29,000 rpm in a Sorvall TH-641 swinging bucket rotor) for 1 hr at 4° to pellet the secretory vesicles. The supernatant is decanted carefully and the pellet may be resuspended in either lysis buffer or another buffer containing osmotic support. We then use this material directly to assay water permeability as we and others have found that freezing the vesicles may result in a loss of enzyme activity,[16,39] presumably because vesicle lysis ensues. Alternatively, the resuspended pellet may be recentrifuged after resuspension in ~10 ml of lysis buffer to obtain a cleaner vesicle fraction.[25] *Note:* The vesicle pellet can be difficult to resuspend and may require vigorous pipetting or rehomogenization.

[39] R. K. Nakamoto and C. W. Slayman, personal communication, 1995.

Electron Microscopy

Vesicle preparations are mixed with an equal volume of 4% glutaraldehyde, 2% paraformaldehyde in 200 m*M* cacodylate buffer, pH 7.4, and centrifuged at 100,000*g* (29,000 rpm in a Sorvall RP45A rotor) for 45 min at 4°. The pellet is dislodged carefully from the side of the tube and incubated for 60 min at room temperature in 1% OsO_4 (w/v), 200 m*M* sodium cacodylate, pH 7.4, 1% $K_4Fe(CN)_6$ (w/v). After several rinses with H_2O the samples are stained *en bloc* overnight at 4° with 0.5% uranyl acetate (w/v) in H_2O. The samples are dehydrated in a graded series of ethanol, embedded in epoxy resin LX-112 (Ladd, Burlington, VT), and sectioned with a diamond knife (Diatome, Fort Washington, PA). Sections, 75–90 nm thick (as determined by their interference colors), are contrasted with uranyl acetate and lead citrate, mounted on Butvar-coated nickel grids, and viewed at 80 kV in a Jeol 100 CX electron microscope.

Results

To examine the biophysical properties of two human water channels (AQP1 and AQP2) *in vitro,* we transformed the yeast *sec6-4* mutant strain with galactose-regulatable plasmids encoding one of the two genes. After the synthesis of the protein was induced by the addition of galactose, the temperature of the culture was shifted to 37° so that secretory vesicles containing the water channel proteins accumulated. Once the aquaporin-containing vesicles were isolated, they were prepared for electron microscopy as described earlier. Figure 1 shows a representative field of view from an electron micrograph. The average diameter of the secretory vesicles determined from the micrographs was 188 ± 13 nm, which differs somewhat from the 100-nm vesicle size reported using electron microscopy measurements within whole yeast cells.[40] This difference could be due to the isolation procedure; vesicles could fuse to produce larger diameter structures. By comparison, when the vesicle diameter was measured by quasi-elastic light scattering, the average particle diameter was 572 nm and the median diameter was 429 nm. This suggests that the vesicles aggregate in solution, even after vigorous resuspension, and agrees with our own experience that the vesicle pellet is very difficult to resuspend (see Methods).

In order to measure the activities of the AQP1 and AQP2 water channels in the secretory vesicles, we used a "shrinkage assay," which has been described previously.[16,22] Briefly, the vesicles are loaded with a fluorescent

[40] E. Harsay and A. Bretscher, *J. Cell Biol.* **131,** 297 (1995).

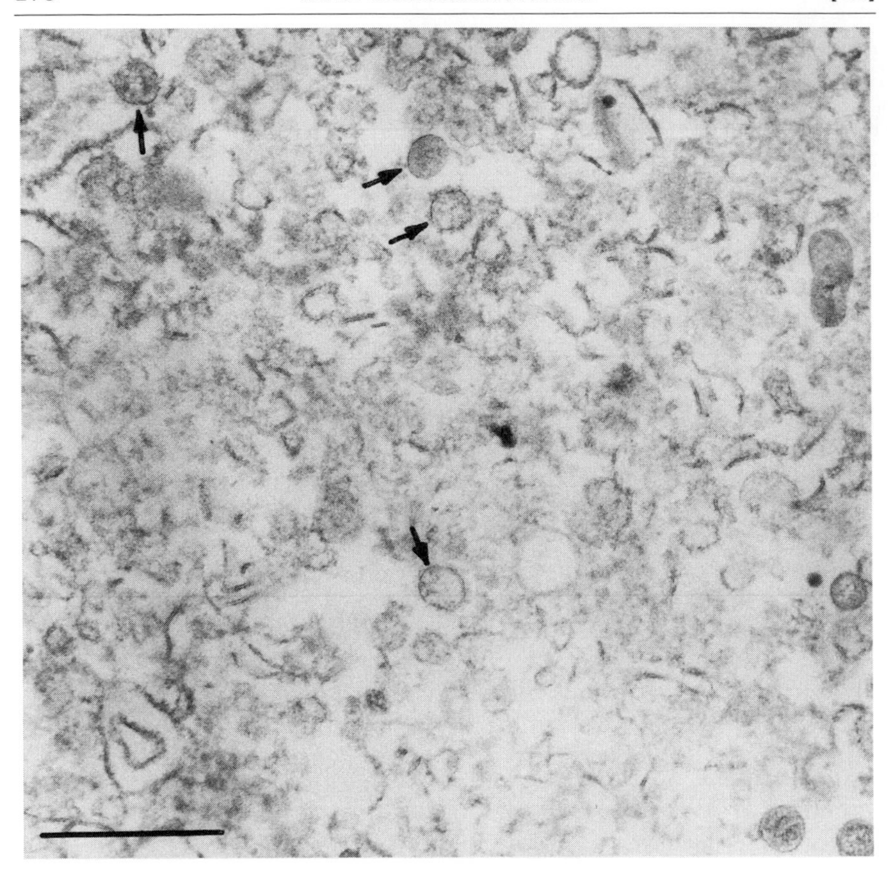

FIG. 1. Electron micrograph showing the crudely fractionated secretory vesicles used in this study (courtesy of Gerard Apodaca, University of Pittsburgh). Other membrane fragments are evident, but only the water permeability of sealed vesicles (marked with arrowheads) can be measured in the assay. Bar: 0.5 μm.

dye, a mixture of 5- or 6-carboxyfluorescein, which exhibits concentration-dependent fluorescence; at high concentrations, the dye self-quenches, causing a decrease in fluorescence.[41] Therefore, vesicle shrinkage, which results in decreased fluorescence, or vesicle expansion, which results in increased fluorescence, can be monitored readily. In practice, the vesicles are shrunk rapidly (instead of expanded, as expansion may result in lysis) by exposure to hypertonic solution on a stopped-flow apparatus, and the change in fluorescence is monitored. This fluorescence change can be converted easily

[41] P. Y. Chen, D. Pearce, and A. S. Verkman, *Biochemistry* **27,** 5713 (1988).

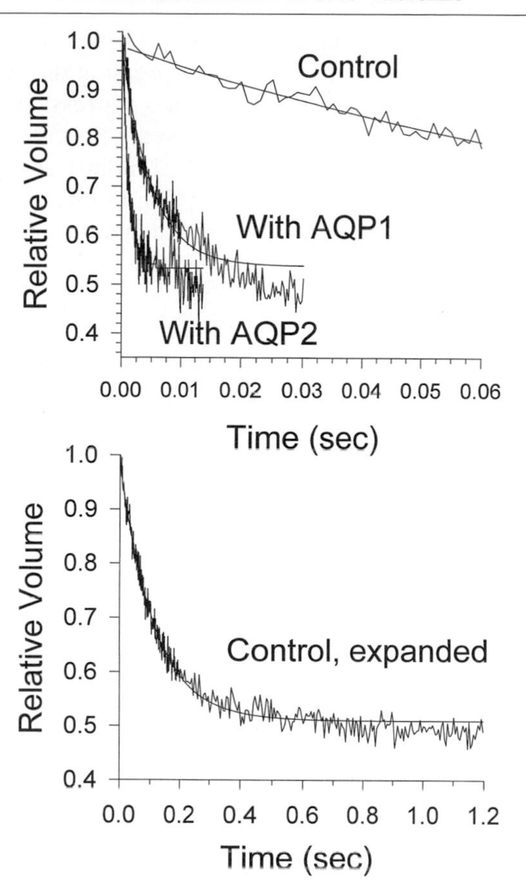

FIG. 2. Water transport in yeast secretory vesicles. (Top) Traces show the change in volume as a function of time for *sec6-4* vesicles containing either AQP1 or AQP2 or no heterologous protein (control). The smooth curve is the computerized fit used for water permeability calculations. (Bottom) Identical curve for the control vesicles on an expanded time scale.

to a change in vesicle volume. The changes in fluorescence (and therefore volume) occur on a millisecond time scale and require a stopped-flow fluorimeter for the measurements. The stopped-flow apparatus allows very rapid mixing of solutions and simultaneous measurements of fluorescence. Quantitation of vesicle fluorescence and details of the shrinkage measurements using the stopped-flow apparatus have been described.[16,22,42,43]

Figure 2 shows the volume changes as a function of time experienced

[42] M. L. Zeidel, S. V. Ambudkar, B. L. Smith, and P. Agre, *Biochemistry* **31,** 7436 (1992).
[43] M. L. Zeidel, A. Albalak, E. Grossman, and A. Carruthers, *Biochemistry* **31,** 589 (1992).

by secretory vesicles containing either no heterologous protein (control) or AQP1 or AQP2. It is clear that vesicles expressing the AQP1 and AQP2 water channels shrink much faster than vesicles expressing no heterologous protein. Water permeabilities for control vesicles and those expressing AQP1 and AQP2 are (in cm/sec, mean ± standard deviation) 0.0022 ± 0.0004, 0.030 ± 0.010 (before optimizing conditions), and 0.14 ± 0.03, respectively. Water permeabilities were calculated as described previously.[22] When conditions described in this article were later used to optimize the expression of AQP1 in *sec* vesicles, a water permeability of 0.087 cm/sec was obtained. Increases in water permeability indicate the presence of aqueous pores through the membrane like those found in functional water channels, as water flow is higher through an aqueous pore than through the lipid bilayer. Values for the water permeability of AQP1 range from 0.0032 cm/sec[44] and 0.019 cm/sec[45] following expression in *Xenopus* oocytes to 0.472 cm/sec for two-dimensional AQP1 crystals.[46] A water permeability for AQP1 expressed in yeast *sec6-4* vesicles of 0.00415 cm/sec has been reported.[17] The water permeability of AQP2 expressed in *Xenopus* oocytes was found to be 0.010 cm/sec,[45] and the water permeability of rat kidney endosomes that are rich in AQP2 was 0.14–0.16 cm/sec.[47] Therefore, our results compare favorably with these values and demonstrate that our secretory vesicles express AQP1 and AQP2 abundantly.

We also calculated the activation energy for water transport for control vesicles and those expressing AQP1 and AQP2 using described methods.[22] The values are (in kcal/mol, mean ± standard deviation) 13.2 ± 1.2, 6.2 (for a single preparation after optimizing conditions), and 4.0 ± 0.5 for control, AQP1, and AQP2 vesicles, respectively. Lower activation energies, such as those obtained for AQP1 and AQP2-containing vesicles, indicate aqueous pores in the membrane as the energy barrier to water flow across a membrane is decreased substantially when water flows through a pore instead of across the hydrophobic bilayer. The expression of AQP1 and AQP2 in these vesicles was confirmed by immunoblotting using affinity-purified antibodies prepared against C-terminal peptides from AQP1[42] and AQP2.[48]

Protein from the secretory vesicles can also be extracted and reconstituted into proteoliposomes. AQP1 from yeast *sec* vesicles is solubilized with

[44] F. Le Caherec, P. Bron, J. M. Verbavatz, A. Garret, G. Morel, A. Cavalier, G. Bonnec, D. Thomas, J. Gouranton, and J. F. Hubert, *J. Cell Sci.* **109,** 1285 (1996).

[45] B. Yang and A. S. Verkman, *J. Biol. Chem.* **272,** 16140 (1997).

[46] T. Walz, B. L. Smith, M. L. Zeidel, A. Engel, and P. Agre, *J. Biol. Chem.* **269,** 1583 (1994).

[47] M. B. Lande, I. Jo, M. L. Zeidel, M. Somers, and H. W. Harris, Jr., *J. Biol. Chem.* **271,** 5552 (1996).

[48] M. A. Knepper, *Am. J. Physiol.* **272** (*Renal Physiol.* **41**), F3 (1997).

TABLE I
WATER PERMEABILITIES (P_f) AND ACTIVATION ENERGIES (E_{act}) FOR *sec6-4* VESICLES OR RECONSTITUTED PROTEOLIPOSOMES[a]

Source	P_f (cm/sec)	E_{act} (kcal/mol)
Control vesicles	0.0022	13.2
Control proteoliposomes	0.0038	12.1
AQP1 vesicles	0.087	6.2
AQP1 proteoliposomes	0.010	3.7
AQP2 vesicles	0.14	4.0

[a] Prepared from *sec6-4* vesicles containing the indicated heterologously expressed proteins.

octyl glucoside and reconstituted using the identical methods employed for the reconstitution of AQP1 from human red blood cells.[42,43] These proteoliposomes exhibit properties similar to those of functional water channels in other systems (data not shown).[16] Water permeabilities and activation energies for proteoliposomes either containing or lacking AQP1 prepared from the secretory vesicles are shown in Table I, along with summarized data for AQP1- and AQP2-containing secretory vesicles. If desired, the amount of reconstituted membrane protein from reconstituted vesicles may be quantitated using a sensitive amido black staining procedure, as has been described.[49] Such measurements are vital if the specific activity of a reconstituted enzyme preparation is necessary.

Discussion

Yeast secretory vesicles offer several advantages as an environment for the heterologous expression of many membrane proteins. The yeast system can provide large amounts of protein, and the heterologous protein can be purified partially by isolating the secretory vesicles that are enriched for the membrane protein of interest. The vesicles may be assayed for activity or, if desired, the resident membrane protein may be reconstituted. If protein expression is under control of an inducible promoter, protein synthesis can be initiated concurrently with the accumulation of the secretory vesicles.

Another potentially useful feature of the *sec6* vesicle expression system is that the topology of the protein is inverted relative to the cell as a whole, i.e., the cytoplasmic side of the protein faces out of the vesicle whereas the extracellular portion of the protein faces into the vesicle lumen. Because

[49] J. L. Brodsky, *Anal. Biochem.* **246,** 262 (1997).

it is likely that the sidedness of the yeast secretory vesicles remains intact throughout the isolation procedure, the sidedness of a given protein could be determined easily. This may provide a facile means to determine the topology of a given membrane protein. For example, the effects of cytoplasmic regulatory factors, addition of antibodies with cytoplasmic epitopes, and channel asymmetries could all be assayed using the yeast *sec6* vesicles.

Although the yeast system has several pitfalls, these pitfalls can usually be circumvented. First, evidence shows that there are significant differences between yeast and mammalian translation machinery. Protein expression in yeast is very sensitive to elements in the 5′-untranslated region of mammalian genes and to yeast genes that control mating type.[27–29] In fact, we have been unable to reproducibly express every protein when the corresponding cDNA is used to transform yeast.

Although we have not systematically investigated the basis of this phenomenon, we have noticed that every gene with less than 20 bases of a 5′-untranslated sequence has been expressed efficiently in yeast. This result is probably due to the lack of secondary structure in the 5′-untranslated regions shorter than 20 bp and not directly due to the length of the untranslated region. Stem–loop structures introduced into the 5′-untranslated region of the *HIS4* gene, for example, can reduce translation efficiency by 95–99%, depending on the exact position of the loop, although duplicating or deleting part of the normal untranslated region has no effect on translation.[50] A similar result was obtained when stem–loops were introduced into a region directly upstream of the *CYC1* translational start site.[51] A 12-bp stem–loop reduced translation by 95% when inserted six nucleotides upstream of the translation start site, and a 42-bp stem–loop at the same position completely obliterated translation of *CYC1*.

There may also be preferred sequences in the 5′-untranslated region that facilitate ribosome binding. Now that the yeast genome has been sequenced, it should be possible to analyze the 5′-untranslated regions of all of the yeast open reading frames to find the most common sequences upstream of the translation start site. One such analysis of the initiator regions of 131 yeast genes has been compiled.[52] This compilation demonstrated that in 95% of the open reading frames studied, translation was initiated from the first AUG in the mRNA; no conserved sequence that might represent a ribosome-binding site was present, and there appears to be no yeast counterpart to the bacterial ribosome-binding consensus

[50] A. M. Cigan, E. K. Pabich, and T. F. Donahue, *Mol. Cell. Biol.* **8,** 1591 (1988).
[51] S. B. Baim and F. Sherman, *Mol. Cell. Biol.* **8,** 1591 (1988).
[52] A. M. Cigan and T. F. Donahue, *Gene* **59,** 1 (1987).

sequence.[53] A more extensive compilation could now provide information about which sequence elements immediately upstream of the translation start site are favorable or unfavorable for efficient translation, as has been done for vertebrate mRNAs.[54,55]

Second, there is a trade-off between increasing the yield of protein and plasmid loss. Typically, selective medium lacking one or more ingredients and growth at high temperature (37°) are required to express proteins in secretory vesicles of the *sec6-4* mutant. Expression from an inducible promoter may require other less than optimal conditions, such as growth in a carbon source other than glucose. Although these conditions guarantee that every cell will contain the plasmid and that every cell will be protected against synthesis of potentially toxic proteins (i.e., proteins that could cause cell death if trafficked to the plasma membrane), these suboptimal conditions can lead to low culture densities even at saturation. For example, we find that the OD_{600} of cultures grown to saturation in galactose medium lacking uracil is more than 10-fold lower than in rich YP medium containing glucose as the carbon source.

Others have reported that the expression of heterologous proteins is improved and that their degradation is reduced by growth in rich media.[56–58] However, we found that a long-term switch to media without selective pressure may lead to an inability to detect the desired protein, presumably due to increased plasmid loss. Others have obtained similar results when selective pressure is lifted.[59,60] Plasmid loss may also depend on other factors, such as the plasmid size.[61] The nature of the gene and promoter may also be a factor because some groups have not observed plasmid loss, despite the absence of selective pressure.[58,62] The method we described in this article, in which selective pressure is lifted just before the initiation of protein expression and the induction of the *sec* phenotype, was developed to minimize each of these potential problems.

Third, improper folding or sorting of proteins may also pose a problem. Several groups have reported that heterologous protein expression in yeast

[53] J. Shine and L. Delgarno, *Proc. Natl. Acad. Sci. U.S.A.* **71,** 1342 (1974).

[54] M. Kozak, *EMBO J.* **16,** 2482 (1997).

[55] M. Kozak, *Biochimie* **76,** 815 (1994).

[56] S. J. Coppella and P. Dhurjati, *Biotechnol. Bioeng.* **33,** 976 (1989).

[57] Z. Wang and N. A. Da Silva, *Biotechnol. Bioeng.* **42,** 95 (1993).

[58] L. A. Castelli, C. J. Mardon, P. M. Strike, A. A. Azad, and I. G. Macreadie, *Gene* **142,** 142 (1994).

[59] R. D. O'Kennedy and J. W. Patching, *J. Indust. Microbiol. Biotech.* **18,** 319 (1997).

[60] Z. Zhang, J. M. Scharer, and M. Moo-Young, *J. Biotechnol.* **55,** 31 (1997).

[61] F. Storici, J. Oberto, and C. V. Bruschi, *Plasmid* **34,** 184 (1995).

[62] T. Hottiger, J. Kuhla, G. Pohlig, P. Furst, A. Spielmann, M. Garn, S. Haemmerli, and J. Heim, *Yeast* **11,** 1 (1995).

(but not necessarily in the *sec6-4* mutant) results in protein accumulation in the endoplasmic reticulum.[63,64] Apparently, space at the plasma membrane is limited, so expression of large amounts of a particular protein results in an overloading of the secretory pathway and an intracellular buildup of protein. Overexpression of at least one endogenous yeast protein even led to an intracellular buildup of endoplasmic reticulum.[65] General problems associated with both homologous and heterologous membrane protein expression have been reviewed.[66]

Fourth, the default targeting pathways between mammalian cells and yeast differ. The vacuole is the destination for membrane proteins lacking specific targeting signals in yeast,[67] whereas the plasma membrane is the default destination for membrane proteins in mammalian cells.[68–70] The default pathway to the vacuole for membrane proteins in yeast does not require prior delivery to the plasma membrane because proteins in the late secretory pathway, such as Sec 1p, are not required.[67,71,72] Furthermore, no specific sequences are required to traffic membrane proteins to the vacuole in yeast,[67,73] although positive sequence information *is* required to traffic *soluble* proteins to the vacuole.[74] This strongly suggests that the vacuole, not the plasma membrane, is the default sorting compartment for membrane proteins in yeast and that these proteins do not travel through plasma membrane-targeted vesicles *en route* to the vacuole.

Because the exact peptide sequences required to target membrane proteins to the plasma membrane in mammalian or yeast cells are unknown, there is currently no way to predict the final destination of membrane proteins in yeast solely from their amino acid sequence. In the case of at least one yeast membrane protein, Ste6p, specific information is almost certainly required for plasma membrane expression. A significant amount of this protein, which is required for the transport of the yeast **a** factor mating pheromone across the plasma membrane, moves from intracellular

[63] Villalba, M. G. Palmgren, G. E. Berberian, C. Ferguson, and R. Serrano, *J. Biol. Chem.* **267,** 12341 (1992).
[64] P. Supply, A. Wach, D. Thines-Sempoux, and A. Goffeau, *J. Biol. Chem.* **268,** 19744 (1993).
[65] R. Wright, M. Basson, L. D'Ari, and J. Rine, *J. Cell Biol.* **107,** 101 (1988).
[66] G. F. X. Schertler, *Curr. Opin. Struct. Biol.* **2,** 534 (1992).
[67] C. J. Roberts, S. F. Nothwehr, and T. H. Stevens, *J. Cell Biol.* **119,** 69 (1992).
[68] M. R. Jackson, T. Nilsson, and P. Peterson, *EMBO J.* **9,** 3153 (1990).
[69] C. E. Machamer and J. K. Rose, *J. Cell Biol.* **267,** 832 (1987).
[70] C. E. Machamer, *Trends Cell Biol.* **1,** 141 (1991).
[71] C. A. Wilcox, K. Redding, R. Wright, and R. S. Fuller, *Mol. Biol. Cell* **3,** 1353 (1992).
[72] A. Cooper and H. Bussey, *J. Cell Biol.* **119,** 1459 (1992).
[73] D. J. Klionsky and S. D. Emr, *J. Biol. Chem.* **265,** 5349 (1990).
[74] C. K. Raymond, C. J. Roberts, K. M. Moore, I. Howald, and T. H. Stevens, *Int. Rev. Cytol.* **139,** 59 (1992).

compartments to the plasma membrane when the other mating pheromone, α factor, binds to its receptor at the plasma membrane.[75] This suggests that transport of at least one protein to the plasma membrane is regulated in yeast.

In sum, because the default targeting pathways for membrane proteins in yeast and mammalian cells differ, the fate of some heterologous membrane proteins in yeast may be uncertain. Specifically, mammalian proteins that traffic to the plasma membrane by default in mammalian cells may not be expressed in *sec6-4*-derived secretory vesicles, but could instead be targeted to the vacuole (for reviews of yeast membrane protein sorting, see Nothwehr and Stevens[76] or Conibear and Stevens[77]). When the sequences required for plasma membrane targeting are better understood, it may be possible to ensure the delivery of all heterologous proteins to the plasma membrane in yeast. Until then, we believe that our and others' success supports the feasibility of using this system to express recombinant membrane proteins.

Finally, yeast show extreme codon bias, using only 25 of the possible 61 codons more than 90% of the time.[78] The frequency at which specific codons in the mRNA are used mirrors the abundance of the corresponding tRNAs.[79] Specific replacement of infrequently used codons with more frequently used codons may result in a large increase in translation efficiency. When 115 of 215 codons (53%) were replaced with yeast-preferred codons in the mouse immunoglobulin κ chain, a 5-fold increase in the rate of protein synthesis resulted.[80] Perhaps more importantly, a 50-fold increase in the steady-state level of the Ig κ chain protein resulted, although there was no increase in the steady-state level of mRNA.[80] This effect was also observed when the reciprocal experiment was performed and preferred codons were substituted for ones that are used infrequently. When 164 codons (39% of all codons) in the yeast phosphoglycerate kinase (PGK) gene were replaced with the least preferred yeast codons for the same amino acids, the steady-state level of the PGK protein decreased by 84%.[81] Because the steady-state level of mRNA also decreased, mRNA stability in this experiment was tied to efficient protein translation. Therefore, expression in yeast of heterologous genes containing infrequently used codons may be diminished as a result of both fewer available tRNAs and decreased

[75] K. Kuchler, H. Dohlman, and J. Thorner, *J. Cell Biol.* **120,** 1203 (1993).
[76] S. F. Nothwehr and T. H. Stevens, *J. Biol. Chem.* **269,** 10185 (1994).
[77] E. Conibear and T. H. Stevens, *Cell* **83,** 513 (1995).
[78] J. L. Bennetzen and B. D. Hall, *J. Biol. Chem.* **257,** 3026 (1982).
[79] T. Ikemura, *J. Mol. Biol.* **158,** 573 (1982).
[80] L. Kotula and P. J. Curtis, *Biotechnology* **9,** 1386 (1991).
[81] A. Hoekema, R. A. Kastelein, M. Vasser, and H. A. deBoer, *Mol. Cell. Biol.* **7,** 2914 (1987).

stability of the mRNA. It should be noted, however, that others have reported little effect from codon optimization in yeast.[82,83]

In conclusion, the yeast *sec6-4* mutant offers a number of advantages as a heterologous protein expression system for membrane proteins. Large amounts of yeast can be produced easily and inexpensively, yeast contain many of the posttranslational processing enzymes of mammalian cells, gene expression can be controlled using inducible promoters, and the expressed protein is ultimately inserted into a physiological vesicle that can be purified and functionally characterized. The pitfalls of yeast, such as differences in the translation machinery, differences in the default targeting pathways, and limited codon usage, may be circumvented in many cases. Undoubtedly, as more genes encoding putative membrane proteins of unknown function are uncovered, powerful systems in which the corresponding gene products may be examined will become vital. We believe that protein expression in the yeast *sec6-4* strain is one such system.

Acknowledgments

This work was supported by National Institutes of Health Grant DK43955 (M. L. Zeidel). J. L. Brodsky is supported by National Science Foundation Grant MCB9506002 and American Cancer Society Fellowship JFRA602. L. A. Coury is supported by National Institutes of Health NRSA 1 F32 DK09485-01.

[82] J. K. Ngsee and M. Smith, *Gene* **86,** 251 (1990).
[83] M. Egel-Mitani, M. T. Hansen, K. Norris, L. Snel, and N. P. Fiil, *Gene* **73,** 113 (1988).

[11] Use of Yeast Artificial Chromosomes to Express Genes in Transgenic Mice

By KENNETH R. PETERSON

Introduction

The establishment of transgenic mice technology as a routine method in many laboratories has facilitated the study of gene function in development and disease.[1–5] First, transgene expression in mice allows the com-

[1] J. W. Gordon, G. A. Scangos, D. J. Plotkin, J. A. Barbosa, and F. H. Ruddle, *Proc. Natl. Acad. Sci. U.S.A.* **77,** 7380 (1980).
[2] R. L. Brinster, Y. Chen, E. Trumbauer, A. W. Senera, R. Warren, and R. D. Palmiter, *Cell* **27,** 223 (1981).

0076-6879/99 $30.00

plementation of existing murine mutants and the assessment of other phenotypic effects. Second, gene expression can be studied throughout development and tissue-specific regulatory elements can be analyzed. Historically, the expression of transgenes was erratic due to position effects and copy number-independent expression. In part, these problems were due to the nature of the transgene constructs. Typically, they were limited in size due to constraints on how large a DNA fragment could be stably cloned into plasmid or cosmid vectors and isolated without degradation for introduction into the mouse. Genes with exons spanning several hundred kilobases or loci containing multiple genes could not be used as transgenes. Thus, potentially important or unidentified *cis*-acting sequences could not be included during the synthesis of transgenes. cDNAs were often substituted for large genes, or individual genes from a multigene cluster were utilized. However, expression from these constructs was subject to the effects of surrounding chromatin into which they were integrated. Some improvement was achieved when additional *cis* elements were included, such as enhancers, introns, and polyadenylation signals.[6–9] These additional sequences boosted the level and reproducibility of transgene expression, but these truncated constructs lacked their natural regulatory elements and thus developmental studies did not necessarily indicate how the native gene might be regulated. Full-size, intact genes or loci used as transgenes might improve the utility of transgenic studies; inclusion of distal regulatory elements as part of the native locus might validate developmental studies and insulate the construct from position effects. To test this possibility, large DNA constructs, in particular yeast artificial chromosomes (YACs), have been utilized to generate transgenic mice.[10–12] Several techniques have been described that are suitable for introducing YACs into transgenic mice.[10,13–17]

---

[3] K. Harbers, D. Jahner, and R. Jaenisch, *Nature* **293,** 540 (1981).

[4] T. E. Wagner, P. C. Hoppe, J. D. Jollick, D. R. Scholl, R. L. Hodinka, and J. B. Gault, *Proc. Natl. Acad. Sci. U.S.A.* **78,** 6376 (1981).

[5] E. F. Wagner, T. A. Stewart, and B. Mintz, *Proc. Natl. Acad. Sci. U.S.A.* **78,** 5016 (1981).

[6] R. L. Brinster, J. M. Allen, R. R. Behringer, R. E. Gelinas, and R. D. Palmiter, *Proc. Natl. Acad. Sci. U.S.A.* **85,** 836 (1988).

[7] R. Jaenisch, *Science* **240,** 1468 (1988).

[8] T. Choi, C. Huang, C. Gorman, and R. Jaenisch, *Mol. Cell. Biol.* **11,** 3070 (1991).

[9] J. W. Gordon, *Methods Enzymol.* **225,** 747 (1993).

[10] B. T. Lamb and J. D. Gearhart, *Curr. Opin. Genet. Dev.* **5,** 342 (1995).

[11] K. R. Peterson, C. H. Clegg, Q. Li, and G. Stamatoyannopoulos, *Trends Genet.* **13,** 61 (1997).

[12] K. R. Peterson, *in* "Genetic Engineering" (J. K. Setlow, ed.), Vol. 19, p. 235. Plenum Press, New York, 1997.

[13] C. Huxley and A. Gnirke, *in* "YAC Libraries, a User's Guide" (D. L. Nelson and B. H. Brownstein, eds.), p. 143. Freeman, New York, 1994.

This article focuses on the purification of YAC DNA for the production of transgenic mice by microinjection of fertilized oocytes and the analysis of YAC transgene structure in mice. In addition, manipulations of YACs in yeast will be outlined.

Manipulation of YACs Prior to Purification/Microinjection

Maintenance of Large DNA Sequences in YAC Vectors

Although YAC vectors are the primary focus here, other vector systems are available that accommodate large DNAs. The large capacity cloning vectors include bacteriophage P1,[18] P1 artificial chromosomes (PACs),[19] fosmids,[20] bacterial artificial chromosomes (BACs),[21] and yeast artificial chromosomes.[22] Of these, YACs have several distinct advantages. First, insert sizes of approximately 2 Mb can be contained in YACs, compared to BACs, in which a maximum of 350 kb can be maintained. Thus, the study of intact genes, multigenic loci, distant regulatory sequences, and higher order genomic structure can be performed in the context of native sequences. Because extensive sequences, including coding regions and flanking genomic DNA, can be maintained in an inert state in YAC clones, all sequences having potential regulatory relevance, as well as distances between genes and control elements, are maintained. Second, site-specific mutagenesis can be performed *in vivo* readily and efficiently using the

---

[14] C. Huxley, *in* "Genetic Engineering" (J. K. Setlow, ed.), Vol. 16, p. 65. Plenum Press, New York, 1994.

[15] N. P. Davies and C. Huxley, *in* "Methods in Molecular Biology" (D. Markie, ed.), Vol. 54, p. 281. Humana Press, Totowa, 1996.

[16] A. Schedl, B. Grimes, and L. Montoliu, *in* "Methods in Molecular Biology" (D. Markie, ed.), Vol. 54, p. 293. Humana Press, Totowa, 1996.

[17] W. M. Strauss, *in* "Methods in Molecular Biology" (D. Markie, ed.), Vol. 54, p. 307. Humana Press, Totowa, 1996.

[18] N. Sternberg *Proc. Natl. Acad. Sci. U.S.A.* **87,** 103 (1990).

[19] P. A. Ioannou, C. T. Amemiya, J. Garnes, P. M. Kroisel, H. Shizuyu, C. Chen, M. A. Batzer, and P. J. de Jong, *Nature Genet.* **6,** 84 (1994).

[20] U.-J. Kim, H. Shizuyu, P. J. de Jong, B. Birren, and M. I. Simon, *Nucleic Acids Res.* **20,** 1083 (1992).

[21] H. Shizuyu, B. Birren, U.-J. Kim, V. Mancino, T. Slepak, Y. Tachiiri, and M. Simon, *Proc. Natl. Acad. Sci. U.S.A.* **89,** 8794(1992).

[22] D. T. Burke, G. F. Carle, and M. V. Olson, *Science* **236,** 806 (1987).

homologous recombination system of the yeast host.[23–26] Point mutations, deletions, insertions, and replacements can be introduced easily into a YAC without leaving behind foreign DNA, such as selectable marker cassettes. These mutant YACs can then be used to generate transgenic mice and the effect of the mutation on transgene expression studied. However, a system allowing homologous recombination in BACs has been developed.[27]

Utilization of YACs as transgenes requires the implementation of several methods, both prior to and after transgenesis. A flow chart summarizing the necessary techniques and steps for the successful production and analysis of YAC transgenics is shown in Fig. 1. Not all of the methods are applicable in every instance, many are optional, and some are alternatives to one another.

Optimization of YAC Vectors

Most YAC clones have been constructed in YAC vectors such as pYAC4, which has the selectable markers TRP1 and URA3 (Fig. 2).[14,28,29] These clones may be used directly to generate transgenic mice, but their versatility is limited for the production of transgenics containing mutant YACs. To enhance their utility, three problems must be rectified by modification of the YAC vector arms. First, mutagenesis by homologous recombination in yeast is best documented using the URA3 marker in a two-step differential selection process, but the presence of the URA3 gene on the arm of the YAC vector precludes its use for this purpose. Second, structural analysis to determine the integrity of YAC transgenes is difficult without a comprehensive restriction map of the cloned insert and due to the lack of unique rare-cutting restriction enzyme sites in the YAC vector. Third, YACs may also be introduced into embryonic stem (ES) cells for the generation of chimeric mice or into other established cell lines, but a se-

[23] P. Hieter, C. Connelly, J. Shero, M. K. McCormick, S. Antonarakis, W. Pavan, and R. Reeves, *in* "Genome Analysis" (K. E. Davies and S. Tilghman, eds.), Vol. I, p. 83. Cold Spring Harbor Laboratory Press, Cold Spring Harbor, NY, 1990.

[24] F. Spencer, G. Ketner, C. Connelley, and P. Hieter, *Methods* **5,** 161 (1993).

[25] R. Rothstein, *in* "Guide to Yeast Genetics and Molecular Biology" (C. Guthrie and G. R. Fink, eds.), p. 281. Academic Press, San Diego, 1995.

[26] K. Duff and C. Huxley, *in* "Methods in Molecular Biology" (D. Markie, ed.), Vol. 54, p. 187. Humana Press, Totowa, 1996.

[27] X. W. Yang, P. Model, and N. Heintz, *Nature Biotechnol.* **15,** 859 (1997).

[28] D. R. Smith, *in* "YAC Libraries, a User's Guide" (D. L. Nelson and B. H. Brownstein, eds.), p. 1. Freeman, New York, 1994.

[29] L. L. Ling, D. R. Smith, and D. T. Moir, *in* "Methods in Molecular Biology" (D. Markie, ed.), Vol. 54, p. 231. Humana Press, Totowa, 1996.

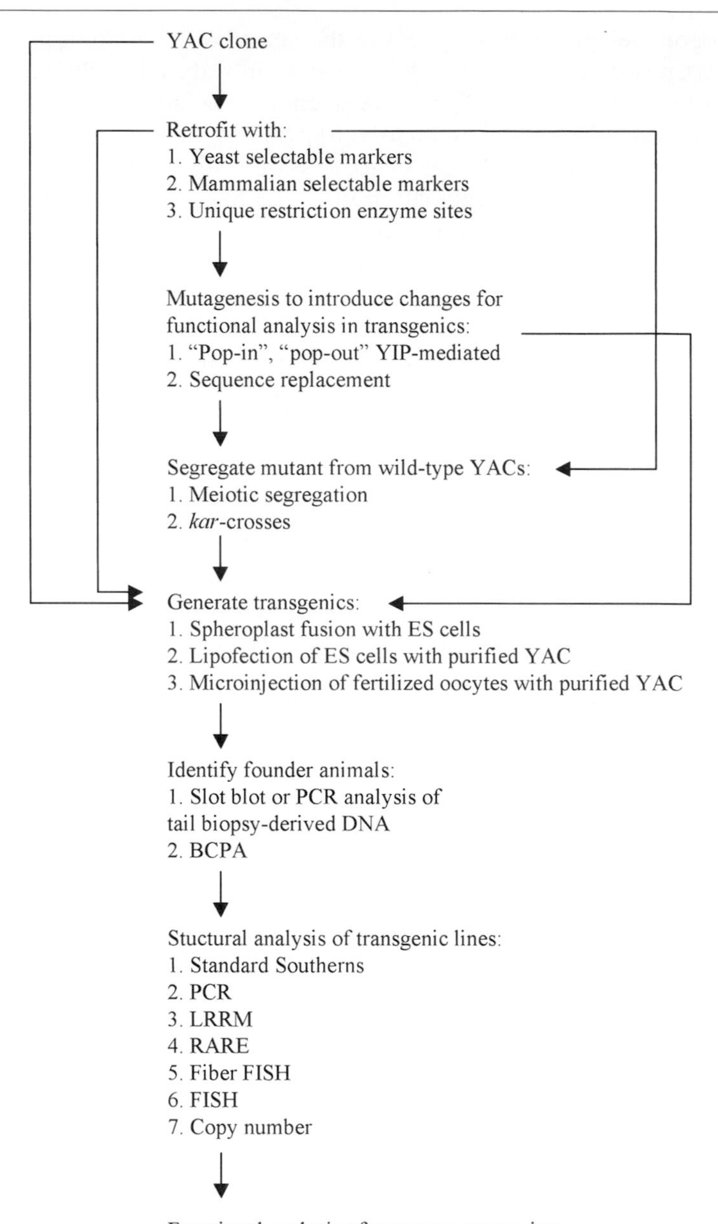

FIG. 1. Flow chart depicting manipulation of YAC clones before and after production of transgenic mice. Each of the steps or methods is outlined in the text. Isolation of purified YAC DNA for the production of transgenics by microinjection is detailed in this article. BCPA, buffy coat plug analysis; LRRM, long-range restriction enzyme mapping; RARE, RecA-assisted restriction endonuclease cleavage.

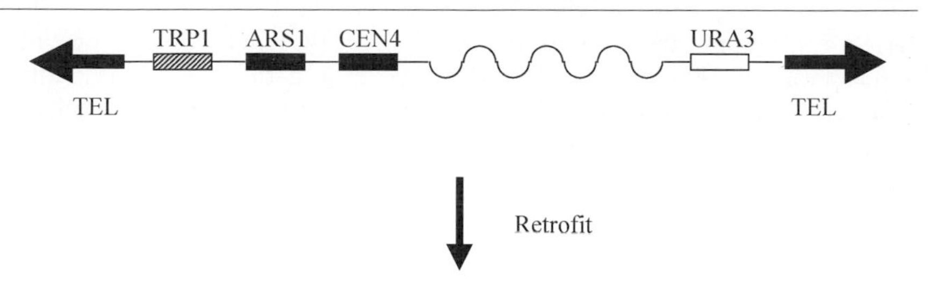

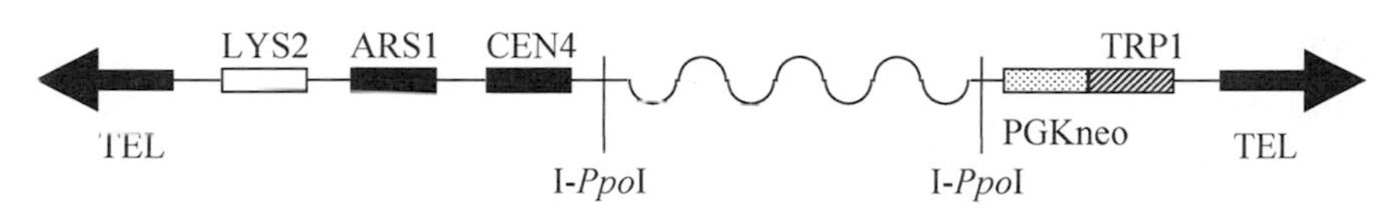

FIG. 2. Schematic representation of a YAC clone. (Top) Unmodified YAC clone based on the pYAC4 vector. (Bottom) YAC clone in which the pYAC vector arms have been modified by retrofitting to enhance the versatility of the YAC. The genomic insert DNA is shown as a wavy line. The yeast chromosomal elements on the YAC vector arms are displayed as variously shaded boxes or arrows. TEL, telomeres; ARS1, autonomous replicating sequence; CEN4, centromere; TRP1, yeast selectable marker for tryptophan prototrophy; URA3, yeast selectable marker for uracil prototrophy; LYS2, yeast selectable marker for lysine prototrophy; PGKneo, mammalian selectable marker for resistance to G418; I-*Ppo*I, rare-cutting restriction enzyme sites introduced into the YAC vector arms for YAC transgene structural analysis.

lectable marker such as a neomycin resistance gene is necessary to select YAC-bearing cell clones. To incorporate these sequence motifs, the YAC vector sequences can be modified by homologous recombination, a process called retrofitting. The details of this process have been described elsewhere and many constructs are available for retrofitting YACs.[12–14,28–30] One series of retrofitting constructs alters the YAC vector such that the three problems just described are solved (Fig. 2).[30] URA3 is made available for mutagenesis of the YAC insert DNA, structures of YAC transgenes can be determined unambiguously, and selection of YAC-containing mammalian cells can be performed, if desired. The TRP1 gene on the left arm of the YAC is replaced with the yeast LYS2 gene and a rare-cutting I-*Ppo*I restriction enzyme site is introduced near the insert–vector junction. The URA3 gene on the right arm of the yeast vector is replaced with the yeast TRP1 gene, a PGKneo gene, and another I-*Ppo*I site near the insert–vector junction. Selection for the YAC in yeast is growth in the absence of tryptophan and

[30] C. Fairhead, E. Heard, D. Arnaud, P. Avner, and B. Dujon, *Nucleic Acids Res.* **23,** 4011 (1995).

lysine, rather than tryptophan and uracil. The PGKneo cassette provides a selectable marker for transfected ES cells, and the I-*Ppo*I sites allow structural analysis of intact YAC inserts without the need for a restriction enzyme map of the cloned insert.

Mutagenesis and Segregation of YACs

Homologous recombination in the yeast host can be used to introduce almost any mutation into the YAC DNA prior to transgenesis. Two methods for this type of manipulation have been described in detail.[12,25,26] The "pop-in," "pop-out" method has been used most extensively and requires the production of only one construct in a yeast integrating plasmid (YIP).[31–36] The YIP integration event requires only a single crossover, as does the excision of YIP sequences. Both steps can be selected. The sequence replacement method is advantageous for the introduction of large deletions, but is also applicable for recombining point mutations, small deletions, or insertions into the target YAC sequence.[25,37] The sequence replacement method has the disadvantage of requiring two constructs. Thus, two transformations of yeast are required for producing a mutant YAC. In addition, each recombination event is a less frequent double crossover, necessitating the analysis of more isolates. However, this approach may be used to introduce mutations into the YACs when a convenient scheme cannot be devised using the YIP-mediated "pop-in, pop-out" method of mutagenesis.

The presence of both mutant and wild-type YACs is observed frequently in the same yeast isolate after mutagenesis.[38] YACs are not replicated and segregated meiotically in strict concordance with yeast chromosomes. Over time, even under selection, daughter cells arise that may carry one or more YAC copies. During mutagenesis via recombination, the targeting vector need integrate in only one of these YAC copies to establish selection through the marker on the vector. Thus, in one yeast cell, mutant and

[31] M. C. Barton, M. F. Hoekstra, and B. M. Emerson, *Nucleic Acids Res.* **18,** 7349 (1990).

[32] K. Duff, A. McGuigan, C. Huxley, F. Schulz, and J. Hardy, *Gene Ther.* **1,** 70 (1994).

[33] J. Bungert, U. Dave, K.-C. Lim, K. H. Lieuw, J. A. Shavit, Q. Liu, and J. D. Engel, *Genes Dev.* **9,** 3083 (1995).

[34] K. R. Peterson, Q. Li, C. H. Clegg, T. Furukawa, P. A. Navas, E. J. Norton, T. G. Kimbrough, and G. Stamatoyannopoulos, *Proc. Natl. Acad. Sci. U.S.A.* **92,** 5655 (1995).

[35] K. R. Peterson, C. H. Clegg, P. A. Navas, E. J. Norton, T. G. Kimbrough, and G. Stamatoyannopoulos, *Proc. Natl. Acad. Sci. U.S.A.* **93,** 6605 (1996).

[36] Q. Lui, J. Bungert, and J. D. Engel, *Proc. Natl. Acad. Sci. U.S.A.* **94,** 169 (1997).

[37] S. P. A. McCormick, K. R. Peterson, R. E. Hammer, C. H. Clegg, and S. G. Young, *Trends Cardiovasc. Med.* **6,** 16 (1996).

[38] K. R. Peterson, C. H. Clegg, Q. Li, P. A. Navas, E. J. Norton, K. A. Leppig, and G. Stamatoyannopoulos, *in* "Hemoglobin Switching" (G. Stamatoyannopoulos, ed.), p. 45. Intercept, Ltd., Andover, 1995.

normal YACs may coexist. These YACs should be segregated from one another to avoid coinjection of the normal and modified YACs using one of two schemes. The mutant and wild-type YACs may be segregated meiotically[25] or, alternately, the "kar-crossing" method may be utilized.[39–41]

Generation of Transgenic Mice by Pronuclear Microinjection of Purified YACs

Three methods have been reported for the production of YAC transgenic mice; yeast spheroplast–ES cell protoplast fusion,[42–45] lipofection of purified YAC DNA into ES cells,[46–51] and direct microinjection of purified YAC DNA into mouse oocytes (Fig. 1).[33–36,52–69] The first two of these

[39] Y. Hugerat, F. Spencer, D. Zenvirth, and G. Simchen, *Genomics* **22,** 108 (1994).

[40] F. Spencer, Y. Hugerat, G. Simchen, O. Hurko, C. Connelly, and P. Hieter, *Genomics* **22,** 118 (1994).

[41] F. Spencer and G. Simchen, *in* "Methods in Molecular Biology" (D. Markie, ed.), Vol. 54, p. 239. Humana Press, Totowa, 1996.

[42] N. P. Davies, I. R. Rosewell, J. C. Richardson, G. P. Cook, M. S. Neuberger, B. H. Brownstein, M. L. Norris, and M. Brüggemann, *Biotechnology* **11,** 911 (1993).

[43] A. Jakobovits, A. L. Moore, L. L. Green, G. J. Vergara, C. E. Maynard-Carrie, H. A. Austin, and S. Klapholz, *Nature* **362,** 255 (1993).

[44] L. L. Green, M. C. Hardy, C. E. Maynard-Currie, H. Tsuda, D. M. Louie, M. J. Mendez, H. Abderrahim, M. Noguchi, D. H. Smith, Y. Zeng, N. E. David, H. Sasai, D. Garza, D. G. Brenner, J. F. Hales, R. P. McGuinness, D. J. Capon, S. Klapholz, and A. Jakobovits, *Nature Genet.* **7,** 13 (1994).

[45] M. J. Mendez, H. Abderrahim, M. Noguchi, N. E. David, M. C. Hardy, L. L. Green, H. Tsuda, S. Yoast, C. E. Maynard Curie, D. Garza, R. Gemmill, A. Jakobovits, and S. Klapholz, *Genomics* **26,** 294 (1995).

[46] T. K. Choi, P. W. Hollenbach, B. E. Pearson, R. M. Ueda, G. N. Weddell, C. G. Kurahara, C. S. Woodhouse, R. M. Kay, and J. F. Loring, *Nature Genet.* **4,** 117 (1993).

[47] B. T. Lamb, S. S. Sisodia, A. M. Lawler, H. H. Slunt, C. A. Kitt, W. G. Kearns, P. L. Pearson, D. L. Price, and J. D. Gearhart, *Nature Gene.* **5,** 22 (1993).

[48] B. E. Pearson and T. K. Choi, *Proc. Natl. Acad. Sci. U.S.A.* **90,** 10578 (1993).

[49] W. M. Strauss, J. Dausman, C. Beard, C. Johnson, J. B. Lawrence, and R. Jaenisch, *Science* **259,** 1904 (1993).

[50] J. T. Lee and R. Jaenisch, *Nucleic Acids Res.* **24,** 5054 (1996).

[51] J. T. Lee, W. M. Strauss, J. A. Dausman, and R. Jaenisch, *Cell* **86,** 83 (1996).

[52] A. Schedl, F. Beermann, E. Thies, L. Montoliu, G. Kelsey, and G. Schütz, *Nucleic Acids Res.* **20,** 3073 (1992).

[53] K. M. L. Gaensler, M. Kitamura, and Y. W. Kan, *Proc. Natl. Acad. Sci. U.S.A.* **90,** 11381 (1993).

[54] K. R. Peterson, C. H. Clegg, C. Huxley, B. M. Josephson, H. S. Haugen, T. Furukawa, and G. Stamatoyannopoulos, *Proc. Natl. Acad. Sci. U.S.A.* **90,** 7593 (1993).

[55] A. Schedl, Z. Larin, L. Montoliu, E. Thies, G. Kelsey, H. Lehrach, and G. Schütz, *Nucleic Acids Res.* **21,** 4783 (1993).

[56] A. Schedl, L. Montoliu, G. Kelsey, and G. Schutz, *Nature* **362,** 258 (1993).

[57] L. Montoliu, A. Schedl, G. Kelsey, H. Zentgraf, P. Lichter, and G. Schutz, *Reprod. Fertil. Dev.* **6,** 577 (1994).

methods introduce YACs into ES cells with the subsequent production of YAC-bearing chimeric mice. Details of these methods are reviewed elsewhere.[10,12,17] The purification of YACs for microinjection is the focus of this article.

Microinjection of murine pronuclei to produce transgenic mice requires purification of YAC DNA. However, *in vitro* manipulation of DNA causes physical damage to DNA molecules larger than 40–50 kb. In early studies the major technical difficulty was stabilizing the YAC DNA *in vitro* to minimize shear and denaturation. Gnirke *et al.* found that solutions of high ionic strength could be used as a protective agent and that intact YAC DNA could be obtained[54,70]; other groups found that polyamines or high salt plus polyamines had a similar effect.[56,71] It is imperative that high ionic strength be maintained to prevent breakage of YAC DNA at the high temperatures required for agarase treatment and during passage of the DNA through the microinjection needle.[54,70] YAC DNA prepared in the presence of 100 m*M* NaCl is sufficiently resistant to shear; not enough data exist to suggest that the addition of polyamines affords any further degree of resistance to shearing than with high salt alone. The second problem to be overcome in early studies was the concentration of DNA solutions within a useful range for microinjection. Current protocols circumvent this problem by using a final concentration step.

---

[58] K. A. Frazer, G. Narla, J. L. Zhang, and E. M. Rubin, *Nature Genet.* **9,** 424 (1995).

[59] S. P. A. McCormick, J. K. Ng, S. Taylor, L. M Flynn, R. E. Hammer, and S. G. Young, *Proc. Natl. Acad. Sci. U.S.A.* **92,** 10147 (1995).

[60] D. J. Smith, Y. Zhu, J. Zhang, J.-F. Cheng, and E. M. Rubin, *Genomics* **27,** 425 (1995).

[61] E. Heard, C. Kress, F. Mongelard, B. Courtier, C. Rougeulle, A. Ashworth, C. Vourc'h, C. Babinet, and P. Avner, *Hum. Mol. Genet.* **5,** 441 (1996).

[62] J. G. Hodgson, D. J. Smith, K. McCutcheon, H. B. Koide, K. Nishiyama, M. B. Dinulos, M. E. Stevens, N Bissada, J. Nasir, I. Kanazawa, C. M. Disteche, E. M. Rubin, and M. R. Hayden, *Hum. Mol. Genet.* **5,** 1875 (1996).

[63] C. Huxley, E. Passage, A. Manson, G. Putzu, D. Figarella-Branger, J. F. Pellissier, and M. Fontés, *Hum. Mol. Genet.* **5,** 563 (1996).

[64] S. Matsuura, V. Episkopou, R. Hamvas, and S. D. M. Brown, *Hum. Mol. Genet.* **5,** 451 (1996).

[65] A. Schedl, A. Ross, M. Lee, D. Engelkamp, P. Rashbass, V. van Heyningen, and N. D. Hastie, *Cell* **86,** 71 (1996).

[66] J. F.-X. Ainscough, T. Koide, M. Tada, S. Barton, and M. A. Surani, *Development* **124,** 3621 (1997).

[67] H. Hiemisch, G. Schutz, and K. H. Kaestner, *EMBO J.* **16,** 3995 (1997).

[68] A. L. Manson, A. E. Trezise, L. J. Mac Vinish, K. D. Kasschau, N. Birchall, V. Episkopou, G. Vassaux, M. J. Evans, W. H. Colledge, A. W. Cuthbert, and C. Huxley, *EMBO J.* **16,** 4238 (1997).

[69] S. P. McCormick, J. K. Ng, C. M. Cham, S. Taylor, S. M. Marcovina, J. P. Segrest, R. E. Hammer, and S. G. Young, *J. Biol. Chem.* **272,** 23616 (1997).

[70] A. Gnirke, C. Huxley, K. Peterson, and M. V. Olson, *Genomics* **15,** 659 (1993).

[71] L. B. Couto, E. A. Spangler, and E. M. Rubin, *Nucleic Acids Res.* **17,** 8010 (1989).

Briefly, the steps in YAC DNA preparation are as follows. (1) Preparative pulsed-field gel electrophoresis (PFGE) is used to fractionate the yeast chromosomes and YAC. (2) A second electrophoresis step is carried out perpendicular to the PFGE separation in a high percentage low-melting point agarose (LMPA) gel to concentrate the YAC DNA. The DNA migrates the length of the PFGE gel slice into the LMPA gel. Because the rate of migration is much slower in the LMPA gel relative to the PFGE gel slice, the YAC DNA is concentrated in a small volume of LMPA gel. (3) The agarose gel slice containing the YAC is digested enzymatically using agarase or gelase in the presence of high salt and/or polyamines to protect against shearing.[56,71–73] Alternately, the PFGE slice can be agarase digested first and the YAC DNA solution concentrated by low-speed ultrafiltration[47,54,70] or dialysis with sucrose.[53] A minor problem to be overcome during development of this protocol was cleaning the DNA solution to avoid clogging the injection needle. Undigested agarose or insoluble particles are removed from the DNA solution by filtration,[34,35,54] dialysis,[53,56] or centrifugation just prior to injection.[54,71] The efficiency of transgenesis obtained by injecting a 248-kb human β-globin locus YAC using this protocol was 10–14%, a level similar to that achieved with injection of smaller plasmid or cosmid constructs,[54] but only 2–4% contained intact YAC transgenes. Schedl *et al.*[56] reported a 1% yield of transgenic mice containing intact YAC DNA following injection of a 250-kb mouse tyrosinase YAC.

Materials and Reagents

Kilobase to megabase size YAC DNA is susceptible to mechanical shear and enzymatic degradation. To ensure recovery of intact YAC DNA for microinjection, care must be taken to maintain sterility and minimize photodamage and shearing forces. High-quality water (Milli-Q-filtered, Millipore, Bedford, MA) should be used in the preparation of all solutions, buffers, and gels. When possible, solutions should be sterilized by autoclaving or filtration, or made from sterile stock solutions.

Reagents

YPD medium: 1% yeast extract, 2% peptone, 2% dextrose (w/v). Dissolve yeast extract and peptone in water, sterilize by autoclaving. Add appropriate volume of filter-sterilized 20% dextrose.

50 m*M* EDTA, pH 8.0

[72] L. Montoliu, A. Schedl, G. Kelsey, P. Lichter, Z. Larin, H. Lehrach, and G. Schütz, *Cold Spring Harb. Symp. Quant. Biol.* **58,** 55 (1993).

[73] J. C. Maule, D. J. Porteous, and A. J. Brookes, *Nucleic Acids Res.* **22,** 3245 (1994).

Solution 1: 1 *M* sorbitol, 20 m*M* EDTA, pH 8.0, 14 m*M* 2-mercaptoethanol (2-ME), 2 mg/ml zymolyase-20T (ICN, Costa Mesa, CA). Make fresh just prior to use.

Solution 2: 1 *M* sorbitol, 20 m*M* EDTA, pH 8.0, 2% SeaPlaque GTG low-melting point agarose (LMPA; FMC, Rockland, ME), 14 m*M* 2-ME. Heat sorbitol, EDTA, and LMPA to a boil and equilibrate at 50°, add the 2-ME, and keep solution at 50° until needed.

Solution 3: 1 *M* sorbitol, 20 m*M* EDTA, pH 8.0, 10 m*M* Tris, pH 7.5, 14 m*M* 2-ME, 2 mg/ml zymolyase-20T. Make fresh.

LDS: 1% lithium dodecyl sulfate (Sigma, St. Louis, MO), 100 m*M* EDTA, pH 8.0, 10 m*M* Tris, pH 8.0. Filter sterilize; store at room temperature for several months.

1× NDS: 0.5 *M* EDTA, 10 m*M* Tris, 1% *N*-laurylsarcosine, pH 9.0. Mix EDTA and Tris base in water, adjust pH to greater than 8.0 with NaOH pellets. Add *N*-laurylsarcosine (predissolved in a small volume of water) and pH to 9.0 with 5 *M* NaOH solution. Bring up to final volume with water and filter sterilize. Store at 4° for up to 1 year. Working stock is 0.2× NDS; dilute to this concentration with sterile water.

TE: 10 m*M* Tris, pH 8.0, 1 m*M* EDTA, pH 8.0

0.5× TBE: 45 m*M* Tris, 45 m*M* boric acid, 1 m*M* EDTA. Autoclave buffer.

2% Absolve solution (Dupont, Boston, MA). Prepare according to manufacturer's instructions.

Agarose MP (Boehringer-Mannheim, Indianapolis, IN)

SeaPlaque GTG agarose (FMC, Rockland, ME)

Nusieve GTG LMP agarose (FMC, Rockland, ME)

Injection buffer: 10 m*M* Tris, pH 7.5, 250 μ*M* EDTA, pH 8.0, 100 m*M* NaCl. Autoclave solution.

β-Agarase I (New England Biolabs, Beverly, MA)

Seakem GTG agarose (FMC, Rockland, ME)

0.2-μm Acrodisk syringe filters (Gelman, Ann Arbor, MI)

Methods

Preparation of Agarose Plugs Containing High-Mass DNA for Preparative PFGE

1. Inoculate 200 ml YPD with 1 ml of an overnight culture of the YAC-containing yeast strain. Incubate overnight at 30° with shaking (225–250 rpm).

2. Chill plug molds on ice. Bio-Rad (Hercules, CA) plug molds form 10 plugs of approximately 0.25 ml each. Each mold will hold 2.5 ml of a yeast cell–agarose suspension.

3. Count cells with a hemocytometer. The culture should be saturated at approximately 1×10^8 cells/ml.

4. Spin down cells at 600*g* (1900 rpm in a Sorvall GSA rotor) for 5 min in a 250-ml polypropylene centrifuge bottle at room temperature. Pour off media and resuspend pellet in 80 ml 50 m*M* EDTA, pH 8.0.

5. Spin down cells again, pour off supernatant, and resuspend pellet in 20 ml 50 m*M* EDTA, pH 8.0. Transfer suspension into a preweighed 50-ml Falcon tube.

6. Spin down cells at 600*g* (1770 rpm in a Sorvall HS-4 rotor) for 5 min at room temperature and decant off supernatant. Weigh the yeast pellet and tube, subtract the weight of the tube, and assume a density of 1 (weight = volume) for the yeast pellet.

7. Add solution 1 to give 8×10^9 cells/ml. The yeast will comprise 75–90% of the volume.

8. Add an equal volume of solution 2. The yeast concentration in the plug should be 4×10^9 cells/ml. Mix rapidly, without introducing bubbles, and pipette aliquots into the chilled plug molds with a blue pipette tip. Leave on ice for 10 min to allow agarose to solidify.

9. Transfer plugs to 50-ml Falcon tube containing 40 ml solution 3. Adjust the volume of solution, if necessary, so that a ratio of 8 ml solution ml plug is maintained. Incubate at 37° for 2 hr with occasional gentle rocking.

10. Remove supernatant with 25-ml pipette and add 40 ml LDS. Maintain a solution to plug ratio of 8 : 1 ml. Incubate at 37° for 1 hr with occasional gentle rocking.

11. Replace LDS with fresh solution and incubate at 37° overnight without agitation.

12. Remove LDS and add 40 ml 0.2× NDS. Maintain the 8 : 1 solution to plug volume ratio. Incubate at room temperature for 2 hr with agitation.

13. Replace 0.2× NDS with fresh solution and incubate as in step 12.

14. Wash plugs twice with 40 ml TE, pH 8.0, for 30 min at room temperature with agitation. Maintain the 8 : 1 solution to plug volume ratio. Remove the second TE wash and add 20 ml fresh TE, pH 8.0. Plugs may be stored at 4° for at least 1 year.

Purification of YAC DNA for Microinjection

PFGE

1. Prepare 4 liters of sterile 0.5× TBE the day before the PFGE run.
2. Set up the CHEF DR-II gel system (Bio-Rad, Hercules, CA). Make sure that the electrophoresis chamber is level using a bubble level.

Circulate 1 liter of sterile water through the electrophoresis chamber and refrigeration unit for 30 min to 1 hr to clean the unit prior to adding sterile 0.5× TBE.

3. Drain water from PFGE apparatus and fill with 2 liters sterile 0.5× TBE. Equilibrate to 12°.
4. Prepare gel casting stand for PFGE. Soak gel casting stand parts and comb in 2% Absolve for 1 hr. Rinse well with sterile water. Assemble gel casting stand wearing gloves. Tape enough teeth of the comb to obtain a preparative lane of approximately 7 cm [5–6 teeth of Bio-Rad 15 tooth comb or use a preparative comb (Bio-Rad)]. The preparative well should be placed as near to the center of the gel as possible to avoid anomalous migration. If DNA migration is uneven, incomplete YAC DNA excision will result, reducing DNA concentration for microinjection. Two YACs of the same or different size may be purified on one gel using separate preparative wells and the PFGE conditions described here. The casting stand should be level.
5. Prepare 200 ml 0.5% agarose MP solution: 1 g agarose in 200 ml sterile 0.5× TBE. Melt by boiling on hot plate or in microwave oven. Cool to 58° and pour. After the gel has solidified and reached room temperature, refrigerate for 15–30 min. This will prevent the wells from collapsing in this low percentage gel when the comb is removed.
6. Load preparative yeast plugs cut to 7 mm height next to one another in the preparative well. In the standard size wells flanking the preparative well, load a small amount of the preparative plug, yeast chromosome markers (New England Biolabs or Bio-Rad), and λ midrange marker I (New England Biolabs). Seal all wells with sterile 0.8% SeaPlaque LMPA in 0.5× TBE.
7. Run the PFGE using the following conditions: 200 V, 60-sec switch, 12° 20 hr. We have used these conditions to successfully isolate 155-, 248-, 450-, and 650-kb YACs. The PFGE conditions given in this protocol are generally applicable for the purification of most YACs; 0.5% agarose MP will fractionate DNAs in the 150-kb to 1.6-Mb size range.[73] Alternate agaroses, gel concentrations, and electrophoresis conditions may be utilized to optimize separation and resolution of YACs from yeast chromosomes. These conditions may be determined empirically or Bio-Rad technical services can suggest conditions tailored to maximally fractionate a given size range for isolation of a given size YAC. For example, we routinely use 1% Seakem GTG agarose, 0.5× TBE, 14°, 200 V, 14-sec switch, 24 hr to purify 155- and 248-kb YACs.

Concentration of YAC DNA by Second-Dimension Agarose Gel Electrophoresis

1. Locate YAC band. When the gel run is completed, cut off the marker lanes flanking the preparative lane and stain them on a rotating platform in 0.5× TBE containing 0.5 μg/ml ethidium bromide for 30 min. Destain in 0.5× TBE for 30 min. Locate the YAC band under UV illumination (long or short wave) and mark the position by cutting out a notch encompassing the YAC band with a sterile scalpel, razor blade, or glass coverslip.
2. Reassemble the gel and cut out the unstained region of the preparative lane containing the YAC DNA. Cut out a second region of the preparative lane encompassing a yeast chromosome of approximately the same size as the YAC to serve as a marker lane for the second gel run. Stain and destain the remaining gel as described to ensure that the YAC band was excised properly.
3. Presoak a minigel apparatus and gel casting stand in 2% Absolve for at least 1 hr.
4. Prepare 50 ml 4% Nusieve GTG agarose solution: 2 g agarose in 50 ml sterile 0.5× TBE. Melt by boiling on hot plate or in microwave oven. Cool to 68–72° before pouring.
5. Place YAC slice and marker slice in the minigel casting stand at a 90° angle to the PFGE run so that the DNA will migrate the length of the PFGE gel slice into the Nusieve gel. Pour the gel around them.
6. Electrophorese in sterile 0.5× TBE at 2.4 V/cm (48 V for 20-cm minigel apparatus) for 16 hr. If other PFGE conditions are utilized, it may take longer for the YAC DNA to completely migrate from the PFGE gel slice into the Nusieve gel. Use of multiple marker lanes is recommended in this case to monitor the run and to determine when it is complete.

Agarase Digestion of LMPA Plug Containing Concentrated YAC DNA

1. Excise the marker lane and stain as described to localize the DNA within the Nusieve gel.
2. Locate the corresponding position in the YAC-containing lane and cut out a Nusieve gel plug containing the YAC (usually 0.5 cm). Do not include any of the PFGE gel slice. Agarase will not digest standard (non-low-melting point) agarose. Stain the remaining YAC lane to demonstrate excision of the YAC plug.
3. Place YAC slice in a preweighed 50-ml Falcon tube and weigh to determine the weight of the plug (in milligrams).

4. Add 40–50 ml injection buffer and equilibrate for 1–2 hr at room temperature with occasional rocking.
5. Transfer the gel slice to a sterile, DNase-free 1.5-ml microfuge tube with sterile forceps.
6. Melt the agarose at 68° for 10 min, transfer rapidly to 42.5°, and equilibrate for 5 min at this temperature.
7. Add 2 units β-agarase I per 100-mg gel slice using a wide-bore yellow pipette tip. Prewarm the appropriate volume of agarase to room temperature in the yellow tip prior to addition. Mix by gently releasing air bubbles into the solution three times from a wide-bore yellow pipette tip. Incubate for 3 hr (or up to overnight) at 42.5°.
8. Place on ice for 10 min to check for undigested agarose. Only if many lumps are present should agarase treatment be repeated. Small amounts of undigested agarose are removed by filtration just prior to microinjection. Alternate methods of YAC DNA concentration following agarase digestion of the PFGE gel slice include low-speed ultrafiltration[47,54,70] and dialysis with sucrose.[53]
9. To determine DNA concentration, check 5 μl of the agarased solution on a 0.8% Seakem GTG agarose gel using λ *Hin*dIII markers of known concentration as a standard.
10. Check the integrity of the YAC DNA by running 20–25 μl of purified DNA on a PFGE gel followed by Southern blot analysis. The agarose and conditions should be chosen to allow optimum resolution of the YAC to be microinjected.

Microinjection into Fertilized Mouse Oocytes

A detailed description of the steps involved in the production of transgenic mice is beyond the scope of this article, but the reader is referred to other sources.[9] Several parameters in the handling of YAC DNA before microinjection must be kept in mind. The purified DNAs should be used as soon as possible, but can be stored for 2–4 weeks at 4° without a detectable increase in degradation. Our YAC DNA preparations generally have concentrations of 5–10 ng/μl. Injections at this concentration reduce embryonic survival and produce a greater frequency of transgenics with deleted YAC copies (unpublished observations, 1993).[74] The latter problem may be due to shearing of the YAC DNA as this higher viscosity DNA solution passes through the needle. Thus, we dilute our DNAs to 1–2 ng/μl in injection buffer and filter the solution through a 0.2–μm Acrodisk. The viscosity of the DNA solution is reduced at this concentration, thus facilitating flow

[74] R. L. Brinster, H. Y. Chen, M. E. Trumbauer, M. K. Yagle, and R. D. Palmiter, *Proc. Natl. Acad. Sci. U.S.A.* **82,** 4438 (1985).

through the injection needle. Filtration removes any undigested agarose, preventing blockage in the needle, and does not seem to have a detrimental effect on YAC integrity.

Transgenic animals are identified by hybridization of slot blots of tail-derived DNA. Our efficiency of transgenesis is 10–14% for 155- and 248-kb YACs. Up to 25% of these contain only intact YAC copies and up to 60% contain both intact and deleted copies (unpublished data, 1998).[12]

Identification and Structural Analysis of YAC Transgenics

Demonstration of YAC integrity in transgenic mice is necessary because fragmentation of YACs occurs during *in vitro* manipulation and microinjection. Incomplete YAC copies may affect spatial and temporal patterns of transgene expression due to the loss of gene structural or regulatory sequences, resulting in inaccurate or misleading data. Only transgenic mice with intact copies of the transgene contained within the YAC should be utilized for functional studies. Generally, four types of structural analysis should be completed prior to beginning functional analysis. These include (1) preliminary identification of transgenic founder animals that contain complete YACs for establishment of transgenic lines using buffy coat plug analysis (BCPA),[12] (2) detailed structural analysis of individual YAC copies in each established line, (3) copy number determination, and (4) determination of site of integration in the murine genome. The second of these is the most important structural parameter for the assessment of transgenic lines and is summarized later. Methodology for the other three protocols is reviewed elsewhere.[12]

Deletions of the 5′ and 3′ ends of YAC clones occur during manipulation of YAC DNA *in vitro,* presumably due to shear. This problem has been minimized through technical advances in the preparation of YAC DNA, but nonetheless, it is still necessary to determine the structures of YAC transgenes in each mouse line.[12] Several methods have been applied to determine the continuity of individual YACs within the murine genome. Standard Southern blot analysis of restriction enzyme-digested transgenic genomic DNA or polymerase chain reaction (PCR) analysis will show the presence of YAC sequences of limited length (~0.1–20 kb). PCR also may be used to demonstrate the presence of the left and right YAC insert–vector junctions.[54,70] The detection of these two junctions suggests that the entire YAC insert may be integrated into the mouse genome, although YAC vector sequences flanking the insert may have been lost. However, none of these methods unambiguously demonstrates that YAC clone sequences are contiguous on one or more molecules.

Three methods have been described to determine the continuity of sequence within individual YAC transgene copies; RecA-assisted restriction endonuclease cleavage (RARE),[75,76] fiber fluorescent *in situ* hybridization (fiber-FISH),[77] and long-range restriction enzyme mapping (LRRM).[12,34,35,38] RARE allows detection of the integrity of the entire insert cloned into the YAC vector, whereas LRRM can be used to analyze the entire insert as a single fragment or internal fragments that encompass all or most of the transgene contained in the YAC insert. RARE and fiber FISH are technically difficult, time-consuming, and expensive to use for structural analysis of many transgenic animals. These methods have been described previously and will not be discussed in this article.[75–77]

LRRM is useful when unique restriction enzyme sites flank the transgene within the YAC insert or exist at the YAC vector–insert junctions. The best example of this method is the structural analysis of individual YAC copies in human β-globin locus YAC (β-YAC) transgenics.[12,34,35] However, this approach may be applied to YAC transgenics in general, particularly if the YAC clone was retrofitted with unique rare-cutting restriction enzyme sites, such as I-*Ppo*I sites, at the insert–vector junctions prior to transgenesis.[12–14,28,30,78]

Integrity of β-YACs in transgenic mice is determined by a detailed examination of a 140–Kb *Sfi*I fragment encompassing most of the locus.[12] Briefly, agarose plugs are prepared from mouse liver cell suspensions and plug slices are digested overnight with *Sfi*I. Wells of an agarose gel are loaded with identically digested plug slices, the DNA is fractionated by PFGE, the gels are capillary blotted, and individual lanes are cut from the membrane. Each lane is hybridized to a different probe spanning the β-globin locus. After washing, the membrane is reassembled and an autoradiograph or a phosphorimage is made. Individual YAC copies are visualized. If they are deleted, the approximate 5′ and/or 3′ breakpoint locations can be observed. Occasionally, all sequences probed for are detected, but the size is altered from the original 140 kb. In this case, one or both of the *Sfi*I sites has been lost, but sequence just internal to these sites is retained. The size of the *Sfi*I fragment then becomes contingent on the juxtaposition of the nearest *Sfi*I site(s) in the murine genome. The importance of detailed structural analysis cannot be overemphasized; in the absence of complete

[75] L. J. Ferrin and R. D. Camerini-Otero, *Science* **254,** 1494 (1991).

[76] S. P. Iadonato and A. Gnirke, *in* "Methods in Molecular Biology" (D. Markie, ed.), Vol. 54, p. 75. Humana Press, Totowa, 1996.

[77] C. Rosenberg, R. J. Florijn, F. M. Van de Rijke, L. A. Blonden, T. K. Raap, G. J. Van Ommen, and J. T. Den Dunnen, *Nature Genet.* **10,** 477 (1995).

[78] J. W. McKee-Johnson and R. H. Reeves, *in* "Methods in Molecular Biology" (D. Markie, ed.), Vol. 54, p. 167. Humana Press, Totowa, 1996.

copies, incorrect conclusions about expression may be drawn from functional data.

Conclusion

Generation of transgenic mice with YACs and other large molecules has proved to be a valuable system for studies of gene structure–function relationships, production of mouse models of human disease, complementation of mouse mutants, use of mice as bioreactors, and synthesis of *in vivo* YAC libraries. Continued refinements of current techniques and development of new protocols should encourage widespread adaptation of this strategy for the aforementioned applications. Use of whole loci as transgenes is an important improvement since the advent of murine transgenesis, one that may allow a more realistic assessment of gene regulatory mechanisms during development.

Acknowledgments

I thank those who have assisted in the development of these protocols: C. H. Clegg, A. Gnirke, H. S. Haugen, C. Huxley, and P. A. Navas. I especially thank G. Stamatoyannopoulos for support and motivation. This work was supported by National Institutes of Health Grants DK45365, DK53510, and HL53750.

Section IV

RNA-Based Control of Recombinant Gene Expression

[12] Use of Ribozymes to Inhibit Gene Expression

By MARTIN KRÜGER, CARMELA BEGER,
and FLOSSIE WONG-STAAL

Introduction

Gene therapy, an approach to cure a disease state by the introduction of therapeutic genes, has been a dream ever since the discovery of genetic alterations in human diseases.[1] Originally, gene therapy was aimed at treating genetic diseases where a single mutation is responsible for the loss of a cellular function and the delivery of a normal gene should replace, restore, or at least compensate for the missing gene function. Today, the spectrum of applications for gene therapy ranges from replacement of a missing cellular function (e.g., for inherited single gene mutations) to the block of undesirable gene expression (e.g., viral genes, oncogenes). During recent years, several technologies, such as the use of ribozymes or antisense compounds, have been developed to specifically target mRNA molecules, thereby inhibiting the transfer of information from a particular gene to the protein involved in a disease state. The high sequence specificity of ribozymes has been used successfully to target mRNA molecules, thereby modulating the activity of these genes or their protein products involved in the disease state of the target cell.

Ribozymes are small RNA molecules with endoribonuclease activity that hybridize to complementary sequences of particular target mRNA transcripts through Watson–Crick base pairing and exert catalytic cleavage activity. The development of ribozymes to specifically target a gene involved in a disease state usually includes two main steps: (1) identification of active ribozymes by *in vitro* cleavage assays and (2) ribozyme-mediated inhibition of gene expression *in vivo*. Because our laboratory is working mainly with hairpin ribozymes against human immunodeficiency virus (HIV), we will use hairpin ribozymes as an example to illustrate these processes. In addition, we will demonstrate how hairpin ribozymes have been developed to a clinical application against HIV.

[1] I. M. Verma and N. Somia, *Nature* **389,** 239 (1997).

0076-6879/99 $30.00

In Vitro Identification of Active Ribozymes

Properties of Hairpin Ribozymes and Mode of Action

The hairpin ribozyme represents a class of ribozyme molecules that was originally isolated from the negative strand of the satellite RNA of tobacco ringspot virus.[2] The name is derived from the characteristic two-dimensional "hairpin" structure, consisting of a small catalytic region of four helical domains and five loops. Based on Watson–Crick base pairing, two helices (helix 1 and helix 2) form between the substrate and the ribozyme (Fig. 1). Helix 2 needs to be four nucleotides long, whereas the length of helix 1 can be four or more nucleotides, with a functional ribozyme typically ranging from 6 to 10 nucleotides. Helix 1 and 2 facilitate binding of the substrate and determine specificity of binding for *trans*-acting hairpin ribozymes. They flank the cleavage site within the substrate, consisting of four nucleotides: N*GUC (cleavage occurs at *), where GUC on the 3′ side of the cleavage site is the only sequence required in the substrate for maximal cleavage and N is any nucleotide.[3]

Hairpin ribozymes are usually engineered to cleave an RNA molecule of interest in *trans*. Following successful substrate cleavage, the two products are released and the ribozyme is free to bind and cleave another substrate. *In vitro,* the ribozymes behave truly catalytic in that more than one substrate molecule is processed per ribozyme molecule.[2] This "recycling" of ribozymes is considered to be one of the major advantages over antisense technology. To assess the antisense rather than the cleavage-specific effect of the ribozymes, most researchers use "disabled" ribozymes that contain mutations in the catalytic domain as controls. Such ribozymes should still be able to form stable complexes with the substrate RNA, but would be unable to mediate cleavage. Additional information about structure, properties, and the design of hairpin ribozymes can be found in Yu and Burke.[4]

Given this mode of action, the RNA sequence of interest would first be searched for the presence of a GUC triplet required as a recognition sequence. In general, functionally important and highly conserved regions of the target RNA should be primarily considered for ribozyme targeting. The accessibility of a given target site within a large molecule cannot be

[2] A. Hampel, S. Nesbitt, R. Tritz, and M. Altschuler, *Methods* **5,** 37 (1993).

[3] P. Anderson, J. Monforte, R. Tritz, S. Nesbitt, J. Hearst, and A. Hampel, *Nucleic Acids Res.* **22,** 1096 (1994).

[4] Q. Yu and J. Burke, *in* "Methods in Molecular Biology" (P. C. Turner, ed.), p. 161. Humana Press, Totowa, NJ, 1997.

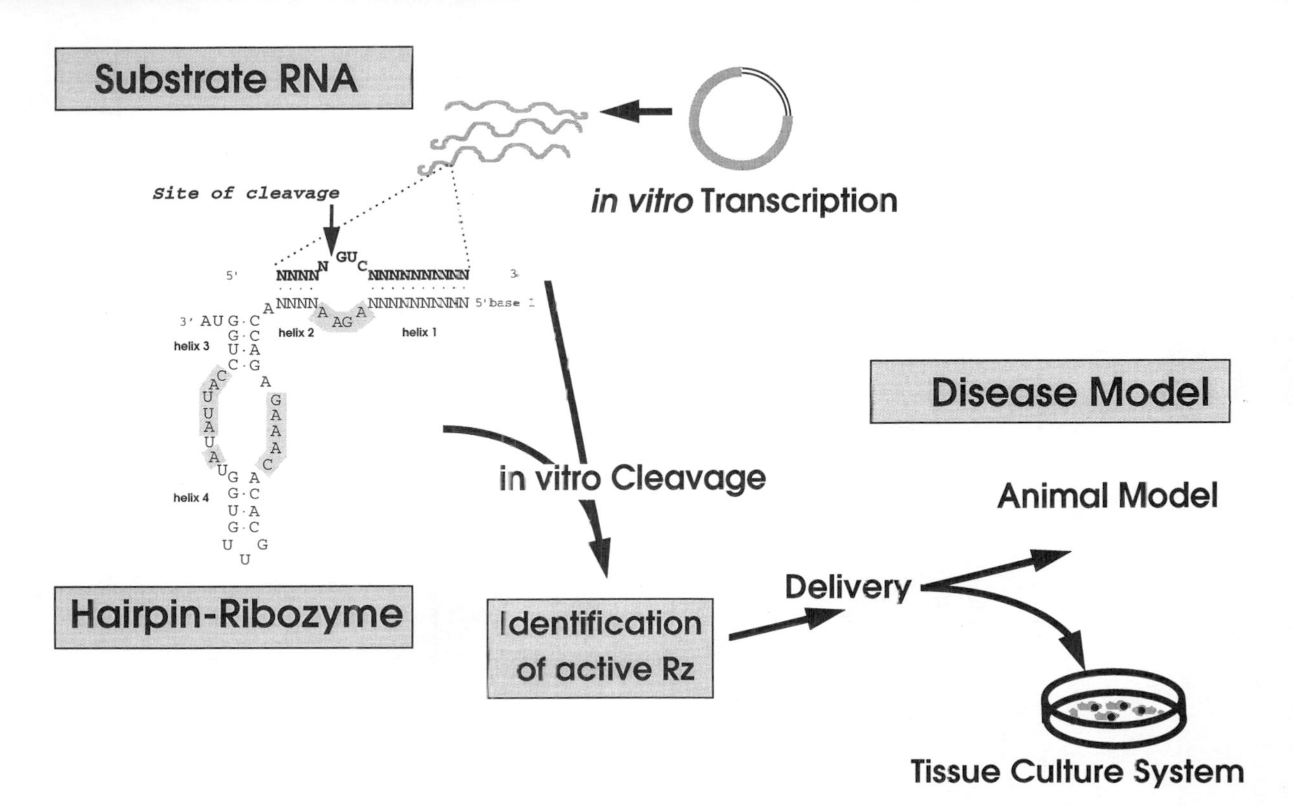

FIG. 1. Sequence and secondary structure of the hairpin ribozyme and strategy for the development of ribozymes to target particular substrate molecules by *in vitro* analysis. An *in vitro* cleavage is performed by combining an *in vitro*-transcribed substrate RNA and ribozyme. Active ribozymes identified by *in vitro* assays are subsequently delivered into a relevant disease model, e.g., animal model or tissue culture system.

determined accurately based on the predicted secondary structure, but must be determined empirically.

Construction and Preparation of Hairpin Ribozyme and Substrate Template DNA

Once a suitable target site has been identified, the ribozyme is engineered so that it base pairs to the nucleotide sequence of the target flanking the GUC triplet. To construct and synthesize a ribozyme, the following technique is used.

A specific oligonucleotide is constructed and synthesized for each ribozyme as a template for *in vitro* transcription.[5] Each ribozyme-specific oligonucleotide shares a complementary sequence to a "universal" oligonucleotide at its 3′ end containing the recognition sequence for the T7 RNA polymerase promoter (see Fig. 2) and further contains unique substrate-binding sequences (helices 1 and 2), indicated by stretches of N′s of 8 and 4 nucleotides in Fig. 2. For example, for a substrate RNA sequence 5′...UGCCGGUCUGUUGUGU...3′ (ribozyme binding regions underlined), the corresponding ribozyme oligonucleotide with the sequence 5′...TGCCTTCTTGTTGTGT...3′ (substrate binding sequences underlined) is constructed. For cloning purposes, unique restriction enzyme sites are included in the specific ribozyme oligonucleotides and the "universal" oligonucleotide. Following conversion to double-stranded DNA, these sites would allow the ribozyme DNAs to be digested with appropriate restriction enzymes and cloned into the vector of choice, such as a ribozyme-expression cassette for intracellular application.

Equimolar amounts of the oligonucleotide specific for a given ribozyme and the "universal" oligonucleotide are combined in 1× Multicore buffer (Promega, Madison, WI) and denatured by heating at 95° for 5 min followed by slowly cooling to room temperature (~60 min) to allow hybridization. The single-stranded regions of the hybrid are converted to double-stranded DNA using exonuclease-free Klenow DNA polymerase (Promega) at a concentration of 1 U/μg DNA (1× Multicore buffer, dNTPs to a final concentration of 200 μM, incubation for 45 min at 37° in an air incubator). The double-stranded DNA template is purified by phenol/chloroform extraction followed by ethanol precipitation according to standard procedures. The pellet is air-dried briefly and resuspended in 20–40 μl diethyl pyrocarbonate (DEPC)-treated nuclease-free H_2O (Ambion, Austin, TX). DNA concentration is determined by analytical agarose gel electrophoresis in

[5] P. J. Welch, R. Tritz, S. Yei, M. Leavitt, M. Yu, and J. Barber, *Gene Ther.* **3,** 994 (1996).

FIG. 2. Oligonucleotides used for the construction of double-stranded DNA templates for *in vitro* transcription of ribozyme (or substrate) RNA. A ribozyme-specific oligonucleotide is annealed to a "universal" oligonucleotide overlapping within the promoter region recognized by bacteriophage T7 RNA polymerase. Single-stranded regions are converted to double-stranded (ds) DNA using Klenow DNA polymerase (fill-in reaction). After purification, the double-stranded template can be used directly for *in vitro* transcription to obtain the ribozyme (or substrate) RNA transcript. For cloning purposes, oligonucleotides contain unique restriction enzyme sites (here *Bam*HI and *Mlu*I). Specific ribozyme oligonucleotides are constructed by changing the substrate-binding sequences (indicated by a stretch of 8 N in helix 1 and 4 N in helix 2, respectively).

comparison to a standard DNA marker with a known concentration and is adjusted to a convenient value, e.g., 0.5 μg/μl. This double-stranded DNA template can be used directly for *in vitro* transcription.

A short substrate template can be generated by the same technique as described for ribozyme template preparation. However, for the transcription of long substrates (>60 nucleotides), the substrate DNA is cloned into an *in vitro* transcription vector [e.g., pGEM-7Zf(+/−) (Promega)], and linearized immediately downstream of the substrate sequence to minimize the amount of vector-derived flanking sequences in the transcripts. Alternatively, a substrate DNA template can be generated by the polymerase chain reaction (PCR), including a phage promoter sequence appended to one of the PCR primers and incorporated into the PCR product.[6] However, for short substrates (<60 nucleotides) and for ribozymes (50–80 nucleotides) we recommend using synthetic oligonucleotides, thereby eliminating time and labor involved in cloning and preparing plasmid DNA.

In Vitro Transcription of Hairpin Ribozyme and Substrate RNA

To obtain a substrate or ribozyme RNA transcript of appropriate length and quantity, bacteriophage T7 RNA polymerase is used according to the manufacturer's instructions (Promega or MEGAShortScript T7 *in vitro* transcription kit, Ambion). Usually, we do not label the ribozymes radioactively, whereas substrate RNA molecules are synthesized enzymatically in the presence of [α-^{32}P]UTP.

The *in vitro* transcription reaction is carried out by adding the following amounts of the indicated reagents (Promega) to an autoclaved 1.5-ml microcentrifuge tube at room temperature: 2 μl (1 μg) template DNA (at 0.5 μg/μl), 4 μl 5× transcription buffer, 2 μl dithiothreitol (DTT) 100 m*M*, 20 U (= 0.6 μl) RNasin ribonuclease inhibitor, 1 μl rGTP 2.5 m*M*, 1 μl rATP 2.5 m*M*, 1 μl rUTP 2.5 m*M*, 1 μl rCTP 2.5 m*M*, 2 μl rCTP P^{32}-labeled (20 μCi) (only for substrate), 2 μl T7 RNA polymerase (at 15–20 U/μl), and RNase-free water to a total volume of 20 μl. For applications where high sensitivity is required, the amount of cold rCTP (2.5 m*M*) can be reduced to 0.5 μl per 20-μl transcription reaction. The components are mixed well by pipetting the reaction mixture up and down several times. The tube is centrifuged briefly to collect the reaction mixture at the bottom of the tube. The reaction is incubated at 37° for 1 to 4 hr, depending on the size and intrinsic transcription efficiency of the template.

The template DNA is removed by adding 1–2 μl of RNase-free DNase I (2 U/μl, Promega), centrifuged briefly, and incubated at 37° for 30 min. The transcript is purified by polyacrylamide electrophoresis under denaturing

[6] K. B. Mullis and F. Faloona, *Methods Enzymol.* **155,** 335 (1987).

conditions: 1 volume of 2× loading buffer (80% formamide, 0.1% bromophenol blue, 0.1% xylene cyanol, 2 m*M* EDTA) is added to the samples, which are denatured at 65–70° for 5 min prior to loading. Following denaturation, the samples are transferred on ice for 2 min and loaded onto a 5–15% denaturing polyacrylamide gel (depending on the size of the expected transcript) containing 7 *M* urea. Electrophoresis is conducted at 250 V for 10 min, then increased to 300–450 V until bromophenol blue dye reaches the bottom of the gel. For nonradiolabeled transcripts, the RNA bands are visualized with a handheld short-wave UV lamp and bands of interest are excised with a clean razor blade. For radiolabeled transcripts, the bands of interest are identified by brief exposure to Kodak X-OMAT film and the corresponding bands are subsequently excised from the gel. The excised bands are crushed with a pestle and shaken for 1 hr at room temperature in a solution containing 0.5 *M* ammonium acetate, 2 m*M* NaEDTA, and 0.01% sodium dodecyl sulfate (SDS). The eluted RNA is recovered by collecting the supernatant after brief centrifugation. RNase-free water is then added to a volume of 400 μl. The solution is extracted once with an equal volume of buffer-saturated phenol : chloroform : isoamyl alcohol and once with an equal volume of chloroform–isoamyl alcohol (24 : 1, v/v). The RNA is precipitated by adding 2.5 volumes of 100% ethanol and chilling for at least 15 min at −20°. The precipitate is collected by centrifugation for 15 min at maximum speed, washed twice carefully with 70% (v/v) ethanol, and the air-dried pellet is resuspended in nuclease-free water.

After purification, the size and quantity of the *in vitro*-transcribed ribozyme are assessed by running an aliquot of the reaction on a 10% polyacrylamide gel followed by ethidium bromide staining and comparison with a standardized marker. Alternatively, the ribozyme can be analyzed by measuring the absorbance in ultraviolet light (260 nm) according to standard methods. The purified, radiolabeled substrate RNA is quantified using a liquid scintillation counter/Cerenkov counting (see manufacturer's recommendations).

In Vitro Cleavage of Substrate RNA by Hairpin Ribozymes

An *in vitro* cleavage is performed by combining *in vitro*-transcribed ribozyme and ^{32}P-labeled substrate RNA molecules under appropriate conditions. Once the ribozyme recognizes the substrate through hybridization with sequences flanking the GUC site (helices 1 and 2), the ribozyme can cleave the target RNA enzymatically (indicated by an arrow in Fig. 1), thereby inactivating the substrate. On completion of cleavage, the ribozyme dissociates from the cleaved substrate, and according to its enzymatic nature, turns over to new target RNA molecules. Typically, the *in vitro* cleavage reactions are performed as follows.

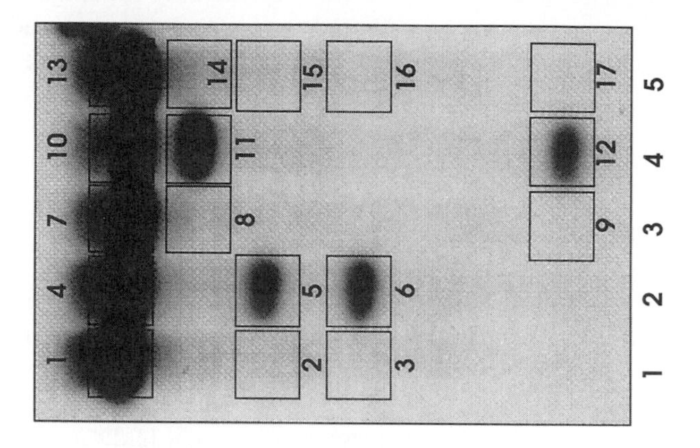
1
2
3
4
5
6
7
8
9
10
11
12
13
14
15
16
17
1 2 3 4 5

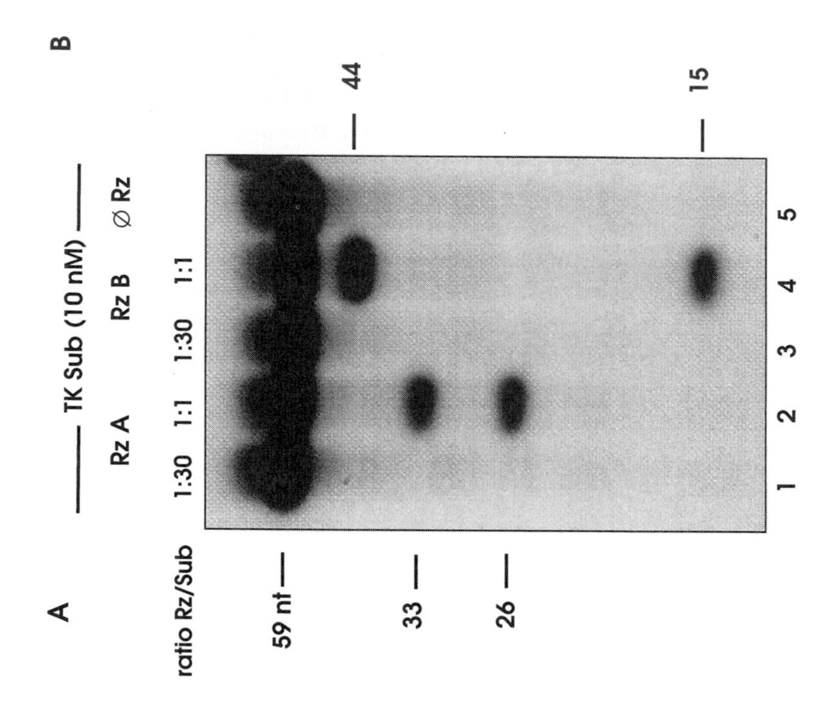
A
B
ratio Rz/Sub
TK Sub (10 nM)
Rz A
Rz B
∅ Rz
1:30 1:1 1:30 1:1
59 nt
33
26
44
15
1 2 3 4 5

Various ratios of substrate to ribozyme are combined in an autoclaved 1.5-ml microcentrifuge tube. For initial analyses, we typically use a substrate concentration of 10 n*M* and the following ratios of substrate to ribozyme: 10:1, 1:1, 1:10, 1:100, and 1:1000. Ribozyme RNA and ^{32}P-labeled substrate RNA molecules are mixed in the desired ratio and the total volume is adjusted to 9 μl by adding nuclease-free H_2O. The samples are mixed by gently pipetting up and down. Substrates with high secondary structure (e.g., long substrates) are denatured briefly by heating at 75° for 5 min followed by chilling on ice for 2 min. Then, 1 μl of 10× cleavage buffer (120 m*M* $MgCl_2$/20 m*M* spermidine/400 m*M* Tris–HCl, pH 7.5) is added and the reaction mixture is incubated at 37° for 1–3 hr (preferably in an air incubator to prevent evaporation of the reaction). For incubation times longer than 3 hr, we recommend including 0.5 μl 0.1 *M* DTT and 0.5 μl RNasin ribonuclease inhibitor (Promega) in the reaction mixture. Reactions are terminated by the addition of 1 volume of formamide gel loading buffer (80% formamide, 0.1% bromophenol blue, 0.1% xylene cyanol, 2 m*M* EDTA) and are denatured by heating at 70° for 5 min followed by immediate transfer on ice. The cleavage products are resolved by electrophoresis on a 5–15% denaturing polyacrylamide gel as described earlier and analyzed by autoradiography and phosphorimaging.

Characterizing Ribozyme Cleavage Reactions

The products of a typical *in vitro* cleavage for a hairpin ribozyme directed against a sequence within the herpes simplex virus thymidine kinase gene (HSV-TK) are shown in Fig. 3A. The substrate is 59 nucleotides long and has two GUC cleavage sites. Ribozyme-mediated cleavage with ribozyme A leads to a 5′ product of 33 nucleotides, and a 3′ product of 26 nucleotides, as visible on the gel (lane 2). Cleavage with ribozyme B leads to products of 44 nucleotides (5′ product) and 15 nucleotides (3′ product) (lane 4).

FIG. 3. Typical *in vitro* cleavage of herpes simplex virus thymidine kinase substrate RNA by specific hairpin ribozymes. Two specific hairpin ribozymes (Rz) were incubated separately with their corresponding substrate (59 nucleotides in length) containing two GUC cleavage sites. Reactions were performed with 10 n*M* substrate (Sub) and incubated at 37° for 3 hr without initial denaturation. (A) Cleavage reactions were resolved on a 15% denaturing polyacrylamide/7 *M* urea gel. Uncleaved substrate RNA and the 5′ and 3′ cleavage products are indicated. Lane 1, ribozyme A, ratio Rz/Sub 1:30; lane 2, ribozyme A, ratio Rz/Sub 1:1 (5′ cleavage product 33 nucleotides, 3′ product 26 nucleotides); lane 3, ribozyme B, ratio Rz/Sub 1:30; lane 4, ribozyme B, ratio Rz/Sub 1:1 (5′ cleavage product 44 nucleotides, 3′ product 15 nucleotides); lane 5, uncleaved substrate (59 nucleotides). (B) Quantitation of cleavage products in relation to uncleaved substrate performed on a Molecular Dynamics phosphorimager.

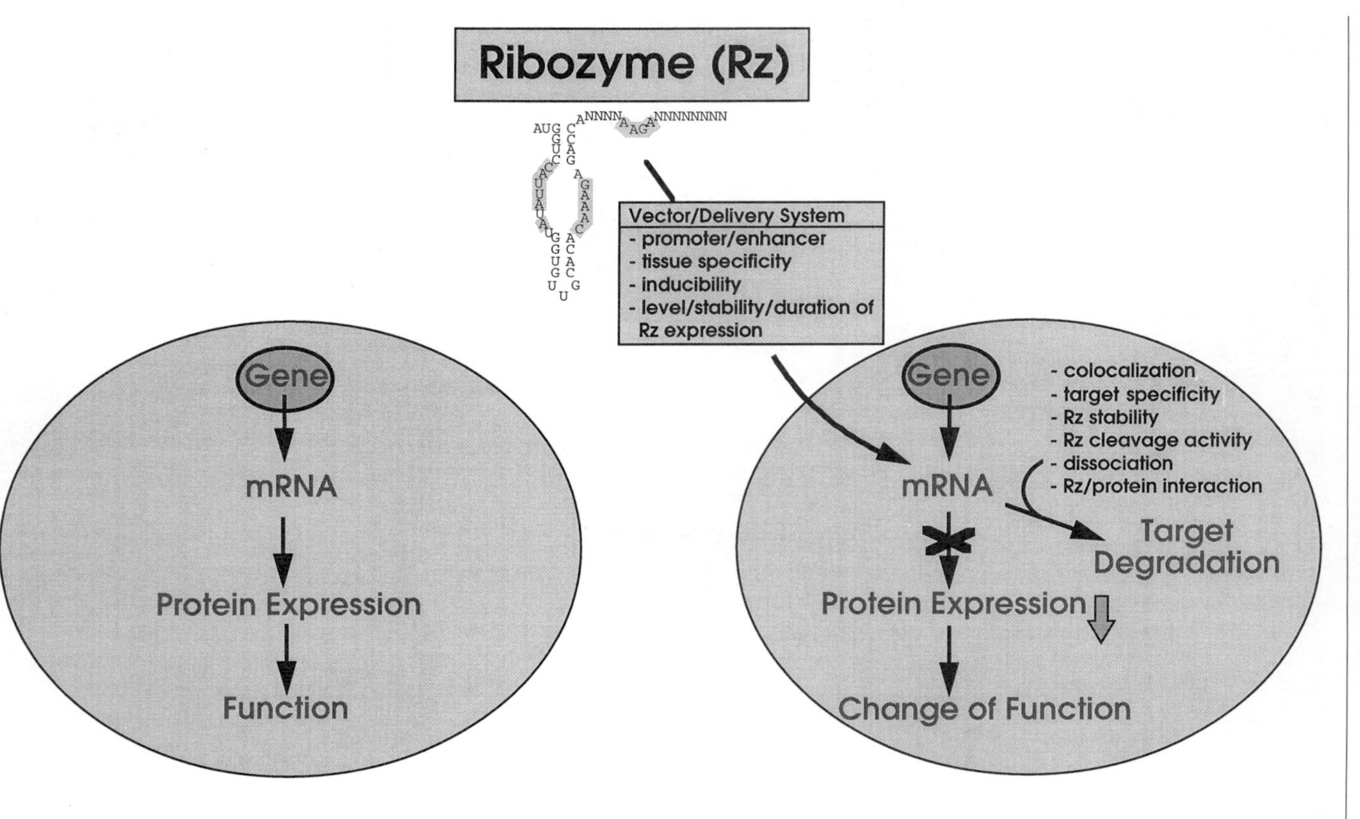
Ribozyme (Rz)
Vector/Delivery System
- promoter/enhancer
- tissue specificity
- inducibility
- level/stability/duration of Rz expression
Gene
mRNA
Protein Expression
Function
Gene
mRNA
- colocalization
- target specificity
- Rz stability
- Rz cleavage activity
- dissociation
- Rz/protein interaction
Target Degradation
Protein Expression
Change of Function
Cell without Ribozyme
Cell with active Ribozyme

Cleavage products are quantitated by phosphorimager analysis (Molecular Dynamics) as shown in Fig. 3B as a typical example. Constant-sized rectangular boxes are drawn around each of the bands in every lane. The boxes in the lane where the uncleaved substrate was loaded (lane 5) serve to determine the background for these areas. To analyze the percentage cleavage, the pixel intensity within each box covering the 5′ and 3′ product bands was calculated and the background pixel intensity for each band is subtracted. Percentage cleavage is defined as (radioactivity of the products)/[(radioactivity of the products) + (radioactivity of remaining substrate)]. Additional informations about the characterization and analysis of ribozyme cleavage reactions can be found in recent publications.[7]

Once active ribozymes and accessible cleavage sites within the substrate RNA are identified by these cleavage assays, the ribozymes are delivered into an appropriate cell culture system or animal model. Although labor-intensive and time-consuming, this approach often leads to the identification of active ribozymes that can suppress a known, undesirable gene function. However, some ribozymes that are active *in vitro* are not effective *in vivo,* and vice versa. This could be due to the myriad of cellular proteins or nucleic acids that could interact with the ribozyme substrate complex and affect various kinetic parameters of the reaction (e.g., association or dissociation rate). Furthermore, colocalization of ribozyme and substrate is another important determinant, as discussed later. Therefore, the activity of ribozymes must be ultimately evaluated in the cell or higher organisms.

In Vivo Inhibition of Gene Expression by Ribozymes

After delivering ribozyme genes into the cellular environment, active ribozymes lead to destabilization of the target RNA molecule and subsequent intracellular degradation (Fig. 4). Consequently, the expression of this specific gene and the synthesis of the encoded protein are prevented. The development of an ideal strategy for ribozyme-mediated inhibition of

[7] P. Hendry, M. J. McCall, and T. J. Lockett, *in* "Methods in Molecular Biology" (P. C. Turner, ed.), p. 221. Humana Press, Totowa, NJ, 1997.

FIG. 4. Ribozyme-mediated *in vivo* gene inhibition following ribozyme delivery into a cellular system. After successful delivery of ribozyme genes into the cellular environment of a target cell, active ribozymes lead to destabilization of the target RNA molecule and subsequent intracellular degradation (right-hand side). Consequently, the expression of this specific gene and the synthesis of the encoded protein are prevented, thereby leading to functional changes. Crucial factors of the delivery system and factors influencing intracellular gene inhibition are mentioned.

gene expression is dependent on several essential components and can be envisioned as follows: an appropriate delivery system facilitates ribozyme entry into the cell, a promoter/enhancer combination that allows high inducible and tissue-specific intracellular ribozyme expression, and optimal conditions for colocalization of ribozyme and target sequences. In recent years, these various parameters have been studied by several groups. Again, using hairpin ribozymes as an example, this section outlines important steps of our current approach toward an *in vivo* application of ribozymes.

In general, two different strategies can be used for delivery and expression of ribozymes in mammalian cells: exogenous and endogenous delivery. For an exogenous delivery system, the ribozyme is presynthesized as RNA or chimeric RNA/DNA molecules *in vitro* and is delivered to the cells via cationic lipids and liposomes, where a variety of products are available commercially. This strategy is suitable for local application of ribozymes, in settings where long-term activity of the ribozyme is not required. An example is the prevention of restenosis of coronary vessels following angioplasty, using ribozymes against cellular genes involved in cell proliferation (e.g., cell cycle genes).[8,9] However, many disease models require long-term continuous activity of the ribozymes to suppress the expression of undesirable genes (e.g., those of an infectious agent such as HIV). In such cases, the investigators would choose the endogenous delivery and expression of ribozymes via a viral or nonviral vector. A variety of nonviral gene transfer techniques are receiving increasing attention, among these are microinjection of DNA, particle bombardment, receptor-mediated gene delivery, and liposomes.[10] In addition, replication-defective viral vector systems can be used to express ribozymes *in vivo,* including retroviral vectors, adenoviral vectors, adeno-associated viral (AAV) vectors, vaccinia vectors, and herpes simplex vectors (review by Verma and Somia;[1] for more detailed information on ribozyme gene delivery, see Birikh *et al.*[11]). Each vector system has certain advantages and disadvantages, which relate to host cell range and toxicity, preferred intracellular location, immunogenicity of the vector, and stable versus transient gene expression, as well as production, titer, and purification features. Murine-based amphotropic retroviral vectors are widely used in current ribozyme approaches, as they provide efficient transduction of a wide spectrum of cell types and result

[8] T. C. Jarvis, L. J. Alby, A. A. Beaudry, F. E. Wincott, L. Beigelman, J. A. McSwiggen, N. Usman, and D. T. Stinchcomb, *Rna* **2,** 419 (1996).

[9] A. Frimerman, P. J. Welch, X. Jin, N. Eigler, S. Yei, J. Forrester, H. Honda, R. Makkar, J. Barber, and F. Litvack, *Circulation* **99,** 697 (1999).

[10] S. A. Brown and T. C. Jarvis, *in* "Methods in Molecular Biology" (P. C. Turner, ed.), p. 441. Humana Press, Totowa, NJ, 1997.

[11] K. R. Birikh, P. A. Heaton, and F. Eckstein, *Eur. J. Biochem.* **245,** 1 (1997).

in stable integration and expression of genes delivered to the host cell. However, retroviral vectors are usually produced in moderate titers and require actively dividing cells for efficient targeting. New generation adenovirus and AAV vectors are currently explored for ribozyme gene delivery and might overcome some drawbacks of retroviral vectors. Titers of adenovirus vectors usually range significantly higher compared with retroviral and AAV vectors.[12] The ultimate choice of gene delivery system depends on the experimental strategy and the target cell used in a particular application. Besides the choice of vectors, the choice of an appropriate promoter to drive the expression of the ribozyme gene is also important. Both RNA pol II (e.g., CMV) and pol III (such as the cellular $tRNA^{Val}$ promoter) promoters have been used successfully to express ribozymes *in vivo,* but the latter seems to yield more consistent results in different cell types.

Design and Construction of Ribozyme Expression Vectors

A minimal ribozyme expression vector contains the ribozyme expression cassette, in which transcription of the ribozyme gene is driven by an appropriate promoter, and a selectable marker for the selection of stable expressers in eukaryotic cells. Because a ribozyme embedded in a long molecule may be impeded in its activity, an autocatalytic cassette is sometimes included in the constructs to allow the precise 3′ termination of the ribozyme transcript.[2] Once an appropriate ribozyme expression system is selected and the ribozyme is cloned into the expression vector of choice between an appropriate promoter and the corresponding termination signal, the exact sequence of the ribozyme vector construct should be verified by standard DNA sequencing techniques.

Intracellular Delivery and Stable Expression of Ribozymes

The choice of the best methods to deliver the ribozyme expression vector into cells depends mainly on the cell type and the particular application. Evaluation can be facilitated by using a reporter gene [such as green fluorescent protein (GFP)], which can be analyzed easily for expression [e.g., by FACS (fluorescence activated cell sorting) analysis]. Most cells can be transfected using commercially available lipids or standard methods, such as DEAE-dextran or calcium phosphate. Transfection of the ribozyme expression vector is performed according to standard protocols, including plating of a certain number of cells (e.g., $2 \times 10^4/cm^2$) in appropriate media on day one in a cell culture dish (e.g., 10-cm^2 dish) and transfection on day

[12] N. Mittereder, S. Yei, C. Bachurski, J. Cuppoletti, J. A. Whitsett, P. Tolstoshev, and B. C. Trapnell, *Hum. Gene Ther.* **5,** 225 (1994).

2 or 3 after seeding. Forty-eight hours after transfection, the cells can be analyzed for RNA or protein expression. If necessary, the selection of stable transfectants can be started by refeeding with cell culture media containing an appropriate concentration of the selecting drug encoded by the expression vector (such as an antibiotic resistance gene, e.g., G 418). If the sensitivity of the cells to the selecting drug is not known, different concentrations of the drug within a usual range (e.g., 400, 600, and 800 μg/ml G 418) can be applied to optimize conditions. Therefore, the cells need to be seeded into multiple smaller dishes/wells (e.g., six-well plates). In parallel, the same concentration(s) of the selectable drug should be applied on nontransfected cells as a negative control. The duration of selection depends on particular cell lines and can be terminated once all negative control cells are killed and resistant colonies are visible in transfected cells. These colonies can be either pooled together or expanded separately for further analysis.

Detection of Expression of Ribozyme or Substrate in Targeted Cells

Once the transfectants have been expanded, expression of ribozyme (RNA) and target gene (RNA and protein) are verified. Total cellular RNA and proteins need to be isolated from harvested cells using standard techniques. Expression of ribozyme and target RNA can be determined by reverse transcription polymerase chain reaction (RT-PCR) or ribonuclease protection assay (RPA). For RT-PCR, specific primers need to be designed to reverse transcribe and amplify the ribozyme or substrate. For reverse transcription, we routinely use MMLV (Moloney murine leukemia virus) reverse transcriptase according to the manufacturer's recommendations (GIBCO-BRL). The level of RNA expression can be quantitated, using (QC) RT-PCR, by measuring expression of the RNA of interest relative to known amounts of a competitor template added to the same amplification reaction.[13] For RPA, a radiolabeled probe specific for the ribozyme or substrate sequence is transcribed *in vitro* and is hybridized to the RNA sample. RPA is performed using reagents that are available from various suppliers (such as RPA II kit from Ambion, Austin, TX). After hybridization, the sample is digested with a mixture of RNase A and T1 to remove the unhybridized, single-stranded probe. The sample is then analyzed by denaturing polyacrylamide electrophoresis. Relative ribozyme expression levels are determined by quantifying the amount of protected product(s) by phosphorimager analysis, and the catalytic activity of the ribozyme is evaluated directly by measuring the reduction of target mRNA in the target

[13] A. A. Beaudry and J. A. McSwiggen, *in* "Methods in Molecular Biology" (P. C. Turner, ed.), p. 325. Humana Press, Totowa, NJ, 1997.

cells. In addition to RT-PCR and RPA, both of which are sensitive and specific for the detection of ribozyme and substrate within cultured cells, Northern blotting can also be used for RNA detection, although this is less sensitive. For experiments performed with ribozymes targeting viral gene expression (e.g., HIV, hepatitis C virus), assays based on viral titer, reduction of infectivity, and proviral DNA (HIV) are good indicators for ribozyme-mediated inhibition of viral gene expression. RNA analyses should reveal a reduced amount of target RNA, providing indirect evidence of ribozyme activity. Direct evidence of cleavage by detection of cleavage products in cells was difficult until recently, when a new technique using reverse ligation-mediated PCR allowed the detection of such products.[14]

For these *in vivo* analyses, various controls are always performed, including mock-transfected cells as well as inactive ribozymes containing mutations in the catalytic domain to establish that any effect observed *in vivo* is a result of specific ribozyme-mediated cleavage.

Determination of Phenotypic Changes

Once a reduction of the target gene expression has been verified, it might be interesting to analyze further the ribozyme effect on the expression level of other genes that are regulated by the target gene. The phenotypic changes that can be detected after the successful intracellular expression of active ribozymes depend on the function of the target gene. If the gene is involved in the regulation of proliferation of the cell, a change in the proliferative capacity can be monitored. If the gene is responsible for certain cellular characteristics, such as the suppression of apoptosis, resistance, or susceptibility to a viral infection or metabolism of a specific agent, a change in this feature can be analyzed in transfected cells.

Ribozyme Gene Therapy against HIV: From Bench to Bedside

Although much progress has been made in the treatment of HIV infection, AIDS continues to be a major threat to global health, especially in developing countries. Even in developed countries, sobering realizations have dawned in the afterglow of the success of combination drug therapies: virus eradication is probably impossible once a person is infected, thus making therapy a lifelong requirement; treatment has to be initiated as early after infection as possible to forestall immune destruction by the virus, and multiple antiviral drugs are necessary to suppress virus replication

[14] E. Bertrand, M. Fromont-Racine, R. Pictet, and T. Grange, *in* "Methods in Molecular Biology" (P. C. Turner, ed.), p. 311. Humana Press, Totowa, NJ, 1997.

and resistance. There is already panic over the lack of alternatives once the current three or four drug regimens fail. Therefore, there is a continued need for the development of new, alternative approaches that would offer long-term antiviral effects. Gene therapy seems to be the tailor-made solution to these challenges. If HIV target cells can be engineered to express a combination of antiviral genes continuously and without toxicity, such cells may escape virus destruction and immune impairment and preferentially expand *in vivo*.

Sarver and co-workers[15] described the first use of a hammerhead ribozyme as an anti-HIV agent in 1990. In this study, HeLaCD4$^+$ cells stably expressing a ribozyme targeting the gag sequence expressed reduced p24 levels compared with the nonribozyme expressing cells on infection with HIV. Since then, a number of groups have utilized ribozymes to inhibit HIV in various preclinical studies (see review by Poeschla and Wong-Staal[16] and references contained therein). Because ribozymes utilize the specificity of Watson–Crick base pairing to interfere with gene expression in a sequence-specific fashion, it is simple to design ribozymes against highly conserved target sites on the virus genome as well as potential mutations that would confer viral resistance. Hence, a cocktail of ribozymes against wild-type and potential mutant viruses can be administered to preempt any possibility of the emergence of resistant viruses. This is a unique advantage of the ribozyme approach.

Our laboratory has been developing gene therapy for HIV infection. We have shown that hairpin ribozymes targeting three different sites on the HIV genome, namely U5,[17] Pol,[18] and Rev/RRE,[19] could effectively inhibit virus replication in a variety of primary and cultured hematopoietic cells against both laboratory and primary virus isolates. To further increase antiviral potency and minimize the chance of viral resistance, we cloned the stem–loop II sequences (SL II) of the HIV-1 Rev response element (RRE) into the ribozyme transcription cassettes.[20] The SL II sequence may further serve to stabilize the ribozyme as well as increase colocalization with HIV mRNA intracellularly, particularly during nuclear export. The

[15] N. Sarver, E. M. Cantin, P. S. Chang, J. A. Zaia, P. A. Ladne, D. A. Stephens, and J. J. Rossi, *Science* **247,** 1222 (1990).

[16] E. Poeschla and F. Wong-Staal, *in* "AIDS Clinical Review" (P. Volberding and M. Jacobson, eds.), p. 1. Dekker, New York, 1995.

[17] J. O. Ojwang, A. Hampel, D. J. Looney, F. Wong-Staal, and J. Rappaport, *Proc. Natl. Acad. Sci. U.S.A.* **89,** 10802 (1992).

[18] M. Yu, E. Poeschla, O. Yamada, P. Degrandis, M. C. Leavitt, M. Heusch, J. K. Yees, F. Wong-Staal, and A. Hampel, *Virology* **206,** 381 (1995).

[19] O. Yamada, G. Kraus, M. C. Leavitt, M. Yu, and F. Wong-Staal, *Virology* **205,** 121 (1994).

[20] O. Yamada, G. Kraus, L. Luznik, M. Yu, and F. Wong-Staal, *J. Virol.* **70,** 1596 (1996).

fusion RNA molecules were shown to function both as an RNA decoy and a ribozyme. Stable Molt 4/8 cell lines expressing "fusion RNA" of the SL II and the U5 ribozyme or its disabled counterpart were expressed stably in T-cell lines without toxicity. When these transduced cells were cocultivated with chronically infected Jurkat cells, much greater protection was observed in cells expressing the functional ribozyme linked to SL II. In a subsequent study,[21] this fusion molecule was expressed in retroviral-based double copy vector to obtain higher expression of this molecule. Furthermore, a sequence was inserted internally to drive the expression of another fusion molecule with a ribozyme targeting the env/rev region linked to SL II to obtain a triple copy vector. These multigene antiviral vectors were subsequently transduced or transfected into human $CD4^+$ T cells (Molt-4). Cell challenge with multiple subtypes of HIV-1 strains (clades A to E) showed increasing levels of virus inhibition with increasing expression of therapeutic genes. This study suggests that the combination of multiple anti-HIV genes, such as ribozymes and decoys, targeting multiple sites of HIV RNA and expressed at high levels, are promising for the treatment of HIV-1 infection.

We also used retroviral vectors to transduce primary human peripheral blood lymphocytes (PBL). Under stimulatory conditions, lymphocytes proliferate and can be transduced by the retroviral viral vector encoding the neomycin resistance gene, nontransduced peripheral blood mononuclear cells (PBMCs) eliminated by culture in the presence of the antibiotic G418. It is possible for the resistant recombinants to be expanded as much as 1000 times if maintained in interleukin 2 (IL-2)-supplemented media. We have successfully transduced human PBMCs with the retroviral vectors pMJT and pLNL-6. Expression of the ribozyme gene had no deleterious effects on T-cell function. pMJT-transduced PBL also resisted infection by HIV-1/HXB-2 over a period of 3.5 weeks in culture, and PBMC isolated from two different donors transduced with the pMJT vector resisted infection by two different HIV-1 clinical isolates.[22] We then explored conditions for optimal transduction and expansion of $CD4^+$ lymphocytes from HIV-1-infected donors, which would be appropriate for subsequent reinfusion.[23] We also determined if transduction by the ribozyme vector would inhibit replication and spread of the endogenous HIV-1 as well as confer preferential survival and expansion of the transduced $CD4^+$ cells. In the absence of selection, about 20% of total lymphocytes were transduced.

[21] A. Gervaix, X. Li, G. Kraus, and F. Wong-Staal, *J. Virol.* **4,** 3048 (1997).

[22] M. C. Leavitt, M. Yu, O. Yamada, G. Kraus, D. Looney, E. Poeschla, and F. Wong-Staal, *Hum. Gene Ther.* **5,** 1115 (1994).

[23] M. C. Leavitt, M. Yu, F. Wong-Staal, and D. J. Looney, *Gene Ther.* **3,** 599 (1996).

Cells transduced with the ribozyme-expressing vector (MJT) exhibited a threefold greater $CD4^+$ cell number after 2 weeks in culture compared to control vector (LNL6)-transduced cells. Viral replication was delayed 2–3 weeks. Both transduced cell populations underwent a 2–3 log expansion in 2 weeks.

Based on these encouraging preclinical data, we initiated the first phase 1 clinical study to test the safety and feasibility of ribozyme gene therapy for HIV-1 patients.[24] Leukapheresed PBMC from three patients were enriched for $CD4^+$ cells and transduced separately with GMP-grade murine retroviral vectors MY-2, expressing two anti-HIV-1 hairpin ribozymes and an autocatalytic hammerhead ribozyme, and LNL-6, the control vector. The cells were expanded *ex vivo* in the presence of antiretroviral drugs (delaviridine, nelfinavir). Equal numbers of the two transduced cell products were combined and assayed for p24, proviral load, replication-competent retrovirus (RCR), endotoxin, and other microbial contamination prior to infusion into the autologous patient. No adverse effects were observed except for transient fever in one patient. Two patients experienced a transient increase in $CD4^+/CD8^+$ ratio, which returned to baseline by 4 weeks. Vector DNA in purified $CD4^+$ PBMCs pre- and postinfusion was quantified by QC-PCR and validated by "sample blinding": control samples spiked with varying amounts of LNL6 and MY-2 transduced T cells were supplied with subject samples as coded pairs. Ribozyme expression was detected by limiting dilution RT-PCR. LNL-6-transduced cells were detected at low levels in all three subjects postinfusion and persisted for 1–24 weeks. The length of persistence of these gene-marked cells was inversely proportional to the viral load of the patients. MY2-transduced cells were detected in only one patient postinfusion. The low levels of detection of MY-2 transduced cells was due to low vector titer because of the presence of the autocatalytic cassette, apoptosis of transduced cells during *ex vivo* expansion, and less efficient detection with MY-2 primers in the QC-PCR assay. An interesting result is that *in vivo* expression of the U5 ribozyme driven by the $tRNA^{Val}$ promoter was detected at 5 and 7 months after reinfusion, whereas the Pol ribozyme driven by the MLV promoter was not expressed. These results suggest that the $tRNA^{Val}$ promoter is a promoter of choice for *in vivo* expression. The emergence of MY-2 expressing cells at 5 months after reinfusion also suggests *in vivo* amplification of these transduced cells. While more patients need to be evaluated with an improved ribozyme vector, the preceeding results at least confirm the safety and feasibility of this approach. This also marks the first clinical study using ribozymes for any disease.

[24] F. Wong-Staal, E. Poeschla, and D. Looney, *Hum. Gene Ther.* **9,** 2407 (1998).

Future Applications of Ribozymes

As discussed earlier, ribozyme technology for expression knock out can potentially be applied to different diseases, including infectious diseases, cancers, and cardiovascular disorders. Despite recent progress in this field, major problems still need to be addressed: the identification of optimal target sites, optimization of ribozymes for intracellular cleavage, the efficiency of gene delivery to appropriate target cells, and persistent gene expression *in vivo,* as well as a relevant tissue culture system or animal models for preclinical evaluation.

In addition to direct therapeutic applications, ribozymes can also be used as a powerful tool in another new frontier in biomedical research. In the decade of the human genome project, new genes are being discovered with tremendous speed, but often little or no information is available about the correlated gene function. In this context, ribozyme technology may allow the direct correlation of sequence and biological function in a cell. Once a ribozyme has targeted a gene successfully, the phenotypic effect of such a "knock out" (gain or loss of function) of a particular gene and the functional effect (change of phenotype) can be studied in a cell culture system or animal model. This new approach of ribozyme technology in functional genomics might substantially speed up gene identification linked to a particular function and might at the same time allow the development of ribozymes as therapeutics targeting the particular genes of interest.

Acknowledgments

The authors thank Richard Tritz, Peter J. Welch, and Jack Barber for helpful discussions and support. Carmela Beger is supported by the Deutsche Forschungsgemeinschaft (Be 1980/1-1).

[13] Translational Control of Gene Expression Using Retroviral Vectors with Iron Response Elements

By Jennifer L. Davis, Paul R. Gross, and Olivier Danos

Introduction

Gene regulatory systems are an important tool to the molecular and cell biologist for a variety of research applications and will be essential to the gene therapist for applications requiring precise expression levels of therapeutic genes. A number of regulated expression vector systems have

0076-6879/99 $30.00

been described to date for both research and gene therapy applications, most of which are regulated transcriptionally. In general, these systems are composed of *cis*-acting control element(s) that are incorporated into the vector expressing the gene of interest and the *trans*-acting control protein(s) that interacts specifically with the *cis*-acting elements. The first vectors developed relied on endogenous systems that are cell type or developmental specific or are responsive to physiological stress. More recent vector systems are hybrid and heterologous in nature and circumvent many of the limitations encountered when using endogenous systems. Because each transcriptionally regulated vector system developed to date has useful properties as well as limitations, it is reasonable to propose that additional regulatory checkpoints will need to be in place to achieve the most optimal and precise level of control. Thus, it is important to also develop vector systems that are regulated posttranscriptionally, e.g., translationally, such that ultimately one can choose the appropriate regulatory system (or combination of systems) for the application at hand. Because our interest is in developing vector systems, including model vector systems, appropriate to the field of gene therapy, we focused our initial efforts on retroviral vectors that are currently widely used in gene therapy. As several groups have already described transcriptionally regulated retroviral vectors[1–4] and because our goal was to expand retroviral gene regulatory capabilities, we investigated the feasibility of a retroviral vector system controlled at the translational level. Given the complexities of the retroviral life cycle, we initially pursued the development of a model vector system employing the well-characterized iron regulatory system to define the basic principles on which a translationally regulated retroviral vector should be built. It is our hope that the principles outlined here, as well as the methodology described, will assist others in the design of additional types of translationally regulated vectors.

In the endogenous iron regulatory system, a 32 nucleotide sequence known as the iron response element (IRE) found in the 5′-untranslated region of ferritin mRNA confers iron-regulated expression at a translational level. Ubiquitous cellular proteins known as iron response binding proteins (IRBPs) bind to this sequence in the absence of iron and prevent translation initiation. In the presence of iron, the IRBPs no longer bind the IRE and

[1] A. Hofmann, G. P. Nolan, and H. M. Blau, *Proc. Natl. Acad. Sci. U.S.A.* **93,** 5185 (1996).
[2] J.-J. Hwang, Z. Scuric, and W. F. Anderson, *J. Virol.* **70,** 8138 (1996).
[3] W. Paulus, I. Baur, F. M. Boyce, X. O. Breakefield, and S. A. Reeves, *J. Virol.* **70,** 62 (1996).
[4] V. M. Rivera, T. Clackson, S. Natesan, R. Pollock, J. F. Amara, T. Keenan, S. R. Magari, T. Phillips, N. L. Courage, F. Cerasoli, Jr., D. A. Holt, and M. Gilman, *Nature Med.* **2,** 1028 (1996).

translation occurs.[5] It has been demonstrated that the IRE can function in a heterologous context when it is placed within the first 67 nucleotides of the message.[6–9] Therefore, we constructed a modified retroviral vector by inserting a synthetic IRE in both the 5′ and the 3′ R regions of the LTRs in the MFG retroviral vector. Because the R region serves important functions in the viral life cycle, there were concerns that the insertion of foreign sequences into this region might negatively impact various aspects of vector performance, such as virus production and genomic stability. We therefore examined these features of vector function in addition to gene regulation. To further address the general utility of this vector, we generated virus particles carrying two different types of transgenes: *lacZ,* a cytoplasmic marker, and human α-interferon (IFNα), a clinically relevant secreted protein. In addition, virus was produced from two different packaging cell systems, transient 293 based or stable NIH 3T3 based, and we examined regulation in cell populations derived from two different lines, C2C12 cells or NIH 3T3 cells.

Experimental Procedures

Design of Iron-Regulated Retroviral Vectors

Vector Design. The parent vector, MFG, has been described in detail elsewhere and contains the complete Moloney murine leukemia virus (MoMLV) LTR driving expression of the transgene.[10] Splicing and packaging signals from the virus are included for efficient virus production and gene expression in transduced cells. To convert MFG into a translationally regulated vector, a synthetic IRE (Fig. 1A) is inserted into MFG at +8 in the R region of both the 5′ and the 3′ LTRs to generate the vector called DG-IRE (Fig. 1C). This location was chosen for two reasons. First, the

[5] T. A. Rouault, R. D. Klausner, and J. B. Harford, *in* "Translational Control" (J. W. B. Hershey, M. B. Mathews, and N. Sonenberg, eds.), p. 335. Cold Spring Harbor Laboratory Press, Plainview, NY, 1996.

[6] N. Aziz and H. N. Munro, *Proc. Natl. Acad. Sci. U.S.A.* **84,** 8478 (1987).

[7] M. W. Hentze, S. W. Caughman, T. A. Rouault, J. G. Barriocanal, A. Dancis, J. B. Harford, and R. D. Klausner, *Science* **238,** 1570 (1987).

[8] S. W. Caughman, M. W. Hentze, T. A. Rouault, J. B. Harford, and R. D. Klausner, *J. Biol. Chem.* **263,** 19048 (1988).

[9] B. Goossen, S. W. Caughman, J. B. Harford, R. D. Klasner, and M. W. Hentze, *EMBO J.* **9,** 4127 (1990).

[10] G. Dranoff, E. Jaffee, A. Lazenby, P. Golumbek, H. Levitsky, K. Brose, V. Jackson, H. Hamada, D. Pardoll, and R. C. Mulligan, *Proc. Natl. Acad. Sci. U.S.A.* **90,** 3539 (1993).

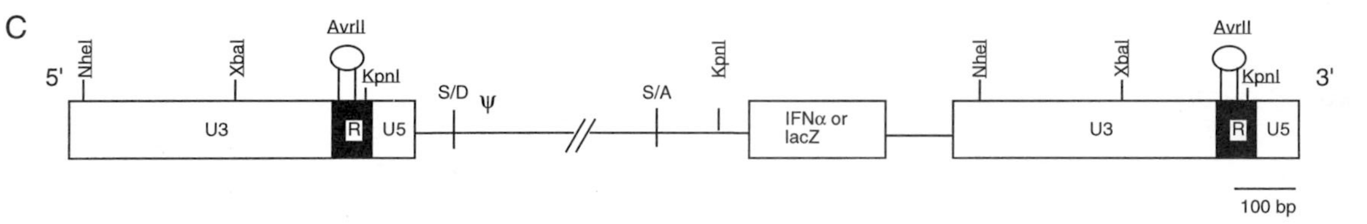

FIG. 1. Schematic vector diagram. (A) Sequence of synthetic iron response element with proposed secondary structure. Location of *Avr*II restriction sites is shown. (B) Sequence of synthetic iron response element with proposed secondary structure. Location of *Sac*II restriction sites is shown. (C) DG-IRE retroviral vector drawn to scale with the exception of the cDNA insert. The approximate insertion site for the IRE sequence in each R region is indicated by the stem–loop diagram. The location of relevant restriction sites is marked. For clarity, only one *Avr*II site is shown; there are, in fact, two sites for each IRE (A). Splice donor (S/D) and acceptor (S/A) sites are shown; ψ, packaging signal.

IRE occurs at +8 in the ferritin gene. Second, it has been shown that the IRE must be within the first 67 nucleotides of the transcript to function.[6,9] The R region is 68 bp in length; therefore, insertion of the IRE had to be within this region. Because perfect homology between the R regions is ideal for faithful and efficient replication[11] and to maintain the integrity of the IRE sequence, it was inserted into both the 5′ and the 3′ R regions.

Plasmid and Vector Constructions. There are three plasmid constructs used in this work that have not been described previously. They are the iron-regulated retroviral vector DG-IRE carrying either IFNα or *lacZ* and a retroviral vector with a different IRE sequence, termed *Sac*II-IRE. For clarity, they are all described here.

To facilitate cloning into each LTR, the MFG vector is subcloned into pBS (Stratagene) as an *Eco*RI to *Spe*I fragment containing the 5′ LTR and into a separate pBS vector as a *Bam*HI to *Hin*dIII fragment containing the 3′ LTR. The DG-IRE vector is then constructed as follows. A two-step polymerase chain reaction (PCR) protocol is used to incorporate the synthetic IRE into either the 5′ or the 3′ LTR. The outside primers used for the 5′ LTR are

5′-GGAATTCAATGAAAGACCCCACCTGTAGGTT-3′

and

5′-GGACTAGTACCGACGCAGGCGC-3′

and the inside primers are

5′-CTCGGCGCGCCAGTCCTAGGTGCTTCAACAGTGCTTGGACC
CTAGGCCTCCGATTGACTG-3′

and

5′-CAGTCAATCGGAGGCCTAGGGTCCAAGCACTGTTGAAGCA
CCTAGGACTGGCGCGCCGAG-3′

For the 3′ LTR, the same inside primers described earlier are used and the two outside primers are

5′-CCGGGATCCGGATTAGTCCAATTTG-3′

and

5′-CCCAAGCTTAATGAAAGACCCCCGCTGACGGGT-3′

PCR reactions are carried out using Vent DNA polymerase (New England Biolabs, Beverly, MA) according to manufacturer's recommendations and

[11] L. I. Lobel and S. P. Goff, *J. Virol.* **53,** 447 (1985).

established procedures.[12] Due to a large region of nonhomology between the internal IRE-containing primers and the template DNA, PCR is performed as follows: 10 cycles of 94° for 1 min, 50° for 1 min, and 72° for 2 min were followed by 20 cycles of 94° for 1 min, 60° for 1 min, and 72° for 2 min. Final PCR products are ligated to the MFG vector in a four-part ligation to generate DG-IRE using standard techniques.[13] Regions of the vector that have been amplified are confirmed by DNA sequence analysis. The *Sac*II-IRE retroviral vector is constructed exactly as described earlier for the DG-IRE vector except that the inside primers are

5′-CTCGGCGCGCCAGTCCGCGGTGCTTCAACAGTGCTTGGACC CGCGGCCTCCGATTGACTG-3′

and

5′-CAGTCAATCGGAGGCCGCGGGTCCAAGCACTGTTGAAGCA CCGCGGACTGGCGCGCCGAG-3′

To generate the IFNα cDNA, synthetic oligonucleotides corresponding to the amino-terminal 30 amino acids of the protein are assembled following standard techniques,[13] and the remainder of the cDNA is generated in a standard PCR reaction using plasmid pRRB201F-23 (ATCC 67979) as template and

5′-GGACCTTGATGCTCCTGG-3′

and

5′GGATCCTCATTCCTTACTTCTTA-3′

as primers. The resulting IFNα coding sequence is confirmed by DNA sequence analysis and is used to construct the MFG-IFNα and DG-IRE-IFNα retroviral vectors.

To construct the DG-IRE-lacZ vector or *Sac*II-IRE-*lacZ* vector, the 3.1-kb *Bam*HI *lacZ* fragment from MFG-*lacZ*[10] is isolated and subcloned into DG-IRE by standard molecular biological techniques.[13]

Iron-Regulated Retroviral Vectors: Assessment of Particle Production and Function

Particle Generation and Cell Transductions. To test whether the DG-IRE vector could generate transducing particles under normal cell culture

[12] R. Higuchi, *in* "PCR Protocols" (M. A. Innis, D. H. Gelfand, J. J. Sninsky, and T. J. White, eds.), p. 177. Academic Press, San Diego, 1990.

[13] J. Sambrook, E. F. Fritsch, and T. T. Maniatis, eds., *in* "Molecular Cloning: A Laboratory Manual." Cold Spring Harbor Laboratory Press, Cold Spring Harbor, NY, 1989.

conditions and to determine whether the presence of the IRE affected the transduction process or basal levels of gene expression, the human interferon α 2 cDNA (IFNα) is subcloned into it and into the MFG vector as a control (see earlier discussion). IFNα was chosen as a marker gene for several reasons, the most important being that it has a very short half-life (10 min in rabbit serum). Thus, it is particularly well suited to a study such as this where we wanted to examine on/off regulation.

Due to difficulties in generating stable producer lines for IFNα in some packaging cell lines, we have produced the virus transiently by cotransfection of DG-IRE-IFNα or MFG-IFNα and a plasmid expressing the VSVG envelope (pMD.G)[14] into a 293 cell line stably expressing MoMLV/*gag-pol* (293gp13)[15] using a calcium phosphate transfection protocol.[16,17] All 293 cells (ATCC, Rockville, MD, CRL1573) and derivatives are maintained in Dulbecco's modified Eagle's medium (DMEM; GIBCO-BRL, Gaithersburg, MD) containing 10% fetal bovine serum (FBS) and antibiotics. As there are numerous variations on the calcium phosphate transfection protocol and as the specifics are critical to optimal virus production, we are elaborating on the protocol here. Cells are seeded 20–24 hr prior to transfection at 7×10^6 cells per 10-cm dish in DMEM plus 10% FBS plus 25 m*M* HEPES, pH 7.0 (IMDM can also be used). The day of transfection, cells are 80–90% confluent. Cells are refed with DMEM plus 10% FBS, 25 m*M* HEPES, pH 7.0, and 25 μM chloroquine 2–3 hr prior to transfection. Plasmid DNA is purified twice over a cesium chloride gradient; a total of 14 μg of DNA (7 μg of retroviral vector plus 7 μg of pMD.G) per 10-cm dish is used for transfection. Reagents and DNA precipitate formation using ice-cold reagents are exactly as described,[16] except that the concentration of sodium phosphate in the 2× HEPES buffered saline (HBS) was 1.5 m*M* and not 2.8 m*M*. Precipitates are immediately added dropwise to the cells while swirling the dish to ensure rapid and even distribution of the DNA. Cells and DNA are incubated for 20–24 hr at 37°. DNA-containing media are then removed, and cells are rinsed gently with Hanks' balanced salt solution (HBSS) (calcium/magnesium free) and refed with normal growth media. It should be noted that there is no sequence homology between the transiently transfected retroviral vector and the envelope vector. This eliminates the possibility of generating replication-competent retrovirus

[14] L. Naldini, U. Blomer, P. Gallay, D. Ory, R. Mulligan, F. H. Gage, I. M. Verma, and D. Trono, *Science* **272,** 263 (1996).

[15] J. L. Davis, R. M. Witt, P. R. Gross, C. A. Hokanson, S. Jungles, L. K. Cohen, O. Danos, and S. K. Spratt, *Hum. Gene Ther.* **8,** 1459 (1997).

[16] C. M. Gorman, D. R. Gies, and G. McCray, *DNA Prot. Eng. Tech.* **2,** 3 (1990).

[17] W. S. Pear, G. P. Nolan, M. L. Scott, and D. Baltimore, *Proc. Natl. Acad. Sci. U.S.A.* **90,** 8392 (1993).

through homologous recombination at the DNA level during the transfection process, a phenomenon that has been observed by others.[17] Because it is important to ensure that replication-competent retrovirus arising through some other mechanism is not contaminating vector preparations, viral supernatants and transduced cells are tested regularly for replication-competent retrovirus using a highly sensitive mobilization assay[15] and are negative. Virus-containing supernatants from transiently transfected cells are harvested every 24 hr for a period of 4 days, filtered through a 0.45-μm filter, pooled, and frozen at $-80°$. Because unconcentrated IFNα viral stocks are consistently of low titer, retroviral particles are pseudotyped with the VSVG envelope such that they could be concentrated easily for experiments where maximum IFNα gene expression and copy number are desired. Virus-containing supernatants are concentrated by ultracentrifugation (50,000*g*, 90 min) followed by resuspension of the pellet at room temperature for 3 hr with intermittent mixing in 1/100th volume of TNE buffer (50 m*M* Tris–HCl, pH 7.8, 130 m*M* NaCl, 1 m*M* EDTA).

In the experiments presented in Fig. 2 and Table I, C2C12 cells are transduced three sequential times with concentrated viral stocks to obtain a high number of integrated copies. C2C12 cells (gift of S. Hardy) and derivatives are maintained as described previously.[18] For each transduction, C2C12 cells in culture are overlayed with virus containing supernatant and 8 μg/ml Polybrene (Sigma, St. Louis, MO) for 20–24 hr. Basal gene expression and genomic copy number in transduced cells are determined (Fig. 2 and Table I). For IFNα expression analysis, 24-hr supernatants are collected (in normal growth media[18]) from the same cells that were used to isolate genomic DNA. For quantitation of IFNα gene expression, an enzyme-linked immunosorbent assay (ELISA) is performed according to the manufacturer's protocol (Endogen, Woburn, MA). The IFNα ELISA assay is performed in duplicate; the mean values and standard deviations are shown in Table I. MFG-IFNα-transduced cells secrete 3.9 ng/ml/10^6 cells/24 hr/genome of IFNα (Fig. 2, Table I) whereas DG-IRE-IFNα-transduced cells secrete 3.2 ng/ml/10^6 cells/24 hr/genome (Fig. 2 and Table I). Thus, expression levels on a per genome basis are equivalent. This result indicates that the presence of IRE in the DG-IRE vector neither enhanced nor hindered the vector production and transduction processes and did not affect basal expression levels relative to the parent MFG vector.

Regulated Expression. To examine whether cells transduced with the DG-IRE vector could regulate expression of a transgene in response to changes in iron concentration in the media, C2C12 cells are transduced with unconcentrated viral stocks of the DG-IRE-IFNα and MFG-IFNα

[18] L. Silberstein, S. G. Webster, M. Travis, and H. M. Blau, *Cell* **46,** 1075 (1986).

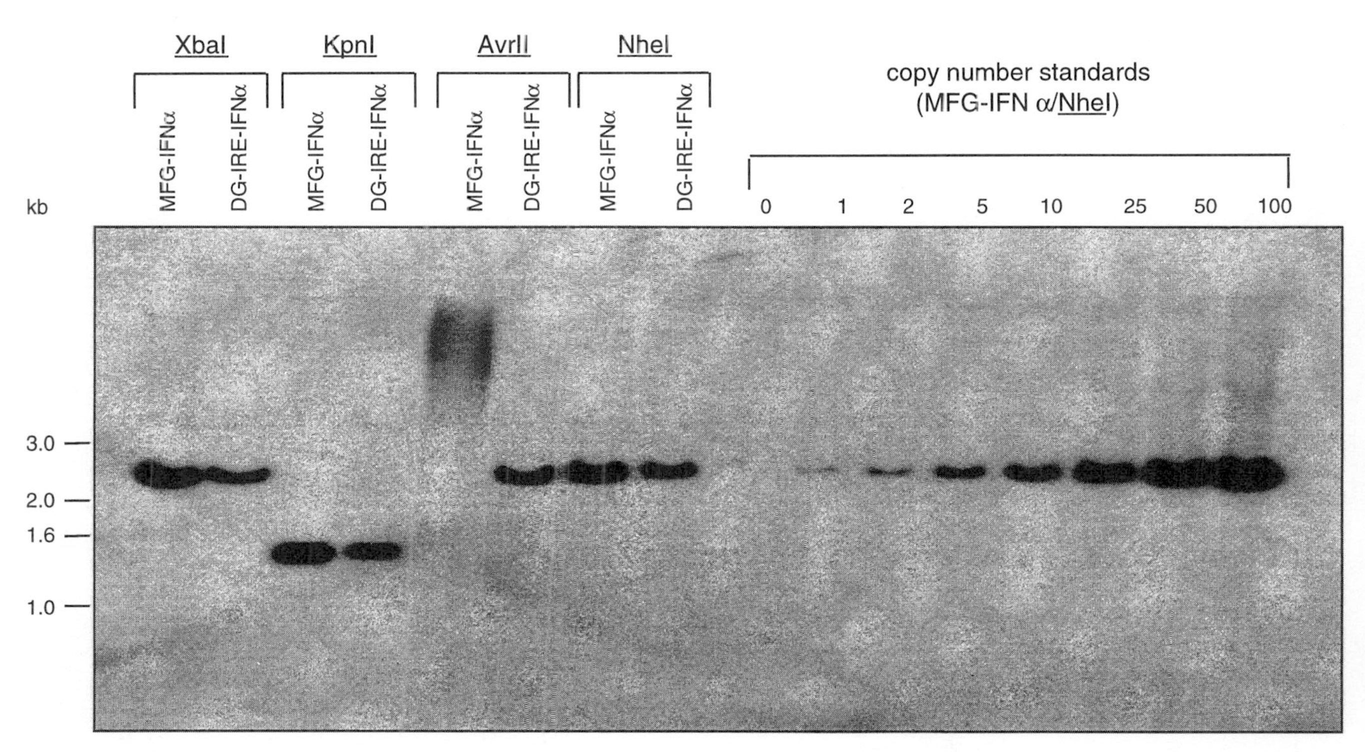

FIG. 2. Southern blot analysis of transduced C2C12 cells. Cells were transduced with either MFG-IFNα or DG-IRE-IFNα as indicated above each lane. Restriction enzymes used are noted above each lane. MFG-IFNα plasmid DNA digested with *Nhe*I was spiked into 10 μg of C2C12 genomic DNA (from untransduced cells) to generate copy number standards with the values indicated. Molecular size standards are noted to the left of the blot.

TABLE I

IFNα SECRETION AND VECTOR COPY NUMBER IN TRANSDUCED C2C12 CELLS[a]

Vector	IFNα secretion (ng/ml/10^6 cells/24 hr)	Copy number	IFNα secretion/genome (ng/ml/10^6 cells/24 hr/copy)
MFG-IFNα	88.6 ± 22.6	23	3.9
DG-IRE-IFNα	22.2 ± 3.7	9	3.2

[a] Twenty-four-hour supernatants collected from C2C12 cells transduced with the indicated vector were harvested and assayed for IFNα levels by ELISA. The same cells were then harvested and genomic DNA isolated for Southern blot analysis (shown in Fig. 2).

vectors in an experiment separate from the one just described. Transduced cell populations are allowed to grow for several days and are then split into low iron containing basal media [MEMα (GIBCO-BRL) supplemented with 10% defined calf serum (Hyclone, Logan, UT)] and antibiotics. Media contain a minimum amount of iron that is derived solely from the added serum and is 58 μg/dl iron, 646 mg/dl transferrin, and ≤10 mg/dl hemoglobin. Changes in gene expression in response to the manipulation of iron in the media are examined by exposing the cells either to extra iron in the form of ferric ammonium citrate (20 μg/ml, Sigma) or to the intracellular iron chelator, Desferal (25 *μM,* Ciba-Geigy, Basel, Switzerland). IFNα expression is quantitated by ELISA assay and the results are presented in Fig. 3. Cells transduced with the MFG vector exhibit similar IFNα secretion rates at all concentrations of iron. This indicates that gene expression in the absence of the IRE is not affected by changes in iron concentration and that the Desferal-induced iron depletion has no secondary effect on IFNα secretion. In contrast, IFNα secretion from cells transduced with the DG-IRE vector is 33-fold higher in the presence of high amounts of iron than in the absence of iron (Fig. 3). Cells grown in low iron containing basal media exhibit an intermediate level of IFNα secretion, indicating a dose–response of gene expression to iron concentration (Fig. 3). Thus, cells transduced with the DG-IRE vector are capable of modulating gene expression in response to changes in the iron content of the surrounding media.

Vector Stability. Because there were concerns that the introduction of foreign sequences into the R region might lead to genomic instability, a Southern blot analysis of cells transduced with the MFG-IFNα or DG-IRE-IFNα vector was carried out (Fig. 2). Genomic DNA is isolated using the Puregene kit (Gentra Systems, Minneapolis, MN) and processed according to standard procedures.[13] Each lane is loaded with 10 μg of DNA and the blot is probed[13] with an IFNα-specific probe. Quantification is achieved by densitometric scanning of the film and comparison to the copy

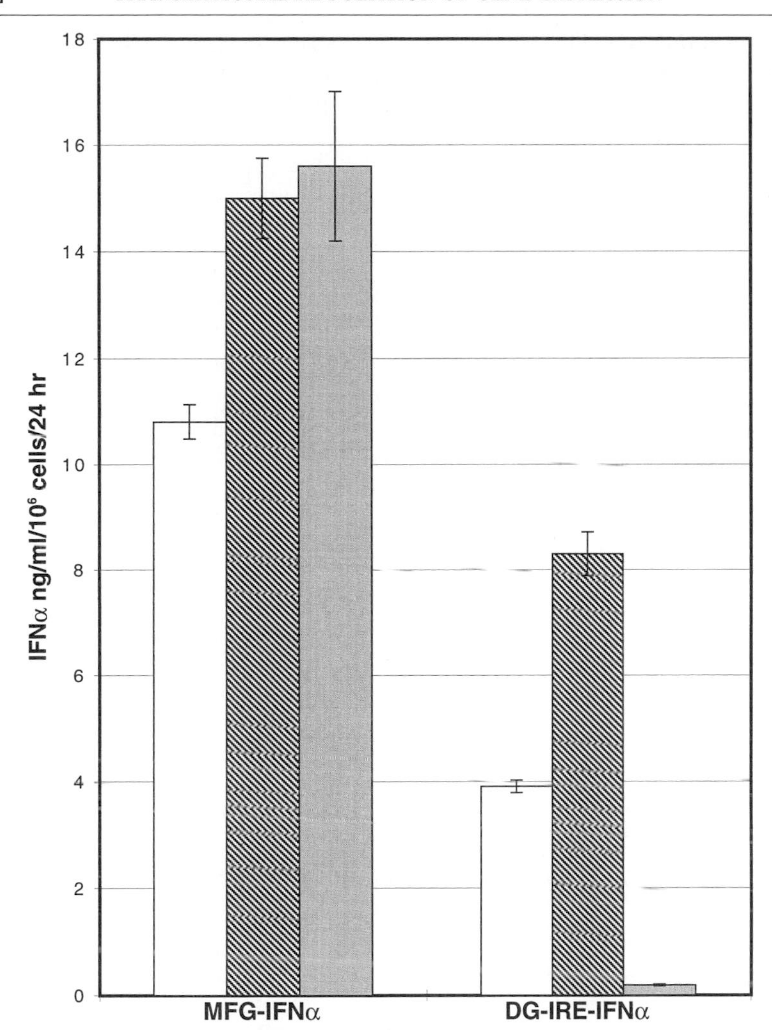

FIG. 3. IFNα secretion from transduced C2C12 cell populations in response to changes in iron concentration. Left-hand side of graph, cells transduced with MFG-IFNα; right-hand side of graph, cells transduced with DG-IRE-IFNα. Transduced cells were seeded in basal media (Materials and Methods) and in triplicate for each media condition. The next day, cells were refed with basal media (open bars), basal media plus 20 μg/ml ferric ammonium citrate (striped bars), or basal media plus 25 μM Desferal (shaded bars). The following day, cells were refed with the same type of media for the collection of 24-hr supernatants. During this time frame, there were no observable phenotypic changes to the cells. IFNα in the media was quantified by ELISA assay (performed in duplicate, error less than 10%). Mean values of triplicate cell platings are plotted and error bars are shown.

number standards. The enzyme *Nhe*I cuts one time in both LTRs and is expected to yield a genomic sized fragment of 2.6 kb when the blot is probed with an IFNα fragment (Fig. 1C). The IRE sequence contains *Avr*II sites (Figs. 1A and C); thus this enzyme is diagnostic for the presence of IRE sequences in the DG-IRE vector and is expected to yield the same genomic-sized 2.6-kb fragment on Southern blot. As can be seen in Fig. 2, both the MFG and the DG-IRE transduced cells contain the 2.6-kb *Nhe*I fragment. In addition, DG-IRE transduced cells yield a 2.6-kb *Avr*II fragment, confirming the presence of the IRE sequences in both LTRs (Fig. 2). MFG transduced cells yield high molecular weight heterogeneous fragments when digested with *Avr*II consistent with the lack of this site in the vector genome (Fig. 2). Two additional enzymes are employed in the analysis to identify gross rearrangements in the genome of cells transduced with the DG-IRE vector: *Xba*I and *Kpn*I, which cut either 5′ or 3′ of the IRE insertion site. There is also a *Kpn*I site 5′ of IFNα (Fig. 1C). As can be seen in Fig. 2, DNA from cells transduced with either the MFG or the DG-IRE vector generate a 2.6-kb *Xba*I fragment and a 1.4-kb *Kpn*I fragment as expected. Thus, at the level of detection afforded by a Southern blot, there are no gross rearrangements in the DG-IRE vector and it appears to be at least as stable as the MFG vector.

Analysis of Stable Producers. To assess the performance of the DG-IRE vector in the context of a stable producer cell as well as to extend our observations to a second marker gene, stable ψCRIP producers were generated with a DG-IRE-*lacZ* vector. The ψCRIP packaging cell line has been described elsewhere and consists of separate *gag-pol* and amphotropic envelope encoding plasmids stably integrated into NIH 3T3 cells.[19] All NIH 3T3 cell lines and derivatives are maintained in DMEM (GIBCO-BRL) containing 10% calf serum and antibiotics. Stable ψCRIP producers are generated by cotransfecting 10 μg vector plus 1 μg pSV2neo (gift of R. Mulligan) in the presence of lipofectamine according to the manufacturer's protocol (GIBCO-BRL). Colonies are selected in 400 μg/ml G418 (active) (Sigma), picked by aspirating into large-bore (P1000) pipette tips and expanded in culture in the absence of selection. Producer clones are screened initially for *lacZ* expression by X-Gal staining. The X-Gal staining protocol has been described previously.[20] Importantly, cells are incubated at 37° for 4–6 hr. They are then either examined immediately under a dissecting scope for the presence of blue cells or are rinsed twice with PBS, air dried, and stored for later viewing. We find that incubation overnight in the X-Gal containing solution increases the nonspecific background staining

[19] O. Danos and R. C. Mulligan, *Proc. Natl. Acad. Sci. U.S.A.* **85,** 6460 (1988).

[20] K. Lim and C. B. Chae, *Bio Techniques* **7,** 576 (1989).

dramatically. Following the initial *lacZ* expression screen, positive producer clones are screened for particle production. For this assay, producer clones are seeded at 2×10^6 cells in a 10-cm dish and refed the next day with normal growth media for the collection of a 24-hr supernatant. Naive NIH 3T3 cells (plated at 2×10^5 cells per 10-cm dish the day prior to transduction) are then exposed to virus containing supernatant with 8 μg/ml Polybrene for 24 hr. Transduced cells are stained with X-Gal 48–72 hr after transduction and examined for the presence of blue cells. For end-point particle determination, 10-fold serial dilutions of virus supernatant in media containing 8 μg/ml Polybrene are generated and used to overlay naive NIH 3T3 cells. Transduced cells are stained with X-Gal 48–72 hr after transduction, and colonies on the plate are counted with the aid of a dissecting scope. End-point particle titers are determined for the best eight clones, and range from 2.5×10^4 to 5.0×10^5 cfu (colony-forming units)/ml. This result is similar to our previous experiences generating stable MFG-*lacZ* ψCRIP producers (data not shown), indicating, as with transient virus production, that the presence of the IRE in the vector does not interfere with virus generation, infection, or basal expression.

To examine regulation in cells transduced with virus from the stable producer, an unconcentrated viral supernatant from the best clone is used to transduce NIH 3T3 cells, and *lacZ* expression in response to changes in iron levels in the media is determined by assaying β-galactosidase activity in cell extracts. As with the IFNα transduced C2C12 cells described earlier, *lacZ*-transduced NIH 3T3 cells are first adapted to grow in low iron basal media [MEMα plus 10% defined calf serum (Hyclone)]. They are then exposed to basal media plus extra iron (75 μg/ml ferric ammonium citrate) or basal media plus Desferal (100 μM) for 2 days before assaying for β-galactosidase expression. We find an approximately fourfold decrease in expression in DG-IRE-*lacZ* transduced cells when they are exposed to Desferal as compared to iron (Table II), whereas the levels of expression in MFG-*lacZ* transduced cells remain the same regardless of the iron content of the media (Table II). The decrease in *lacZ* expression in DG-IRE-*lacZ* transduced cells on the addition of Desferal is not as great as with DG-IRE-IFNα transduced cells, either reflecting a much longer half-life of β-galactosidase or a cell-type-specific difference in iron regulation.

Importance of IRE-Stem Sequence to Function. The IRE sequence is synthetic and its design is based on data indicating that the primary sequence of the loop section is critical for function (binding to the IRBP), whereas the primary sequence of the stem can be variable.[9] We chose to incorporate a unique restriction site into the stem of the IRE to facilitate subsequent molecular analysis of the iron-regulated vector. In addition to the IRE containing an *Avr*II site, which functions appropriately in the DG-IRE vector as described earlier, we constructed a second synthetic IRE that

TABLE II
LacZ Expression in Transduced NIH 3T3 Cells[a]

Vector	β-Gal μU/10^6 cells	
	Iron	Desferal
MFG-*lacZ*	41.9	39.5
DG-IRE-*lacZ*	693	164

[a] In response to iron concentration changes in the media. NIH 3T3 cells transduced with the indicated vector were adapted to grow in low iron containing basal media [MEMα with 10% defined calf serum (Hyclone)]. Cells were then plated into either media containing extra iron (basal media plus 20 μg/ml ferric ammonium citrate) or media containing the iron chelator, Desferal (basal media plus 100 μ*M* Desferal). Three days after plating into the plus or minus iron media, cell extracts were prepared and β-galactosidase activity was determined. For quantitation of β-galactosidase gene expression, extracts of cells were prepared and assayed with a kit according to the manufacturer's protocol (Promega). In this assay, one unit of β-Gal hydrolyzes 1 μmol of *O*-nitrophenyl-β-D-galactopyranoside to *O*-nitrophenol and galactose per minute at pH 7.5 and 37°. β-Galactosidase activity in cell extracts was normalized to the number of cells used in the assay.

incorporates a *Sac*II site, 5′-CCGCGG-3′ into the stem (Fig. 1B). A retroviral vector analogous to DG-IRE was made that incorporates the *Sac*II-IRE into both 5′ and 3′ LTRs, and the *lacZ* gene is subcloned into it (see earlier). This vector (*Sac*II-IRE-*lacZ*) is transfected transiently into NIH 3T3 cells. Cells transfected transiently are then exposed to low iron basal media [MEMα plus 10% defined calf serum (Hyclone)], basal media plus extra iron (75 μg/ml ferric ammonium citrate), or basal media plus Desferal (100 μ*M*). Data are presented in Fig. 4. There is negligible expression from the *Sac*II-IRE-*lacZ* vector under any of the media conditions, whereas the parent MFG vector produces readily detectable levels of β-Gal, which, in agreement with data for the transduced MFG-*lacZ* vector in NIH 3T3 cells (Table II), does not vary significantly with the iron concentration of the media. Because the *Avr*II-IRE does not negatively affect gene expression

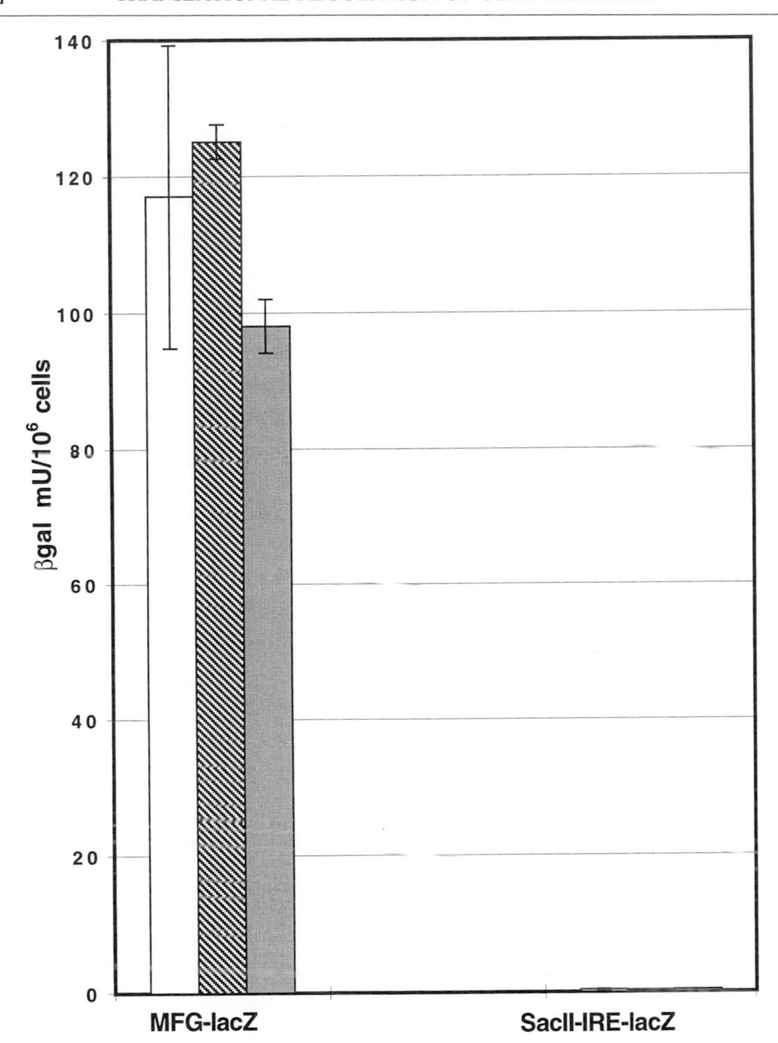

FIG. 4. LacZ expression in transiently transfected NIH 3T3 cell populations and response to changes in iron concentration. Left-hand side of graph, cells transfected with MFG-*lacZ*; right-hand side of graph, cells transfected with *Sac*II-IRE-*lacZ*. Cells were transfected in low iron containing basal media (see text) and in triplicate for each media condition. The following day, cells were refed with basal media (open bars), basal media plus 20 μg/ml ferric ammonium citrate (striped bars), or basal media plus 100 μM Desferal (shaded bars). After 48 hr, cells were lysed and extracts were prepared for β-galactosidase assays. β-Galactosidase activity was quantitated (see text) and normalized to the number of cells assayed. Mean values of triplicate transfections are plotted and error bars are shown.

in basal media and is regulated by the iron concentration of the media (see earlier), the result with the *Sac*II-IRE suggests that this IRE has a very detrimental impact on gene expression that cannot be overcome with the addition of iron to the media. Possibly, the high GC content of the *Sac*II site is problematical, perhaps leading to a very stable RNA secondary structure that inhibits gene expression by itself. Although we have not investigated the molecular mechanism underlying the elimination of gene expression in the *Sac*II-IRE-*lacZ* vector, we felt it was worthwhile to include these data to aid others in the design of similar vectors.

Concluding Remarks

We have described the construction of a translationally regulated retroviral expression vector (DG-IRE) in which the regulatory components were incorporated into the LTR such that the retroviral promoter was utilized to drive transcription in transduced cells. We have shown that the insertion of foreign sequences (specifically, the *Avr*II-IRE) into the R region did not interfere with virus production nor with basal levels of gene transcription. There are two pieces of evidence that underscore this point. One, viral titers from stable ψCRIP DG-IRE-*lacZ* producers were essentially equivalent to those obtained with the parent MFG-*lacZ* vector, reaching at least 5×10^5 infectious particles/ml. In addition, basal expression in cells transduced with DG-IRE-IFNα or with MFG-IFNα was essentially the same on a per genome basis. Adam *et al.*[21] also observed that the R region can accommodate additional nucleotide sequences, in their case an approximately 1-kb cDNA, without severely compromising vector function. However, they found that viral titers were one order of magnitude lower than with the unmodified parent vector. It could be that the small size of our insert (32 nucleotides) accounts for the undetectable impact on virus production in our experiments. In contrast to the results with the *Avr*II-IRE, we observed that the insertion of the *Sac*II-IRE decreased gene expression dramatically from a transiently transfected vector. Clearly, then, while some foreign sequence insertions in R may be well tolerated, it is important to keep in mind that some sequences may not be. This needs to be tested empirically.

We also showed that the level of expression of transgenes in the DG-IRE vector was regulated in a dose-dependent manner by the iron concentration in the media, and the fold regulation that we observed was consistent with what others have described for plasmid-based vectors that contain IRE sequences. Regulation with the retroviral vector presumably occurs

[21] M. A. Adam, W. R. A. Osborne, and A. D. Miller, *Hum. Gene Ther.* **6,** 1169 (1995).

at the translational level, as it is well documented that that is where the IRE functions.[8,9] The DG-IRE vector showed no evidence of genomic instability in transduced cells.

Unfortunately, because the DG-IRE vector uses iron as the regulating agent, it is currently impractical to use the vector in an *in vivo* setting, particularly for any application that would involve the use of human subjects. However, the basic parameters regarding vector function defined in this article, e.g., that some short regulatory sequences can be incorporated into the LTR without compromising vector function, should be important to future endeavors in developing alternatives to the IRE that are clinically appropriate. Ultimately, it will be interesting to combine a translationally regulated vector, such as described here, with a transcriptionally regulated promoter system to achieve an even tighter degree of control than either system alone affords.

Acknowledgments

The authors thank Kaye Spratt, Ya-Li Lee, and David Colvin for help with some molecular constructions and sequence analysis and Steve Hardy for critical reading of the manuscript and helpful discussions.

[14] Strategies to Express Structural and Catalytic RNAs in Mammalian Cells

By JAMES D. THOMPSON

Introduction

RNA is probably most well known for its role in shuttling protein-coding information from the nucleus to the cytoplasm in the form of mRNA. However, the vast majority of the total RNA in mammalian cells—greater than 95%—is composed of structural and functional RNAs that perform a wide variety of functions. Examples of cellular structural/functional RNAs include tRNAs and ribosomal RNAs involved in protein translation, 7SL RNA involved in protein translocation, small nuclear RNAs (snRNAs) such as U1 and U6 involved in pre-mRNA splicing, and the RNA subunit of telomerase that serves as a template for synthesis of telomeric DNA at the ends of chromosomes. The diversity of functions that RNA plays in mammalian cells reflects the ability of RNA to form complex secondary,

0076-6879/99 $30.00

tertiary, and quaternary structures in solution.[1–4] This probably is best illustrated by the existence in nature of RNAs with catalytic activity.[5–7] The discovery of such "ribozymes" supplanted the notion that only proteins can serve as catalysts and enzymes in nature.[8,9]

The structural and functional properties of RNA are finding increasing practical applications in the laboratory and the clinic, where RNA is being used to modulate gene expression and activity in the form of antisense RNA, RNA decoys and aptamers, and ribozymes.[10] Indeed, several clinical trials have been approved or initiated to evaluate the safety and therapeutic benefit of using recombinant effector RNAs to modulate the expression and activity of cellular and viral genes. In addition, there is a significant unmet need in the genomics field for technologies that can be used to determine the function of the thousands of new gene sequences identified each year. The ideal technology will be one that can be designed from raw sequence information alone. Antisense and ribozyme technologies satisfy this criterion and, as such, could prove to be valuable research tools in the functional genomics area.

A variety of strategies have been used to express recombinant structural and functional RNAs in mammalian cells, including the use of classical mRNA promoters and promoters from such structural/functional RNAs as rRNAs, tRNAs, and snRNAs.[11,12] To date, no single expression strategy has emerged as being superior for obtaining maximal intracellular activity. Many factors contribute to achieving optimal activity of recombinant structural/functional RNAs, including the amount of transcript produced in the target cell type, the stability and intracellular localization of the recombinant transcript, and preservation of the biological activity of the effector RNA in the context of other sequences contained in the recombinant transcript. This article reviews the different promoters used to express effector RNAs in mammalian cells, discusses the intracellular localization

[1] H. F. Noller, *Annu. Rev. Biochem.* **53,** 119 (1984).
[2] J. A. Jaeger, D. H. Turner, and M. Zuker, *Proc. Natl. Acad. Sci. U.S.A.* **86,** 7706 (1989).
[3] T. R. Cech, *Gene* **135,** 33 (1993).
[4] L. Gold, *Nucleic Acids Symp. Ser.* **20** (1995).
[5] T. Cech, *J. Am. Med. Assoc.* **260,** 3030 (1988).
[6] R. H. Symons, *Annu. Rev. Biochem.* **61,** 641 (1992).
[7] A. M. Pyle, *Science* **261,** 709 (1993).
[8] K. Kruger, P. J. Grabowski, A. J. Zaug, J. Sands, D. E. Gottschling, and T. R. Cech, *Cell* **31,** 147 (1982).
[9] C. Guerrier-Takada, K. Gardiner, T. Marsh, N. Pace, and S. Altman, *Cell* **35,** 849 (1983).
[10] J. A. Smythe and G. Symonds, *Inflamm. Res.* **44,** 11 (1995).
[11] H. Ilves, C. Barske, U. Junker, E. Bohnlein, and G. Veres, *Gene* **171,** 203 (1996).
[12] P. D. Good, A. J. Krikos, S. X. Li, E. Bertrand, N. S. Lee, L. Giver, A. Ellington, J. A. Zaia, J. J. Rossi, and D. R. Engelke, *Gene Ther.* **4,** 45 (1997).

of the different transcription units, and describes strategies to improve the stability and to preserve the activity of the different types of effector RNAs in the context of recombinant transcripts.

Recombinant Structural and Catalytic RNAs and Their Applications

An understanding of the different types of recombinant structural and functional RNAs, hereafter referred to as "effector RNAs," that are currently available is important in order to identify expression strategies that are compatible with the mechanism of action of the effector RNAs. There are three general classes of effector RNAs predominantly used today to modulate gene expression and activity: antisense RNA, RNA decoys and aptamers, and enzymatic RNAs. The stages in gene expression where each type exerts its activity are illustrated in Fig. 1 and are described in more detail later. Special considerations for expressing each of these different effector RNAs will be mentioned in this section and addressed in more detail in the following section.

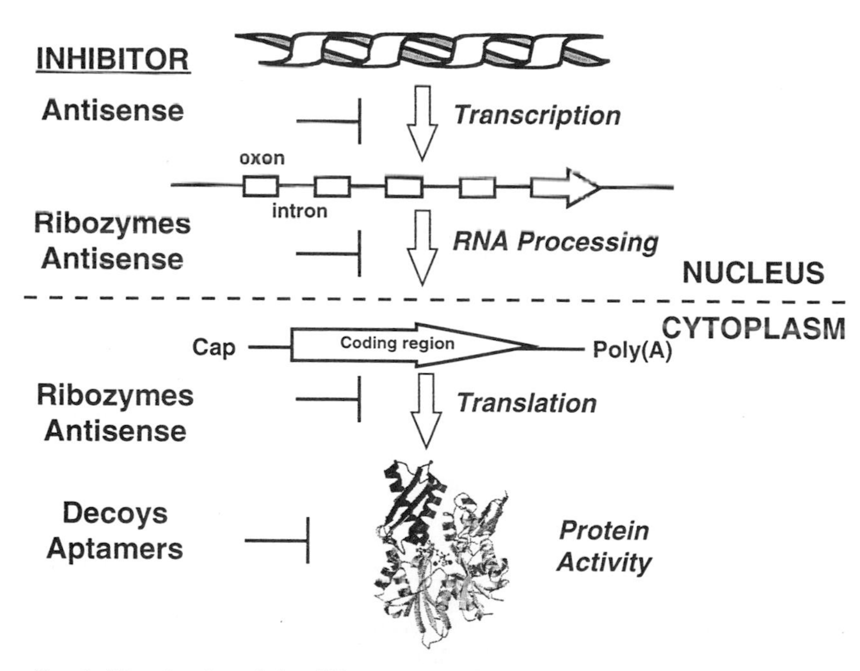

FIG. 1. Site of action of the different types of effector RNAs designed to inhibit gene expression or activity.

Antisense RNAs

Antisense RNAs can be used to inhibit gene expression at the level of transcription, mRNA processing, or translation (Fig. 1).[13–15] The specificity of antisense RNA is determined by Watson–Crick base pairing and Hoogsteen triple-helix pairing to target nucleic acids. Antisense RNAs function by binding to DNA to block transcription, hybridizing to splice donor/acceptor regions of pre-mRNAs to block splicing, or annealing to mRNAs to block translation (Fig. 1). Antisense RNA can also be used to insert missense or nonsense mutations in coding regions of mRNAs via adenosine to inosine conversion.[16]

Ideally, antisense RNAs designed to bind to DNA or pre-mRNAs should accumulate in the nucleus. Alternatively, antisense RNAs designed to inhibit translation should colocalize with their target mRNAs in the cytoplasm. Therefore, the transcription units chosen to express antisense RNAs should be designed to traffic through the relevant intracellular compartments. In addition, because antisense RNAs rely on hybridization to their target nucleic acids for activity, the recombinant transcripts in which they are expressed should be optimized to support this intermolecular interaction.

Decoys and Aptamers

RNA motifs that bind regulatory proteins have evolved in nature to control RNA transcription, trafficking, and stability. For example, the genome of the human immunodeficiency virus (HIV) contains two RNA regulatory elements: TAR, which is recognized by the viral TAT protein and endogenous cellular proteins to regulate viral gene expression, and RRE, which binds the viral REV protein to regulate RNA trafficking.[17] Recombinant versions of these protein-binding RNA motifs can act as decoys to block the binding of cellular or viral regulatory proteins to their normal RNA ligands. In the case of HIV, overexpression of recombinant TAR and RRE decoys in cells leads to the inhibition of viral replication.[18,19] Another class of decoy RNAs being developed is based on recognition and cleavage of pre-tRNAs by RNase P.[9] Recombinant RNAs, termed external guide sequences (EGSs), can be designed to hybridize to target RNAs

[13] H. M. Weintraub, *Sci. Am.* **262,** 40 (1990).
[14] W. Nellen and G. Sczakiel, *Mol. Biotechnol.* **6,** 7 (1996).
[15] R. Knee and P. R. Murphy, *Neurochem. Int.* **31,** 379 (1997).
[16] B. L. Bass and H. Weintraub, *Cell* **55,** 1089 (1988).
[17] B. R. Cullen and W. C. Greene, *Cell* **58,** 423 (1989).
[18] B. A. Sullenger, H. F. Gallardo, G. E. Ungers, and E. Gilboa, *Cell* **63,** 601 (1990).
[19] T. C. Lee, B. A. Sullenger, H. F. Gallardo, G. E. Ungers, and E. Gilboa, *New Biol.* **4,** 66 (1992).

and to form complexes that mimic pre-tRNAs.[20] These complexes are recognized and cleaved by RNase P, leading to inactivation of the target RNA.

While several naturally occurring RNA motifs have been used successfully as decoys, RNA biologists are not limited to RNA motifs found in nature. In fact, it has been proposed that nonnaturally occurring RNA aptamers are present in "RNA sequence space," a term used to describe all possible RNA sequences, that are capable of specifically binding to any given protein.[21] Such RNA aptamers can be selected from pools of random RNA sequences by repeated rounds of binding to the targeted protein, purification of the RNA/protein complexes, and amplification of the bound RNA.[22,23] This selection strategy has been used to identify new RNA motifs that bind to a number of different cellular and viral proteins.[24,25] Such selected RNA aptamers can be used as antagonists to inhibit gene function in a manner similar to the decoy RNAs described earlier.

RNA decoys and aptamers must traffic through the same intracellular compartments as do their target proteins in order to be effective. Therefore, decoys and aptamers that bind to nuclear proteins should accumulate in the nucleus, whereas aptamers targeted to cytoplasmic proteins must gain access to this compartment. Consequently, intracellular localization is a major consideration when choosing transcription units to express particular decoys and aptamers. In addition, RNA decoys and aptamers must fold into complex structures to be recognized by their target proteins. Therefore, recombinant transcripts must be designed to promote proper folding and display of the RNA decoys and aptamers.

Catalytic RNAs

Several different catalytic RNAs have been identified in nature (Table I). Examples include the RNA subunit of prokaryotic RNase P that catalyzes the nucleolytic processing of pre-tRNAs, group I and II self-splicing introns found in ribosomal and tRNA genes of bacteria and lower eukaryotes, and a variety of self-cleaving RNA motifs found in human and plant viral satellite RNAs (Table I). Several of the self-cleaving RNA motifs listed in Table I have been engineered in the test tube to function *in trans*

[20] Y. Yuan and S. Altman, *Science* **263,** 1269 (1994).
[21] L. Gold, D. Brown, Y. He, T. Shtatland, B. S. Singer, and Y. Wu, *Proc. Natl. Acad. Sci. U.S.A.* **94,** 59 (1997).
[22] C. Tuerk and L. Gold, *Science* **249,** 505 (1990).
[23] A. D. Ellington and J. W. Szostak, *Nature* **346,** 818 (1990).
[24] D. P. Bartel, M. L. Zapp, M. R. Green, and J. W. Szostak, *Cell* **67,** 529 (1991).
[25] C. Tuerk and S. MacDougal-Waugh, *Gene* **137,** 33 (1993).

TABLE I
NATURALLY OCCURRING CATALYTIC RNAS

Catalytic RNA	Function	Reaction catalyzed	Approximate size (nucleotides)	Organism
Group I introns	RNA splicing	*cis* RNA cleavage/ ligation	413	*Tetrahymena*
Group II introns	RNA splicing	*cis* RNA cleavage/ ligation		Plants, fungi, bacteria
M1 subunit of RNase P	tRNA processing	*trans* RNA cleavage	375	Eubacteria
Hammerhead	Genome processing	*cis* RNA cleavage	34	Plant pathogen
Hairpin	Genome processing	*cis* RNA cleavage	50	Plant pathogen
Hepatitis delta virus	Genome processing	*cis* RNA cleavage	60	Human pathogen
VS ribozyme	RNA satellite processing	*cis* RNA cleavage	154	*Neurospora*

as true riboendonucleases.[6] Specifically, the hammerhead and hairpin motifs have been engineered to cleave cellular mRNAs in a highly sequence-dependent manner.[26,27] Group I and II self-splicing introns also have been modified in the test tube to catalyze the splicing of new RNA sequences into target mRNAs *in trans*. These *trans*-splicing ribozymes can be used to alter the coding regions of target mRNAs.[28]

In addition to the naturally occurring catalytic RNA motifs described earlier, nonnaturally occurring RNA ribozymes have been selected in the test tube by the *in vitro* evolution of existing ribozyme motifs[29,30] or from RNA sequence space.[31] This strategy has led to the identification of new ribozymes that catalyze other chemical reactions such as cleavage of DNA[29] and amide bond formation.[32]

Similar to protein enzymes, ribozymes must adopt appropriate secondary, tertiary, and quaternary structures for optimal catalytic activity. Catalytic activity of some of the smaller ribozyme motifs, such as the hammerhead, are particularly sensitive to the presence of nonribozyme sequences

[26] J. Haseloff and W. L. Gerlach, *Nature* **334,** 585 (1988).
[27] A. Hampel, R. Tritz, M. Hicks, and P. Cruz, *Nucleic Acids Res.* **18,** 299 (1990).
[28] B. A. Sullenger and T. R. Cech, *Nature* **371,** 619 (1994).
[29] A. A. Beaudry and G. F. Joyce, *Science* **257,** 635 (1992).
[30] J. Tsang and G. F. Joyce, *Methods Enzymol.* **267,** 410 (1996).
[31] D. P. Bartel and J. W. Szostak, *Science* **261,** 1411 (1993).
[32] P. A. Lohse and J. W. Szostak, *Nature* **381,** 442 (1996).

TABLE II
PROPERTIES OF RNA PROMOTERS

RNA polymerase	Promoter	Cell-type restrictions	Expressed in mice and humans
Pol I	rRNA	No	No
Pol II	mRNAs	Cell specific or ubiquitous	Species specific or ubiquitous
	U1	No	Yes
Pol III	tRNA	No	Yes
	Adeno VA	No	No
	U6	No	Yes

contained in recombinant transcripts.[33] Therefore, preserving ribozyme activity in the context of recombinant transcripts is an essential and often complicated task.

Strategies to Express Recombinant Structural and Functional RNAs

Mammalian cells utilize three different RNA polymerases—RNA polymerase I, II, and III (pol I, II, and III)—to produce the variety of structural, functional, and protein-coding RNAs found in cells. Pol I is used exclusively for the transcription of ribosomal RNAs. Pol II is used by cells to produce mRNAs and a subset of the snRNAs such as U1 through U5. Pol III is used to transcribe tRNAs, 5S ribosomal RNA, and snRNAs such as U6. Thus, there are many different RNA promoters to choose from to express recombinant RNAs. These promoters differ in regard to the organization of controlling elements such as promoter and termination elements, and also tissue and species specificity. Often, researchers choose promoters based on the biological properties of the natural RNA produced from the wild-type genes. For example, tRNA promoters are often chosen because tRNAs are abundant in all nucleated cells and accumulate in the cytoplasm. However, recombinant transcripts rarely behave like their cellular counterparts: small changes in the primary sequence of a tRNA affect stability, processing, and intracellular trafficking greatly.[34,35] As such, promoter choice should be based primarily on the desired rate of transcription in the target cell or tissue type (Table II).

[33] J. D. Thompson, D. F. Ayers, T. A. Malmstrom, T. L. McKenzie, L. Ganousis, B. M. Chowrira, L. Couture, and D. T. Stinchcomb, *Nucleic Acids Res.* **23,** 2259 (1995).
[34] M. Zasloff, T. Santos, and D. H. Hamer, *Nature* **295,** 533 (1982).
[35] J. A. Tobian, L. Drinkard, and M. Zasloff, *Cell* **43,** 415 (1985).

For the most part, the behavior of an RNA after it has been transcribed is determined by sequence elements contained in the transcript. Nonnatural sequence elements that direct recombinant transcripts to different intracellular compartments are just beginning to be identified.[36,37] Other elements such as 5′ cap structures, Sm-binding sites, and certain RNA structural motifs contribute significantly to stability, intracellular trafficking, and biological activity of a given effector RNA. Therefore, careful thought must be given to the design of recombinant transcripts. As will be discussed later, sequence elements included in a recombinant transcript should be tailored to match the mechanism of action of the effector RNA being expressed.

To date, pol I promoters have not been utilized extensively to express recombinant transcripts (see later). Currently, the promoters most often used to express recombinant transcripts are pol II mRNA promoters, the pol II U1 snRNA promoter, and pol III tRNA, adenoviral VA, and U6 promoters. Table II summarizes the species and tissue specificity of these different RNA polymerase promoters, with species specificity considerations limited to comparisons between human and mouse.

Pol I rRNA Expression Units

Few groups have published on expression of exogenous RNAs from pol I promoters. At first glance, this seems surprising as pol I-transcribed ribosomal RNA comprise up to 50% of the total RNA transcribed in a cell at any given time. However, there are theoretical constraints that could account for the underutilization of pol I promoters. Mammalian genomes contain hundreds of rRNA genes clustered on several chromosomes, yet only a small percentage of these genes are actively transcribed at any given time. This is because ribosomal genes compete for the limiting transcription factors UBF and SL1.[38] As such, recombinant pol I transgenes also must compete with the endogenous pool of rRNA genes for these limiting factors. In addition, rRNA genes are transcribed in a discrete location in the interphase nucleus, the nucleolus. Therefore, pol I-based expression vectors also may need to localize within the nucleolus for optimal transcription.

Despite these potential limitations, pol I expression vectors have been used successfully to express recombinant proteins in cells.[39] However, the expression levels were no better than those obtained with conventional pol

[36] C. Grimm, E. Lund, and J. E. Dahlberg, *EMBO J.* **16,** 793 (1997).

[37] C. Grimm, E. Lund, and J. E. Dahlberg, *Proc. Natl. Acad. Sci. U.S.A.* **94,** 10122 (1997).

[38] R. H. Reeder, "Transcription Regulation" (K. Yamamoto and S. L. McKnight, eds.). Cold Spring Harbor Laboratory Press, Cold Spring Harbor, NY, 1992.

[39] T. D. Palmer, A. D. Miller, R. H. Reeder, and B. McStay, *Nucleic Acids Res.* **21,** 3451 (1993).

II mRNA promoters. This author is unaware of any reports describing the use of pol I promoters to express effector RNAs in mammalian cells. However, antisense RNAs have been expressed in pol I rRNA transcripts in *Tetrahymena thermophila*,[40] an organism in which all endogenous copies of the rRNA genes can be replaced with recombinant versions. These studies demonstrate that effector RNAs can be active when expressed from pol I promoters.

Pol II mRNA Expression Units

Pol II mRNA promoters represent the most diverse and versatile promoters available. Tissue- and species-specific promoters have been isolated, as well as ubiquitously expressed promoters such as the cytomegalovirus (CMV) immediate early and simian virus 40 (SV40) viral promoters, which can be used in rodent and human cells (Table II). In addition, regulatable promoters have been developed that can be activated or repressed by small molecules that can be added to cell culture media or included in the diets of transgenic animals.[41–43]

mRNAs are transcribed by pol II as hnRNAs (Fig. 2B), which are then processed to mature mRNAs by the removal of introns and the addition of a 5′ cap and a 3′ poly(A) tail (Fig. 2C). Effector RNAs are usually inserted into either the 5′ or the 3′ nontranslated regions (UTRs) of recombinant mRNA transcripts (Fig. 2C). There seems to be no compelling data supporting placement into either UTR for achieving biological activity; hammerhead ribozymes targeting HIV have been shown to be effective when placed either 5′ or 3′ of a protein-coding region.[44,45] However, there are potential size and structural considerations associated with placement into the 5′ UTR of protein-coding mRNAs, as the translational machinery must scan through this region to find the AUG start codon. There is no size constraint when placing effector RNAs into the 3′ UTR. Effector RNAs can also be placed within introns (Fig. 2B). However, in general, introns are not very stable after being spliced out of hnRNAs and thus do not accumulate to appreciable levels in cells. This could be a limiting for effector RNAs such as antisense RNAs that must accumulate to levels

[40] R. Sweeney, Q. Fan, and M. C. Yao, *Proc. Natl. Acad. Sci. U.S.A.* **93,** 8518 (1996).
[41] M. Gossen, S. Freundlieb, G. Bender, G. Muller, W. Hillen, and H. Bujard, *Science* **268,** 1766 (1995).
[42] D. No, T. P. Yao, and R. M. Evans, *Proc. Natl. Acad. Sci. U.S.A.* **93,** 3346 (1996).
[43] Y. Wang, F. J. DeMayo, S. Y. Tsai, and B. W. O'Malley, *Nat. Biotechnol.* **15,** 239 (1997).
[44] C. Zhou, I. C. Bahner, G. P. Larson, J. A. Zaia, J. J. Rossi, and E. B. Kohn, *Gene* **149,** 33 (1994).
[45] L. Q. Sun, L. Wang, W. L. Gerlach, and G. Symonds, *Nucleic Acids Res.* **23,** 2909 (1995).

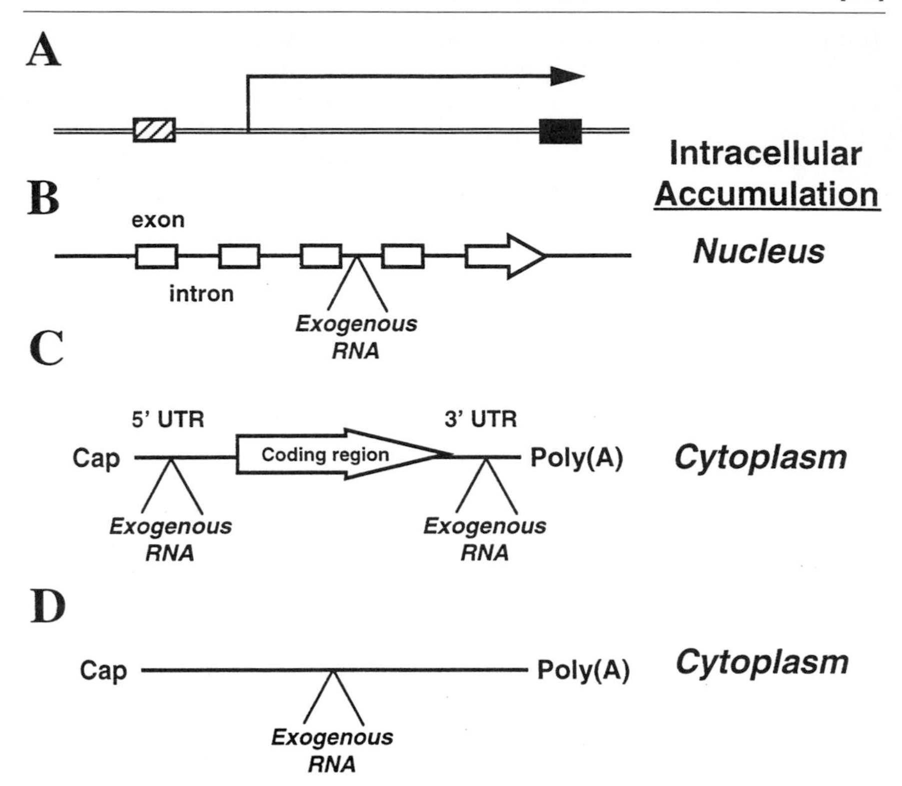

FIG. 2. Strategies to express effector RNAs using the pol II mRNA expression unit. (A) Schematic diagram of the genomic organization of pol II mRNA genes, including the promoter (hatched boxes), primary transcript (arrow), and termination sequences (closed box). Double lines represent DNA. (B) Primary mRNA transcript (hnRNA). Exons are indicated by open boxes; lines separating exons represent introns. Effector RNAs can be placed into introns as shown to obtain nuclear localization. (C) Diagram of the mature mRNA. Open arrow refers to protein coding region. Effector RNAs can be placed in the UTRs. (D) Diagram of a recombinant pol II mRNA where the protein-coding region has been replaced with an effector RNA.

several hundredfold in excess of targets sequences for efficacy.[46] Finally, effector RNAs can be inserted in place of the protein-coding regions of mRNAs (Fig. 2D).

Many factors contribute to the stability of mRNA transcripts.[47] The cap and poly(A) tail are required for mRNA stability[48]; mRNAs lacking either

[46] S. Wang and B. J. Dolnick, *Nucleic Acids Res.* **21,** 4383 (1993).

[47] A. B. Sachs, *Cell* **74,** 413 (1993).

[48] J. Ross, *Trends Genet.* **12,** 171 (1996).

of these modifications as a result of intracellular cleavage by ribozymes or by a RNase H-mediated antisense DNA mechanism are degraded rapidly in mammalian cells. While not essential, the presence of introns in recombinant mRNA transcripts can also improve levels obtained with certain mRNA expression units. Translation has also been linked to mRNA degradation. RNA stem–loop structures placed into the 5′ UTR designed to inhibit translation have been shown to improve the stability of short-lived messages.[49] Protein-coding regions may not be necessary as efficacy with antisense and ribozymes has been reported by groups using mRNA expression vectors lacking coding regions.[50,51]

Capped and polyadenylated mRNAs are transported actively to the cytoplasm whereas introns remain in the nucleus. Therefore, exons are ideal sites for the insertion of effector RNAs to achieve cytoplasmic accumulation, whereas introns are suitable if only nuclear localization is desired.

Pol II U1 Expression Unit

Several of the snRNAs, such as U1 through U5, also are transcribed by pol II.[52] The U1 pol II promoter has been used most often to express recombinant RNAs in mammalian cells. The U1 promoter is active in all cell types in both human and rodent cells (Table II).

The U1 promoter is contained in the 5′-nontranscribed region of the U1 gene (Fig. 3A). Native U1 transcripts are capped at the 5′ end in the nucleus and trimmed at the 3′ end posttranscriptionally to produce the mature U1 snRNA.[53] During this processing, U1 transcripts are transported actively to the cytoplasm. The cap, sequences in the first 124 nucleotides of the transcript, and the terminal stem–loop structure at the 3′ end all contribute to this cytoplasmic transport process.[54] U1 transcripts are then reimported into the nucleus where spliceosome assembly is completed. The U-rich Sm sequence present in the 3′ end of U1 transcripts is required for this nuclear transport event.

Effector RNAs can be inserted anywhere within the U1 transcription unit or inserted in place of wild-type U1 transcript (Fig. 3B).[55] However, the more U1 sequences removed from recombinant transcripts, the less

[49] T. Aharon and R. J. Schneider, *Mol. Cell. Biol.* **13,** 1971 (1993).

[50] J. T. Stout and C. T. Caskey, *Somat. Cell Mol. Genet.* **16,** 369 (1990).

[51] S. Larsson, G. Hotchkiss, M. Andang, T. Nyholm, J. Inzunza, I. Jansson, and L. Ahrlund-Richter, *Nucleic Acids Res.* **22,** 2242 (1994).

[52] J. E. Dahlberg and E. Lund, *Mol. Biol. Rep.* **12,** 139 (1987).

[53] N. Hernandez, *EMBO J.* **4,** 1827 (1985).

[54] M. P. Terns, J. E. Dahlberg, and E. Lund, *Genes Dev.* **7,** 1898 (1993).

[55] E. Bertrand, D. Castanotto, C. Zhou, C. Carbonnelle, N. S. Lee, P. Good, S. Chatterjee, T. Grange, R. Pictet, D. Kohn, D. Engelke, and J. J. Rossi, *RNA* **3,** 75 (1997).

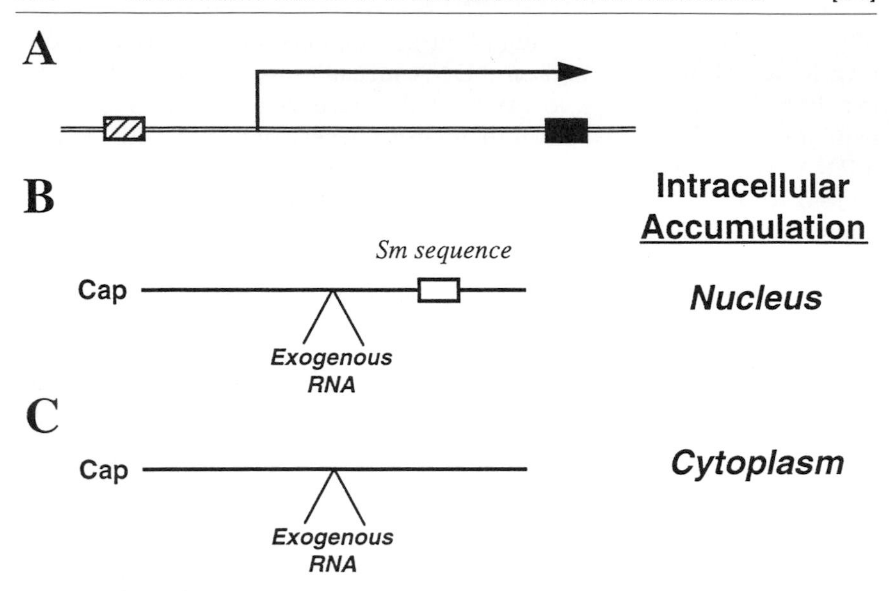

FIG. 3. Strategies to express effector RNAs using the pol II-driven U1 snRNA expression unit. (A) Genomic organization of U1 snRNA regulatory elements, including the promoter (hatched boxes), primary transcript (line with arrow), and termination sequences (closed box). Double lines correspond to DNA. (B) Effector RNAs can be placed anywhere within the U1 transcript. Open box refers to the Sm sequence (consensus = $PuAU_{(4-6)}GPu$). (C) The Sm sequence can be removed from recombinant transcripts, resulting in cytoplasmic localization.

U1-like the transcripts behave in regard to stability and intracellular trafficking. For example, recombinant U1 transcripts lacking the Sm sequence accumulate in the cytoplasm (Fig. 3C). Also, the half-lives of improperly processed U1 transcripts are significantly shorter than wild-type U1 transcripts.

Pol III Expression Units

There are three general classes of pol III promoters, type 1, type 2, and type 3.[56,57] Type 1 promoters are used exclusively for transcription of the ribosomal 5S genes and are not covered in this article. Type 2 promoters are represented by the tRNA and adenoviral VA RNAs. Type 2 promoter elements are contained within the transcription units themselves. Type 3 promoter elements are 5′ of the start site of transcription and are represented by the U6 gene. All three types of pol III expression units utilize

[56] E. P. Geiduschek and G. P. Tocchini-Valentini, *Annu. Rev. Biochem.* **57,** 873 (1988).
[57] I. M. Willis, *Eur. J. Biochem.* **212,** 1 (1993).

the same simple pol III consensus termination sequence consisting of a contiguous stretch of T's followed by a GC-rich region.

Type 2 Pol III Promoters. tRNA and adenoviral VA[58] are the type II promoters that have been utilized most widely to express recombinant RNAs. The promoter elements for this class of pol III transcripts are contained entirely in the 10–12 nucleotide A and B boxes contained within the transcription units themselves (Fig. 4A). Therefore, these promoter sequences must be included within recombinant transcripts. Type 2 tRNA promoters are active in all cell types and function efficiently in both rodent and human cells (Table II). Adenoviral VA promoters may work less efficiently in other species[11]; however, this class has not been studied as extensively as the tRNA promoters. As such, this section focuses on the use of tRNA promoters. Much of the material covered also will apply to adenoviral VA expression units.[59,60]

tRNAs are transcribed as precursor RNAs, which are then subjected to extensive processing to yield mature tRNAs.[61] These modifications include the endonucleolytic removal of 5′ leader and 3′ trailer sequences, posttranscriptional base modifications, and addition of the CCA sequence to the 3′ acceptor stem. During this process, tRNAs are transported actively to the cytoplasm. The RNA and protein enzymes that catalyze these modifications are extremely dependent on the sequence and structure of the pre-tRNA transcripts for recognition. Even as little as a single nucleotide change in a pre-tRNA transcript is sufficient to perturb posttranscriptional modification or transport events.[35] As such, recombinant tRNAs rarely behave as their wild-type counterparts.

The A box promoter element plays a role in the initiation of transcription.[57] The spacing between the A box and the transcriptional start site is only 10–20 nucleotides, making this region prohibitive for the insertion of effector RNAs. Effector RNAs can be placed between the A and the B box promoters (Fig. 4B).[62] However, the distance between the A and the B box promoters is limited to roughly 10–100 bp, with 30–60 bp being optimal for transcriptional activity.[57] Therefore, this region can accommodate only small effector RNAs (Fig. 4B). As such, the region downstream of the B box promoter has been utilized most for the insertion of effector RNAs (Fig. 4C). This expression strategy was developed in the laboratory of M. Zasloff, where it was shown that sequences downstream of the B

[58] Y. Ma and M. B. Mathews, *J. Virol.* **67,** 6605 (1993).
[59] P. A. Jennings and P. L. Molloy, *EMBO J.* **6,** 3043 (1987).
[60] A. Lieber and M. Strauss, *Mol. Cell. Biol.* **15,** 540 (1995).
[61] M. Zasloff, T. Santos, P. Romeo, and M. Rosenberg, *J. Biol. Chem.* **257,** 7857 (1982).
[62] M. Cotten and M. L. Birnstiel, *EMBO J.* **8,** 3861 (1989).

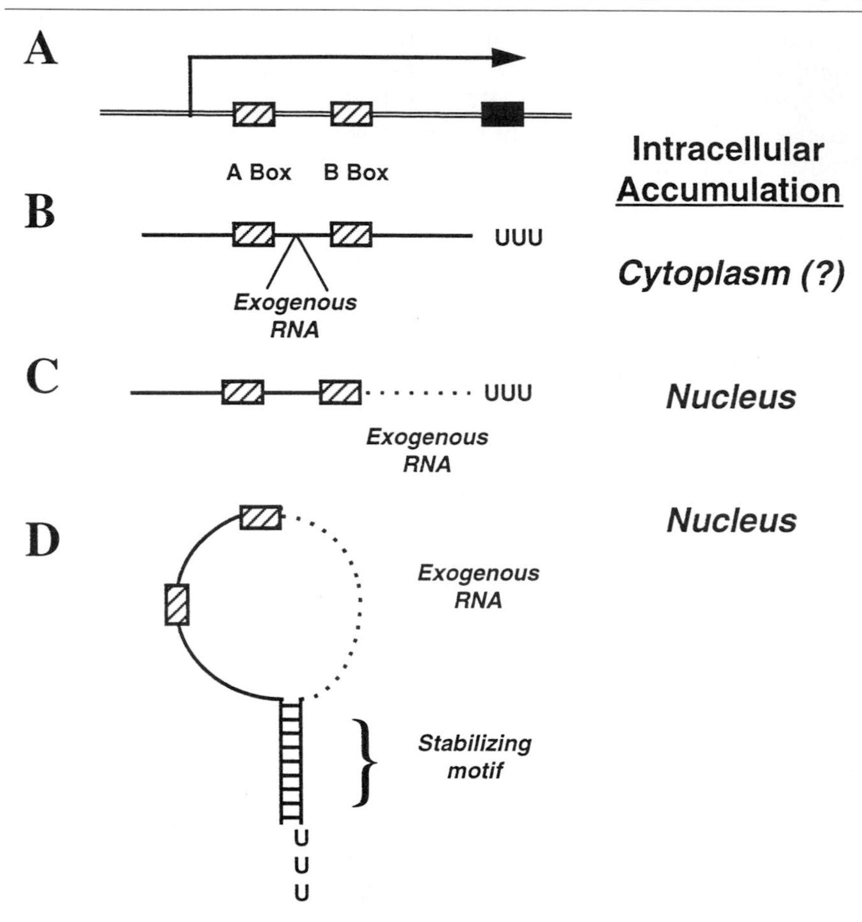

FIG. 4. Type 2 pol III expression units. (A) Type 2 pol III genes contain internal A box and B box promoter elements (hatched boxes) that are transcribed and therefore included in the primary transcript. The consensus pol III termination sequence (closed box) is a stretch of three or more T's followed by a G:C-rich region. An effective pol III termination signal often used in recombinant pol III genes is the simple 5′-GTTTTTG-3′ heptameric sequence.[68] (B) Effector RNAs can be placed between the A and the B box promoters; however, there is a size constraint associated with this region (see text). (C) Effector RNAs can be inserted downstream of the B box promoter element in place of the wild-type sequences without affecting transcription. In addition, such modifications result in the inactivation of 3′ endonucleolytic processing that normally takes place with primary tRNA transcripts. (D) Schematic diagram of a structural motif found to improve the intracellular stability of recombinant pol III transcripts.[33] A sequence complementary to the 5′ portion of the recombinant transcript is appended to the 3′ end of the effector RNA. This additional sequence element hybridizes to the 5′ end of the mature transcript, resulting in increased stability and accumulation of the recombinant transcript.

box in the vertebrate initiator methionine (Met_i) tRNA gene could be deleted and replaced with exogenous sequences without affecting the rate of transcription.[63] In addition, this group found that the 3′ endonucleolytic processing step was abolished in such recombinant transcripts, ensuring that the exogenous sequences replacing this region would not be removed by the normal tRNA-processing machinery. There seems to be no length requirement for insertion downstream of the B box as long as the exogenous sequence lacks a canonical pol III termination signal (see later).

The simple pol III termination sequence can present problems when utilizing pol III expression units. As few as three contiguous T's within a recombinant pol III transcription unit can be sufficient for transcriptional termination. However, because pol III termination is somewhat context dependent, the presence of a stretch of T's within a transcription unit is not an indication that termination will occur efficiently at this site. We have found that prescreening of recombinant pol III expression units by transient transfection followed by Northern blotting is usually sufficient for identifying cryptic pol III termination sequences within recombinant transcription units.

In general, recombinant tRNAs are much less stable than their wild-type counterparts. However, structural elements have been identified that improve the intracellular stability of recombinant tRNA transcripts. For example, the addition of stem–loops in the ends of recombinant transcripts has been shown to improve steady-state accumulation significantly.[64] More dramatic improvements in accumulation have been observed by engineering the 3′ end of the transcript to hybridize to the 5′ end of the pre-tRNA (including the 5′ leader sequence) to form an intramolecular duplex (Fig. 4D).[33] This strategy also has been found to be effective with other pol III expression units such as recombinant U6 and adenoviral VA genes for improving transcript stability (D. Macejak and J. Thompson, unpublished observation, 1995).

One potential disadvantage of using type 2 pol III promoters is the need to include in the recombinant transcripts sequences from the transcriptional start site to the end of the B box. These extra sequences have the potential to interfere with proper folding or function of effector RNAs (Fig. 5A). For example, these sequences have been shown to inhibit the activity of hammerhead ribozymes.[33] However, this interference was subsequently avoided by including additional nonribozyme sequences within the recombinant transcript designed to form stable secondary stem–loop structures

[63] S. Adeniyi-Jones, P. H. Romeo, and M. Zasloff, *Nucleic Acids Res.* **12,** 1101 (1984).
[64] S. W. Lee, H. F. Gallardo, E. Gilboa, and C. Smith, *J. Virol.* **68,** 8254 (1994).

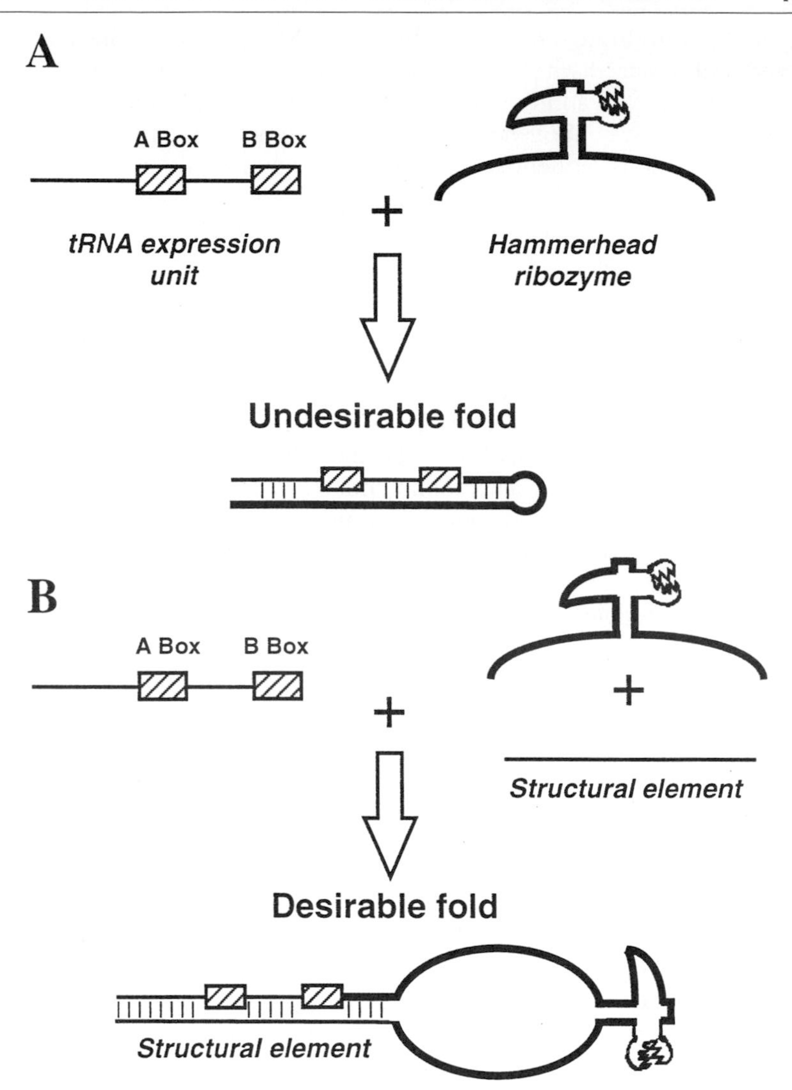

FIG. 5. Preserving the structure/function of effector RNAs in the context of complex transcription units. (A) Many effector RNAs, such as the hammerhead ribozyme, must adopt appropriate secondary and tertiary folding (indicated by the "hammerhead" diagram) for optimal activity. Often, noneffector sequences present in the recombinant transcript, such as the A and B box promoter elements cotranscribed within tRNA expression units, interact with effector RNA sequences, resulting in undesirable folding of the effector RNA (indicated as the absence of the hammerhead structure in the folded molecule). (B) This problem can be alleviated by including additional noneffector "structural elements" designed to form stable secondary structures with the noneffector sequences. In this strategy, noneffector sequences become sequestered from the effector RNA, allowing it to fold into the desired structure (again depicted as the hammerhead diagram in the full structure).

with the wild-type tRNA sequences (Fig. 5B). This strategy to sequester the ribozyme sequences from the tRNA sequences led to recombinant transcripts exhibiting ribozyme activity equivalent to ribozymes lacking any tRNA flanking sequences.[33]

Type 3 pol III Expression Units. The type 3 pol III expression unit used most extensively to express effector RNAs is U6. The U6 promoter is expressed in all cell types and is active in both human and mouse cells (Table II).[11] The U6 transcript spends its entire lifetime in the nucleus, where it is transcribed and then capped at the 5′ end.[52] Type 3 pol III promoters have an advantage over type 2 promoters in that type 3 promoter sequences are outside of the transcription unit and thus are not present in the recombinant transcripts. Effector RNAs can be placed anywhere within the U6 transcript downstream of the obligatory guanine present at the transcriptional start site (Fig. 6B).[11,12,55,65] In fact, the entire U6 portion can be replaced with the effector RNA with the exception of the +1 guanine and the pol III termination sequence (Fig. 6C).[12] This strategy limits the unwanted effects described earlier for type 2 expression units where internal promoter and spacer sequences can affect folding and activity of effector RNAs. However, it is advantageous to include the first 27 nucleotides of the wild-type U6 transcript in recombinant transcripts as this region forms in hairpin structure that is required for capping.[66] The stability of recombinant U6 transcripts is improved significantly when this region is included.[65] Fortunately, this hairpin structure is reasonably stable and thus is less likely to interfere with the folding of effector RNAs.

Recombinant U6 transcripts accumulate in the nucleus[12,65] and, except perhaps during mitosis, never become exposed to the cytoplasmic compartment. Recombinant U6 transcripts are often less stable than their wild-type counterparts that become associated with the spliceosome. In our hands, recombinant U6 transcripts accumulate to lower levels in cells relative to comparable recombinant tRNAs containing the stabilizing domain described earlier. Also, like any pol III transcript, one must be cautious of canonical pol III termination sequences that can be present in effector RNAs, leading to premature termination (see earlier discussion).

Vector Considerations

All of the expression units described earlier can be incorporated into plasmids or viral vectors such as retroviral, adenoviral, or adenoviral-associ-

[65] S. B. Noonberg, G. K. Scott, M. R. Garovoy, C. C. Benz, and C. A. Hunt, *Nucleic Acids Res.* **22,** 2830 (1994).

[66] R. Singh, S. Gupta, and R. Reddy, *Mol. Cell. Biol.* **10,** 939 (1990).

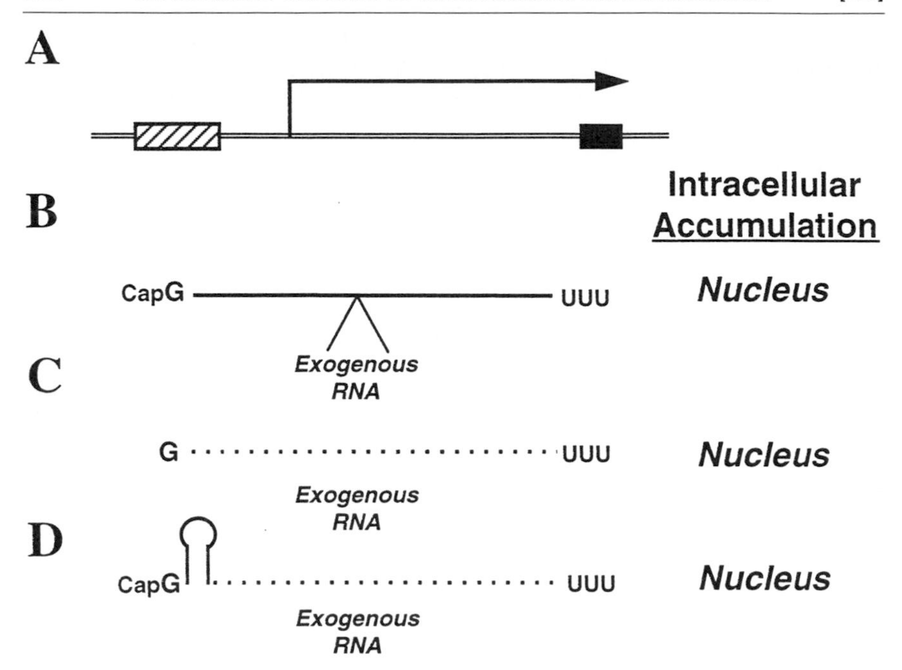

FIG. 6. Placement of effector RNAs in type 3 pol III expression units. (A) In contrast to type 2 pol III genes (Fig. 4), promoter elements for type 3 pol III expression units (hatched box) reside outside of the transcription unit. The termination sequence (closed box) is the same pol III consensus sequence as described in the legend to Fig. 4. Effector RNAs can be placed within the U6 transcription unit (B) or inserted in place of the wild-type U6 sequences with the exception of the +1 G (C). However, the first 27 nucleotides of the U6 transcript form a hairpin structure that is necessary for 5′ capping (D).

ated viral (AAV) vectors.[67] The smaller pol II snRNA and pol III expression units have a size advantage over the pol II mRNA expression units, as they are significantly smaller and can fit more readily into AAV and retroviral vectors that have size constraints.

In the case of retroviral vectors, it should be noted that the site of insertion of certain pol II snRNA and pol III expression units can affect expression levels significantly. Several groups have observed up to a 100-fold reduction in expression when such transcription units were included within the LTR-driven pol II transcription units of retroviral vectors.[11,68] This problem can be avoided by using the double-copy vector strategy developed in E. Gilboa's group, where transcription units are cloned into

[67] R. J. Kaufman, *Methods Mol. Biol.* **62,** 287 (1997).

[68] B. A. Sullenger, T. C. Lee, C. A. Smith, G. E. Ungers, and E. Gilboa, *Mol. Cell. Biol.* **10,** 6512 (1990).

the U3 region of the 3′ LTR.[68] This region is duplicated in the nontranscribed region of the 5′ LTR during conversion of the viral vector into the DNA provirus, producing two copies of the recombinant transcription unit in the provirus. The copy present in the 5′ LTR is outside of the LTR transcription unit and thus is not subject to interference. Interestingly, this interference is not apparent in transient transfection experiments with plasmids (unpublished observation, 1994). This could be due simply to the high copy number achieved with such vectors in transient transfection experiments, where pol II transcription factors may be limiting relative to pol III factors.

Conclusions and Future Directions

The expression strategies described in this article differ in regard to tissue and species specificity, intracellular localization, and ultimately the biological activity of the recombinant transcripts produced. We are just beginning to understand how the unique characteristics of each expression strategy can best complement the mechanisms of action of the different types of effector RNAs described. It will be important to continue to characterize the distinctive properties of each expression system so that the full potential of recombinant structural and functional RNAs can be realized.

The therapeutic potential of several types of recombinant structural and functional RNAs are now being evaluated in the clinic.[69,70] In addition, structural and functional RNAs can serve as important research tools in the genomics field. It is likely that the coding regions for all 100,000 human genes will be known within the next few years. Currently, there is a significant unmet need for technologies that can be used to associate function with the enormous amount of raw sequence information being generated. Several of the effector RNAs described in this article, especially antisense RNAs and ribozymes, are ideally positioned to meet this need because they can be designed from raw sequence information alone. Such effector RNAs can be used in cell culture experiments in a relatively high throughput manner to inhibit gene expression. Global effects on cell function and gene expression patterns can then be measured to determine the function of the targeted gene. In addition, antisense RNA and ribozyme transcription units can be incorporated into transgenic mice, producing "knockdown" mice

[69] M. Yu, E. Poeschla, and F. Wong-Staal, *Gene Ther.* **1,** 13 (1994).
[70] J. J. Rossi, *Br. Med. Bull.* **51,** 217 (1995).

for evaluating function in whole animals.[71] Such studies can lead to the identification and validation of therapeutic gene targets for drug development. Finally, the effector RNAs used in such studies to validate gene targets also represent candidate therapeutics in their own right.

[71] D. L. Sokol and J. D. Murray, *Transgen. Res.* **5,** 363 (1996).

Section V

Small Molecule Control of Recombinant Gene Expression

[15] Regulation of Gene Expression with Synthetic Dimerizers

By ROY POLLOCK and VICTOR M. RIVERA

Introduction

Systems that allow the precise control of gene expression in response to external inducers are valuable tools in basic research and may ultimately find use in the clinic. The ability to regulate levels of a protein can be important when studying its function, particularly when constitutive expression of the protein is toxic to cells or when deletion of its gene is lethal. External control of gene induction can also be applied to the generation of conditional phenotypes both *in vivo* and in cell culture. In addition, precise control of the timing and dosing of therapeutic proteins via an orally bioavailable drug is likely to be an important requirement for certain gene therapy applications.

Ideally, any system for the small molecule control of gene expression should exhibit low or undetectable levels of transcription in the uninduced state, but give a high induction ratio over a broad range in response to the inducer molecule. Furthermore, none of the components of such a system, including the small molecule inducer, should affect normal cellular physiology. Finally, an inducer molecule with high solubility, bioavailability, and good pharmokinetic properties would facilitate its use in animals.

A number of "gene switches" based on the regulation of transcription factor activity by small molecule dimerizers have been described that display many of the desirable features just listed.[1–4] Nevertheless, the small molecule dimerizers utilized to date retain the ability to interact with endogenous proteins and hence the potential to interfere with normal cellular functions. Unlike other small molecule inducers (for review, see Clackson[5]), dimer-

[1] P. J. Belshaw, S. N. Ho, G. R. Crabtree, and S. L. Schreiber, *Proc. Natl. Acad. Sci. U.S.A.* **93,** 4604 (1996).

[2] S. N. Ho, S. R. Biggar, D. M. Spencer, S. L. Schreiber, and G. R. Crabtree, *Nature* **382,** 822 (1996).

[3] V. M. Rivera, T. Clackson, S. Natesan, R. Pollock, J. F. Amara, T. Keenan, S. R. Magari, T. Phillips, N. L. Courage, F. Cerasoli, Jr., D. A. Holt, and M. Gilman, *Nature Med.* **2,** 1028 (1996).

[4] J. F. Amara, T. Clackson, V. M. Rivera, T. Guo, T. Keenan, S. Natesan, R. Pollock, W. Yang, N. L. Courage, D. A. Holt, and M. Gilman, *Proc. Natl. Acad. Sci. U.S.A.* **94,** 10618 (1997).

[5] T. Clackson, *Curr. Opin. Chem. Biol.* **1,** 210 (1997).

0076-6879/99 $30.00

izers play a relatively passive, nonallosteric role in the regulation of transcription factor activity (see later). Because there is no requirement to retain allosteric binding activity, dimerizers should be more amenable to modifications that introduce desirable properties without compromising their induction capabilities. This article illustrates this point by describing the control of target gene expression with a new class of synthetic dimerizer modified to avoid interactions with endogenous cellular proteins.

General Principles of Dimerizer-Regulated Gene Expression

Dimerizer-regulated gene expression takes advantage of the modular nature of eukaryotic transcription factors[6] and the development of cell-permeant chemical inducers of dimerization (for reviews, see Crabtree and Schreiber[7] and Spencer[8]). Eukaryotic transcription factors can be divided into functionally separable DNA-binding and transcriptional activation domains and these domains are able to reconstitute a sequence-specific transcriptional activator even when brought together via a noncovalent interaction.[9] Fusion of each transcription factor "half" to a heterologous ligand-binding domain allows reconstitution of an active transcription factor in response to the addition of cell-permeant dimeric ligands.[1–4]

The prototypical dimeric ligand is a derivative of the natural product FK506, which binds with high affinity to the immunophilin FK506-binding protein (FKBP)-12 (see Schreiber[10] and references therein). A molecule comprising two covalently joined FK506 molecules, termed FK1012, is able to mediate the intracellular dimerization of two FKBP domains.[11] The addition of FK1012 to cells bearing an appropriate reporter gene, together with plasmids expressing DNA-binding domain–FKBP and activation domain–FKBP fusion proteins, has been shown to activate target gene expression over a wide range in a dose-dependent fashion in cell culture. In addition, transcription can be turned off rapidly by the addition of the FK506 monomeric ligand,[2] which acts a competitive inhibitor of dimerization.

Subsequently, a smaller and wholly synthetic dimerizer, termed AP1510, has been developed.[4] AP1510 is composed of linked monomers that mimic only the FKBP-binding interface of FK506. It is therefore much easier to synthesize than FK1012 and is significantly more amenable to modification.

[6] R. Brent and M. Ptashne, *Cell* **43,** 729 (1985).
[7] G. R. Crabtree and S. L. Schreiber, *Trends Biochem. Sci.* **21,** 418 (1996).
[8] D. M. Spencer, *Trends Genet.* **12,** 181 (1996).
[9] S. Fields and O. Song, *Nature* **340,** 245 (1989).
[10] S. L. Schreiber, *Science* **251,** 283 (1991).
[11] D. M. Spencer, T. J. Wandless, S. L. Schreiber, and G. R. Crabtree, *Science* **262,** 1019 (1993).

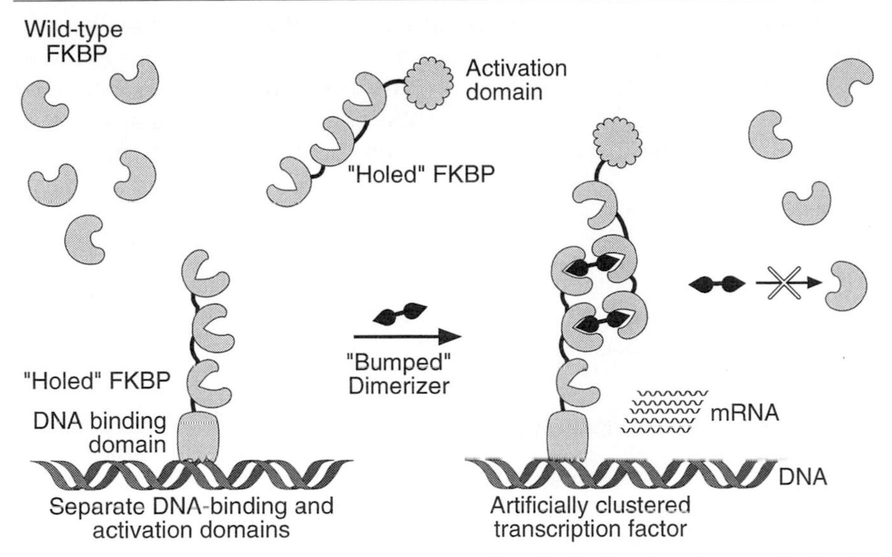

FIG. 1. Regulation of gene expression by "bumped" synthetic dimerizer. See text for details.

Furthermore, AP1510 activates transcription in the dimerizer system more effectively than FK1012, particularly in cell lines where transcription factor components are produced from stably integrated expression vectors.[4,12]

Although AP1510 is a highly effective inducer and does not have any of the immunosuppresive activity of FK506, it interacts with wild-type FKBP protein and therefore retains the potential to interfere with normal cellular functions. This article outlines the use of a new generation of dimerizer modified so that it is no longer able to bind significantly to wild-type FKBP. The inability of the modified dimerizer to interact efficiently with endogenous FKBP reduces potential effects on normal cellular physiology and eliminates non-productive interactions between the transcription factor fusions and the cellular FKBP pool. The modified dimerizer can be thought of as bearing a "bump" at the molecular interface with FKBP that precludes binding to wild-type FKBP protein. Instead, the modified dimerizer is designed to be used in conjunction with FKBP proteins bearing compensatory "holes" generated by amino acid substitutions in the ligand-binding pocket (see Fig. 1).

Overview

This article describes the components that make up the system for dimerizer control of gene expression, the use of dimerizer to regulate

[12] V. M. Rivera, *Methods* **14,** 421 (1998).

expression of a target gene in transiently transfected cells, and the generation and properties of dimerizer-responsive stable cell lines. The goal is to give the researcher an idea of the characteristics of induction with a modified dimerizer, including the effect of the ratio of activation domain to DNA-binding domain, and the dose response and kinetics of induction.

Components of System

To minimize potential complications due to immunogenicity in human gene therapy applications, all transcription factor and dimerizer-binding domains have been derived from human protein sequences.

Transcription Factor Domains

In order to avoid inappropriate regulation of the target gene by endogenous DNA-binding proteins, the DNA-binding domain should bear a sequence specificity distinct from that of known mammalian transcription factors. This requirement precludes the use of a naturally occurring human DNA-binding domain. Instead, we incorporate the artificial DNA-binding domain ZFHD1, described by Pomerantz *et al.*,[13] into the dimerizer system. ZFHD1 is derived from human protein sequences and consists of the Oct-1 homeodomain fused to two zinc fingers from the Zif268 DNA-binding domain. ZFHD1 binds to a novel sequence that is not recognized by known mammalian transcription factors, including Zif268 and Oct-1, and therefore displays a novel DNA-binding specificity.

The activation domain is derived from the carboxy terminus of the human NF-κB p65 subunit (amino acids 361–550).[14] We have found this activation domain to be more potent than the VP16 activation domain, both in the dimerizer system and when fused covalently to a DNA-binding domain.[15]

Ligand-Binding Domain

Configuration

Each transcription factor domain is fused to three tandemly repeated FKBP domains. Although the constructs described here contain FKBP domains at the carboxy terminus of ZFHD1 and the amino terminus of

[13] J. L. Pomerantz, P. A. Sharp, and C. O. Pabo, *Science* **267,** 93 (1995).
[14] M. L. Schmitz and P. A. Baeuerle, *EMBO J.* **10,** 3805 (1991).
[15] S. Natesan, unpublished data (1996).

the p65 activation domain (Fig. 3A), FKBPs can be fused to either end of the DNA-binding or activation domain with similar results. We have found the number of FKBPs fused to each transcription factor domain to be a more critical parameter. In the optimal configuration, each transcription factor domain is fused to three FKBPs and levels of activation drop when only one or two are used. This may reflect the potential for each DNA-binding domain to recruit more than one activation domain to the promoter through multivalent clustering (in theory, activation domain fusions linked directly to the DNA-binding domain might themselves serve to recruit additional activation domain fusions via unliganded FKBP moieties) and/or the role of avidity in assembling stable transcription factor complexes.

Modification

Knowledge of the crystal structure of FKBP bound to a variety of high-affinity ligands was used to alter the ligand-binding interface of FKBP by site-directed mutagenesis.[16] Mutation of FKBP residue 36 from phenylalanine to valine effectively truncates the phenylalanine side chain, creating a hole in the hydrophobic FK506-binding pocket. This results in an FKBP F36V mutant protein ($FKBP_{F36V}$) that can accommodate modified ligands bearing bulky groups at the appropriate position. $FKBP_{F36V}$ can therefore be thought of as a holed FKBP that can be used in combination with a bumped synthetic dimerizer such as AP1889 (see later).

Synthetic Dimerizer

A class of synthetic dimerizers that bind FKBP in the same hydrophobic pocket as FK506 has been described.[4] Their relative simplicity and ease of synthesis has allowed the development of derivatives such as AP1889 (Fig. 2), which bears a bulky ethyl group at a position predicted by structural modeling to cause steric exclusion from the FKBP-binding pocket. Structural modeling also predicts that binding of this modified or bumped compound can be accommodated in the holed $FKBP_{F36V}$ protein. As expected, the monomeric form of AP1889 is unable to bind wild-type FKBP with significant affinity, but binds the holed $FKBP_{F36V}$ protein with nanomolar affinity *in vitro*. AP1889, therefore, effectively dimerizes $FKBP_{F36V}$ but not wild-type proteins.[16,17]

[16] T. Clackson, W. Yang, L. W. Rozamus, M. Hatada, J. F. Amara, C. T. Rollins, L. F. Stevenson, S. R. Magari, S. A. Wood, N. L. Courage, X. Lu, F. Cerasoli, Jr., M. Gilman, and D. A. Holt, *Proc. Natl. Acad. Sci. U.S.A.* **95,** 10437 (1998).

[17] W. Yang, T. Clackson, D. A. Holt, and M. Gilman, unpublished data (1996).

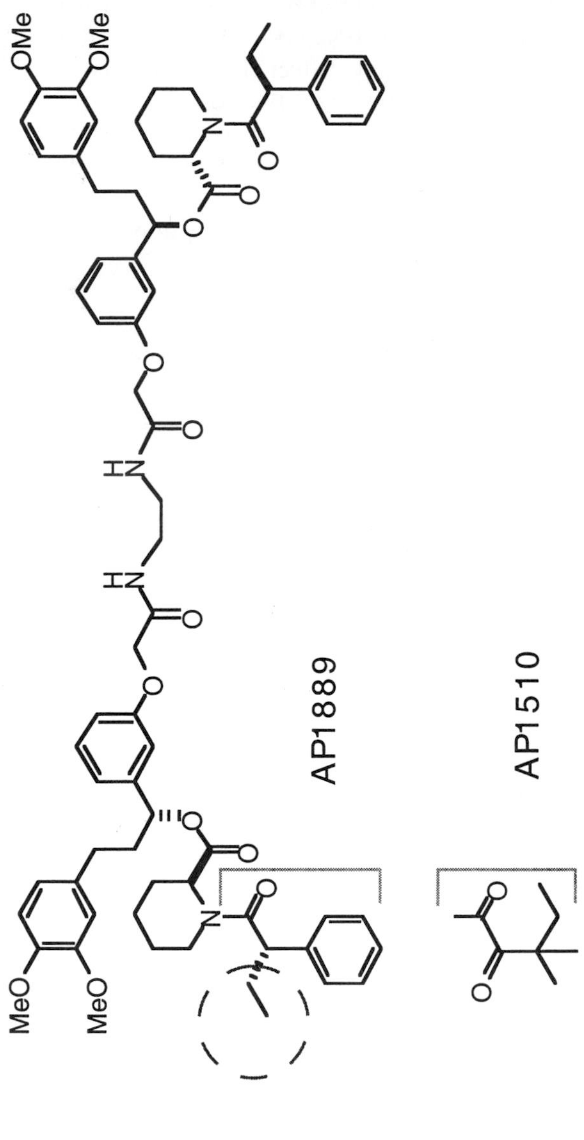

FIG. 2. Synthetic dimerizer AP1889. The structure of AP1889 is shown with a circle surrounding the ethyl group (bump) that sterically hinders binding to wild-type FKBP. The equivalent region of the unbumped dimerizer AP1510 is shown for comparison.

Target gene

To control gene expression with dimerizers, the gene of interest is cloned downstream of a minimal promoter bearing ZFHD1-binding sites (Fig. 3B). The presence of ZFHD1-binding sites upstream of reporter genes appears to have no effect on basal levels of expression in multiple cell types, which is consistent with there being no transcription factors that recognize this site in cell lines tested to date.[13,18] Transient transfection experiments have revealed that increasing the number of ZFHD1-binding sites upstream of the reporter gene leads to increased reporter gene expression on activation, although this effect plateaus at 12 binding sites.[19] Basal transcription or "leakiness" appears to be a function of the minimal promoter itself. We have tested several different minimal promoters, including those from the human interleukin-2 (IL-2), cytomegalovirus (CMV) immediate early, and simian virus 40 (SV40) early genes. The CMV and SV40 early promoters generally support the highest levels of induced transcription, but often produce clones with significant basal expression when introduced stably into the genome.[20] This presumably results from integration near endogenous enhancer elements. However, clones bearing the IL-2 minimal promoter rarely give rise to detectable basal reporter expression and frequently display high levels of dimerizer-induced transcription.[20] We therefore recommend the IL-2 minimal promoter for most applications, although the CMV early promoter may be preferable if basal transcription can be tolerated and high absolute levels of transcription are a more critical requirement.

Materials

Plasmids

Transcription Factor Expression Plasmids. pCGNNZ1F_V3 and pCGNNF_V3p65 (Fig. 3A) contain ZFHD1 and p65 coding sequences, respectively, fused to three tandem copies of the human FKBP12 coding sequence. Each FKBP coding sequence is modified such that it expresses FKBP protein bearing a Phe → Val substitution at residue 36.[16] Transcription factor fusions are expressed from pCGNN.[21] Inserts cloned into pCGNN as *Xba*I–*Bam*HI fragments are transcribed under the control of

[18] S. Natesan, R. Pollock, and V. M. Rivera, unpublished data (1996).
[19] R. Pollock, unpublished data (1996).
[20] V. M. Rivera, unpublished data (1996).
[21] R. M. Attar and M. Z. Gilman, *Mol. Cell. Biol.* **12,** 2432 (1992).

A

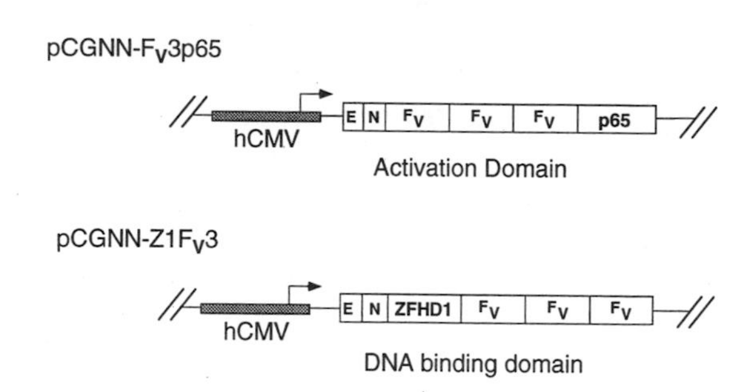

pCEN-F_V3p65/Z1F_V3/Neo

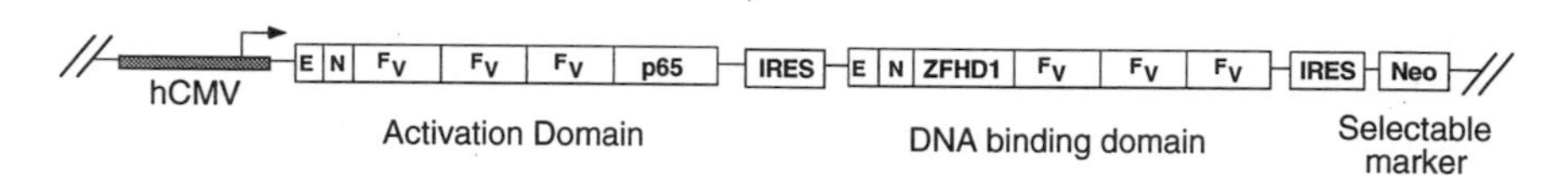

B

LH-Z_{12}-I-PL

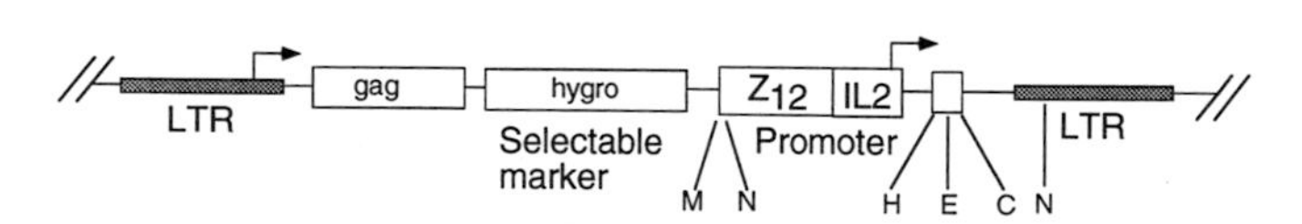

LH-Z_{12}-I-S

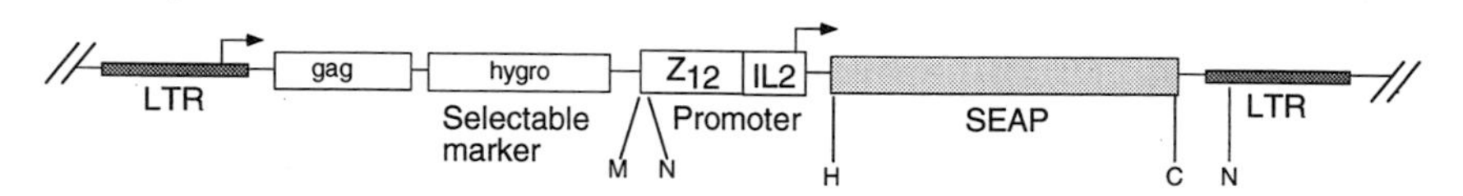

FIG. 3. Plasmids used in dimerizer-regulated transcription system. (A) Schematic diagrams of transcription factor fusion expression plasmids. Transcription factor fusions contain FKBPs bearing the F36V mutation (F_V) and are produced under the control of the human cytomegalovirus (hCMV) immediate early promoter and enhancer. E, an epitope tag, and N, the SV40 T antigen nuclear localization sequence, both fused to the amino termini of the transcription factors. pCEN-F_V3p65/Z1F_V3/Neo contains an internal ribosome entry sequence (IRES) from the encephalomyocarditis virus separating each cistron of the tricistronic transcript. Apart from the FKBP F36V mutation, all transcription factor plasmids are identical to previously

the human CMV enhancer and promoter and are expressed with an amino-terminal epitope tag (a 16 amino acid portion of the *Haemophilus influenzae* hemagglutinin protein) and nuclear localization sequence from SV40 large T antigen.

pCEN- $F_V3p65/Z1F_V3$/Neo (Fig. 3A) is designed to facilitate the construction of stable cell lines expressing transcription factor fusions. A tricistronic mRNA encoding both transcription factor fusions and a Neo selectable marker is expressed from the mammalian expression vector pCEN, a derivative of pCGNN.[21,22] As in pCGNN plasmids, inserts cloned between the *Xba*I and the *Bam*HI sites of pCEN are transcribed under the control of the human CMV enhancer and promoter and each transcription factor fusion is expressed with an amino-terminal epitope tag and nuclear localization sequence. The tricistronic message encodes, in the following order, the activation domain fusion $FKBP_{F36V}3$-p65, the DNA-binding domain fusion ZFHD1-$FKBP_{F36V}3$, and the *neo* gene, each separated by an internal ribosome entry sequence (IRES) from the encephalomyocarditis virus.[23] Because the *neo* gene is encoded by the same message as the transcription factor fusions, neomycin-resistant clones obtained after stable transfection of pCEN-$F_V3p65/Z1F_V3$/Neo should also express both transcription factor fusion proteins.

We have found that the most critical parameter affecting the inducibility of stable cell lines is the level of expression of the activation domain fusion (see later). For this reason, $FKBP_{F36V}3$-p65 is expressed from the 5′-most cistron where translation is most efficient. Nevertheless, levels of $FKBP_{F36V}3$-p65 protein expressed from pCEN-$F_V3p65/Z1F_V3$/Neo are generally lower than those expressed from the monocistronic plasmid.[20]

[22] M. Tanaka and W. Herr, *Cell* **60,** 375 (1990).

[23] S. K. Jang, H. G. Krausslich, M. J. Nicklin, G. M. Duke, A. C. Palmenberg, and E. Wimmer, *J. Virol.* **62,** 2636 (1988).

described versions bearing wild-type FKBP sequences.[3,4,12] Details concerning the generation of the FKBP F36V modification are presented elsewhere.[16] (B) Schematic diagrams of target gene retroviral vectors. LH-Z_{12}-I-PL[12] contains a polylinker with unique *Hin*dIII (H), *Eco*RI (E), and *Cla*I (C) sites that allow target genes to be placed downstream of an IL-2 minimal promoter (IL2) bearing 12 tandem ZFHD1-binding sites (Z_{12}). A unique *Mlu*I site (M) and a nonunique *Nhe*I site (N) are present upstream of the Z_{12}-IL-2 target promoter to facilitate cloning into alternative reporter constructs. The plasmids contain long terminal repeats (LTR) and a hygromycin selectable marker and can be introduced into cells by stable transfection or by retroviral infection following packing in an appropriate cell line. The SEAP reporter plasmid LH-Z_{12}-IL2-S was used in experiments shown in Figs. 4, 5, and 6. LH-Z_{12}-IL2-S is identical to vector LH-Z_{12}-I-PL except that it contains a *Hin*dIII–*Cla*I fragment bearing the SEAP gene in place of the polylinker.

Target Gene Plasmids. Target gene plasmid/retroviral vectors (Fig. 3B) contain 12 ZFHD1-binding sites and a minimal IL-2 promoter (−72 to +45 relative to the start site[24]) upstream of a polylinker (LH-Z_{12}-I-PL) or secreted alkaline phosphatase (SEAP) gene (LH-Z_{12}-I-S).[3,12] The promoter and target gene are flanked by long terminal repeats (LTRs) derived from the Moloney murine leukemia virus (MoMLV), one of which drives expression of the hygromycin resistance gene.[25] Despite the presence of enhancers within the LTRs, these vectors display low or undetectable basal expression. Target genes inserted into LH-Z_{12}-I-PL can be introduced into cells either by transient or stable transfection or by retroviral infection after packaging in an appropriate cell line. Only the coding sequence of the test gene (without introns or a polyadenylation signal) need be inserted into LH-Z_{12}-I-PL. In fact, this is preferable if the retrovirus is to be generated. Alternatively, the ZFHD1-IL-2 target promoter can be removed from these vectors and inserted upstream of the gene of interest (see Fig. 3B).

Dimerizer: Storage and Handling of AP1889

Lyophilized AP1889 (molecular mass 1230 Da) should be reconstituted as a concentrated stock in an organic solvent. We recommend dissolving the lyophilized material in ethanol to make a 1 m*M* stock solution. Once dissolved, the stock solution and any further dilutions in ethanol can be stored at −20 ° indefinitely in a glass or polypropylene vial. Dimerizer can be added to the growth medium immediately before it is added to the cells. Alternatively, dimerizer can be added directly from an ethanol stock to individual wells after medium addition if, for example, a dimerizer titration is being performed. In either case we recommend that the final concentration of ethanol in the media be kept below 0.5% to prevent detrimental effects of the solvent on the cells and that the final concentration of AP1889 in aqueous solution not exceed 5 μM to ensure complete solubility in cell culture medium.

Controlling Target Gene Expression in Transient Transfection Assays

Transient transfections are quick, convenient, and can often provide helpful information about the effect of induction of the gene of interest. It is therefore useful to perform dimerizer induction of a given test gene in transient transfections even if the ultimate goal is the generation of stable cell lines expressing the gene of interest under dimerizer control.

[24] U. Siebenlist, D. B. Durand, P. Bressler, N. J. Holbrook, C. A. Norris, M. Kamoun, J. A. Kant, and G. R. Crabtree, *Mol. Cell. Biol.* **6,** 3042 (1986).

[25] J. P. Morgenstern and H. Land, *Nucleic Acids Res.* **18,** 3587 (1990).

The use of the dimerizer system differs from other methods for the small molecule control of gene expression in that the regulatable transcription factor is composed of two separate polypeptides. The transcription factors can be introduced as two separate plasmids or simultaneously with the pCEN-F_V3p65/Z1F_V3/Neo IRES construct. The relative amount of the ZFHD1-$FKBP_{F36V}$3 and $FKBP_{F36V}$3-p65 encoding plasmids introduced into cultured cells is an additional variable that can affect the level of induction achieved.

In transient transfections, the ratio of activation domain to DNA-binding domain can be controlled and optimized by the cotransfection of different relative amounts of pCGNNZF_V3 and pCGNNF_V3p65 plasmids. An example of such an optimization is shown in Fig. 4. A constant amount of the DNA-binding domain plasmid pCGNNZF_V3 was cotransfected into HT1080 cells together with the SEAP reporter plasmid LH-Z_{12}-I-S and increasing amounts of the activation domain plasmid pCGNNF_V3p65. All wells contained 100 n*M* AP1889, a concentration known to give maximal transcription induction (see later). Increasing the amount of activation domain plasmid increases induced levels of transcription, up to an activation domain to DNA-binding domain plasmid ratio of 2:1. The addition of activation domain plasmid beyond this ratio fails to increase and, in fact, slightly reduces the levels of induced transcription. It is important to note that these results may not hold for all transfection protocols or cell types and should be treated as a reasonable starting point for optimization.

It is also possible to express both transcription factor fusions from the pCEN-F_V3/Z1F_V3/Neo plasmid, in which the ratio of activation domain and DNA-binding domain expressed is fixed by the relative efficiency of translation of the transcription factor cistrons. Although designed to simplify the generation of cell lines stably expressing transcription factor components (see later), pCEN-F_V3/Z1F_V3/Neo performs comparably to the two plasmid system in transient transfections (Fig. 4, IRES column).

The following transient transfection protocol is used for lipofection, although other transfection protocols, such as calcium phosphate, may also be used.

Transient Transfection Protocol

Plate cells (human HT1080 fibrosarcoma cells) into 12-well plates (2×10^5 cells/well). Transfect each well with a total of 400 ng of DNA, including pUC carrier (370 ng), reporter plasmid (100 ng), and transcription factor fusion plasmids [pCGNNZ1F_V3 (10 ng) plus pCGNNF_V3p65 (20 ng) or pCEN-F_V3p65/Z1F_V3/Neo (30 ng)]. Add 86 μl of Lipofectamine/Opti-MEM mixture [6 μl Lipofectamine (GIBCO-BRL, Gaithersburg, MD),

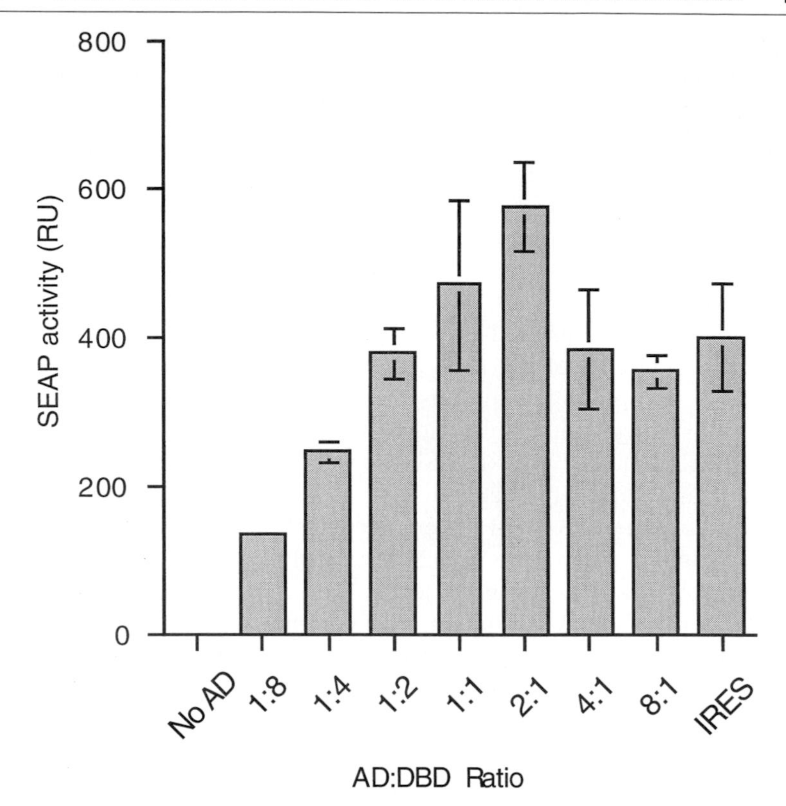

FIG. 4. Effect of activation domain: DNA-binding domain ratio (AD : DBD) on dimerizer-induced gene expression. Plasmids encoding ZFHD1-FKBP$_{F36V}$3 and FKBP$_{F36V}$3-p65 were transfected into HT1080 cells with the LH-Z$_{12}$-IL2-S reporter gene. Wells contained 10 ng pCGNNZ1F$_V$3 with 0 ng (No AD), 1.25 ng (1 : 8), 2.5 ng (1 : 4), 5 ng (1 : 2), 10 ng (1 : 1), 20 ng (2 : 1), 40 ng (4 : 1), and 80 ng (8 : 1) of pCGNNF$_V$3p65 or 30 ng pCEN-F$_V$3p65/Z1F$_V$3/Neo (IRES). SEAP activity secreted into the growth medium was measured after 24 hr as described.[11] Transfections were performed in triplicate and mean values (in relative units) ± SD were plotted.

80 μl Opti-MEM [(GIBCO-BRL)] to the DNA and incubate at room temperature for 20 min, prior to the addition of a further 330 μl of Opti-MEM. Wash cells once with Opti-MEM and add 420 μl of the DNA/Lipofectamine/Opti-MEM transfection mix. Leave the transfection mix on the cells for 4–5-hr and then replace with 1 ml growth medium [minimal essential medium, glutamine, nonessential amino acids (GIBCO-BRL), 10% fetal bovine serum] plus dimerizer.

Incubate cells plus dimerizer for 24 hr before removing 100 μl growth medium for SEAP assay. SEAP activity secreted into the growth medium can be measured in the supernatant using fluorescence or chemiluminescence.[11,26,27]

This transfection protocol can be successfully scaled down to a 24-or 96-well format, which conserves dimerizer, and is particularly useful when screening stable cell line clones (see later).

Generation of Dimerizer Responsive Stable Cell Lines

Many applications of dimerizer technology involve the generation of cell lines containing integrated copies of vectors expressing transcription factor fusions and the target gene of interest. This process involves two sequential transfections: to introduce dimerizer-regulated transcription factor fusions and to introduce the target gene to be regulated. It is usually preferable to introduce the transcription factor fusions first, as the resulting cell line can be used to regulate the expression of any subsequently introduced target gene.

The following procedure is used for the generation of a stable cell line containing transcription factor components and a dimerizer-regulated SEAP reporter.

To Stably Integrate Regulated Transcription Factor Fusions

(1) Transfect cells with pCEN-F_V3p65/Z1F_V3/Neo. Although transcription factors can be introduced successfully sequentially on separate plasmids,[20] this involves two stable transfections, each of which requires cointegration with a separate selectable marker plasmid. It is therefore much more convenient and efficient to introduce both transcription factor fusions simultaneously by transfection with pCEN-F_V3p65/Z1F_V3/Neo (see Plasmids).

(2) Select G418-resistant clones. Transfected HT1080 cells are selected in growth medium containing 500 μg/ml G418 (Mediatech, Herndon, VA). An additional advantage of transfecting with pCEN-F_V3p65/Z1F_V3/Neo is that most, if not all, G418-resistant clones express the transcription factor fusions, as the *neo* gene is expressed from the same tricistronic mRNA as the transcription factor fusions.

(3) Screen for clones with high dimerizer-dependent induction. After transfection of pCEN-F_V3p65/Z1F_V3/Neo, resulting G418-resistant clones are expanded and screened for high dimerizer-dependent induction. The simplest and most direct way to do this is to transiently transfect a reporter

[26] B. R. Cullen and M. H. Malim, *Methods Enzymol.* **216,** 362 (1992).

[27] S. R. Kain, *Methods Mol. Biol.* **63,** 49 (1997).

construct in the presence of 100n*M* AP1889. The reporter gene need not be the same as the target gene to be integrated. Therefore, a sensitive and easily assayable reporter such as SEAP may be used. At this stage it is useful to utilize a scaled down version of the transient transfection protocol to speed the process and conserve dimerizer. In a 96-well format, each well is transfected with 50 ng DNA, including 25 ng pUC carrier and 25 ng LH-Z_{12}-I-S reporter plasmid, in a total of 60.75 μl Lipofectamine/Opti-MEM mix (0.75 μl Lipofectamine plus 60 μl Opti-MEM). It is important that the inducibility of different clones can be compared directly. This can be done by plating all clones at the same cell density or by including an internal transfection control plasmid bearing an alternative reporter gene under control of a constitutive promoter. However, counting cells for each clone can become laborious if many are to be screened, and a transfection control necessitates performing an additional reporter gene assay. An alternative method is to plate cells into additional wells that can be transfected with a constitutively active SEAP reporter such as pSEAP2-control (Clontech, Palo Alto, CA). For each clone, inductions obtained with the LH-Z_{12}-I-S reporter plasmid can then be normalized to SEAP levels in wells transfected with the pSEAP2-control plasmid. We find that dimerizer-responsive clones occur at a high frequency. For example, to generate the 37-10 transcription factor cell line, 16 G418-resistant clones were screened: 11 displayed dimerizer-dependent induction and, of these, 6 gave inductions of 100-fold or greater.[20] The most important parameter affecting the generation of a highly inducible transcription factor cell line appears to be the level of expression of the $FKBP_{F36V}$3-p65 fusion protein. High levels of $FKBP_{F36V}$3-p65 fusion protein as measured by Western blot generally correlate with a highly responsive transcription factor clone. In contrast, even in highly inducible clones, the DNA-binding domain fusion is often expressed at levels undetectable by Western blot.[20] Higher levels of the transcription factor fusion proteins are obtained when expressed from separate plasmids (see Plasmids). Therefore, in situations where high levels of induced transcription are the most important factor, the researcher may wish to forgo the convenience of the pCEN-F_V3p65/Z1F_V3/Neo IRES plasmid and integrate the transcription factors on separate plasmids.

To Stably Integrate Target Plasmid

(1) Introduce the target gene. Once identified, clones displaying high levels of dimerizer-induced reporter gene transcription are used for a second stable transfection in which the target gene is stably integrated. Target genes inserted into LH-Z_{12}-I-PL can be introduced into cells by stable transfection or retroviral infection (see Plasmids). To generate 41-5 cells

LH-Z_{12}-I-S, a derivative of LH-Z_{12}-I-PL bearing the SEAP coding sequence was introduced by retroviral infection into the 37-10 cell line.

(2) Select hygromycin-resistant clones. The HT1080-derived transcription factor cell line is selected in growth medium containing 250 μg/ml hygromycin B (Calbiochem, La Jolla, CA). Because LH-Z_{12}-I-PL contains the hygro gene, hygromycin-resistant clones contain a stably integrated target gene.

(3) Screen for clones with low basal background and high dimerizer-dependent induction of reporter gene expression. Individual clones are incubated in media ±100 n*M* AP1889, and target gene expression is monitored either by immunological detection or by an appropriate biochemical assay. We have found that introduction of LH-Z_{12}-I-S generally gives rise to hygromycin-resistant clones with low or undetectable basal levels of SEAP activity, and cell lines exhibiting several hundredfold induction in response to dimerizer can be obtained relatively easily. For example, in the course of generating 41-5 cells expressing a dimerizer-regulated SEAP reporter, 22 hygromycin-resistant clones were screened by the addition of AP1889. All 22 clones exhibited undetectable basal SEAP expression that could be induced 100-fold or greater by the addition of 100 n*M* to the media.[20] Once identified, highly inducible clones can be investigated in more detail with respect to the kinetics and dose response of dimerizer induction.

Properties of Dimerizer-Responsive Stable Cell Lines

In this section, HT1080 clone 41-5 bearing an integrated LH-Z_{12}-I-S SEAP reporter gene under the control of ZFHD1-$FKBP_{F36V}3$ and $FKBP_{F36V}3$-p65 transcription factor fusion proteins is used to illustrate some of the properties of a AP1889-responsive cell line.

Figure 5A shows a dose–response curve of SEAP expression in response to the addition of increasing amounts of AP1889 to the media. 41-5 cells were exposed to the indicated concentrations of dimerizer for 24 hr. One hundred microliters of media was subsequently removed and SEAP activity assayed. SEAP production was detected at AP1889 concentrations as low as 5 n*M*, reached half-maximal levels at approximately 30 n*M*, and approached maximal levels at 100 n*M*. Higher concentrations of AP1889 up to 1 μM neither increased nor decreased SEAP production. SEAP production in the absence of AP1889 is indistinguishable from untransfected HT1080 control cells. Thus, 41-5 cells exhibit dimerizer-dependent and concentration-responsive SEAP production.

An additional feature of dimerizer-regulated gene expression is the ability to block transcription by competition with monomeric ligand. Figure

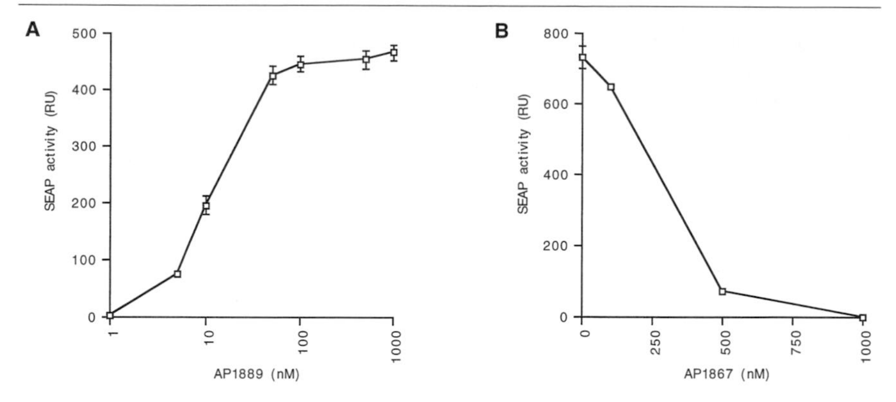

FIG. 5. Dimerizer control of gene expression in a stably transfected cell line. (A) Dose–response activation of SEAP expression by AP1889. 41-5 cells were split into 24–well plates (1×10^5 cells/well) and were treated the next day with the indicated concentration of AP1889. SEAP activity secreted into the growth medium was measured 24 hr after dimerizer addition. (B) Inhibition of dimerizer-induced gene expression by excess monomeric compound. 41-5 cells were plated as in (A) and treated simultaneously with 100 n*M* AP1889 and the indicated concentration of the monomeric ligand AP1867. SEAP activity secreted into the growth medium was measured after 24 hr. Assays were performed in triplicate and mean values (in relative units) ± SD were plotted.

5B shows the effect of adding 100 n*M* AP1889 to 41-5 cells together with increasing concentrations of AP1867, a monomeric ligand that binds tightly to the modified $FKBP_{F36V}$-binding pocket.[16] AP1867 is an effective competitor of AP1889-induced transcription with SEAP expression almost entirely prevented at a 5-fold molar excess. In the case of FK1012, a reduction in the rate of transcription has been shown to occur within 30 min of the addition of the competing monomer FK506.[2]

To assess the kinetics of SEAP induction in response to dimerizer addition, SEAP production was measured at 4-hr intervals after the addition of 100 n*M* AP1889 to 41-5 cells. The results are shown in Fig. 6A. SEAP activity is detectable after 8 hr of incubation with AP1889, and SEAP continues to be secreted into the media at a constant rate throughout the duration of the 32-hr experiment.

Figure 6B shows a time course for the decay in SEAP production following the removal of AP1889. Cells were incubated in media containing 100 n*M* AP1889 for 24 hr. Samples were removed for SEAP assay prior to washing of cells and addition of AP1889-free media. Sample removal and media replacement were repeated at 24-hr intervals for 6 days. Twenty-four hour exposure to AP1889 gave rise to significant levels of SEAP production for a period of 3 days, dropping to undetectable levels after 6 days. Because SEAP production is not a direct measure of transcription,

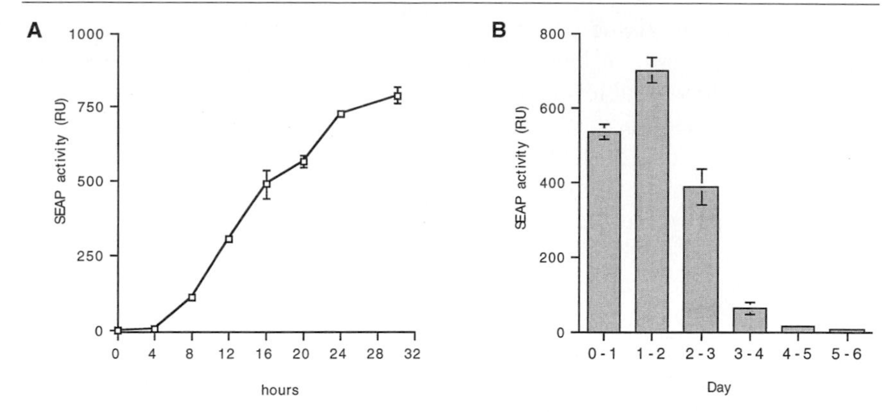

FIG. 6. Kinetics of dimerizer-induced SEAP expression. (A) Time course of SEAP induction in response to AP1889. 41-5 cells were treated with 100 n*M* AP1889, and SEAP activity secreted into the growth medium was measured at 4-hr intervals after dimerizer addition. (B) Time course of the decay in SEAP expression after the removal of AP1889. 41-5 cells were treated with 100 n*M* AP1889 for 24 hr and 100 μl of growth medium was removed and saved. The remaining medium was removed and cells were washed prior to the addition of fresh medium with no dimerizer. This process was repeated every 24 hr for 6 days. SEAP activity secreted into the growth medium during each 24-hr time interval was measured. Assays were performed in triplicate and mean values (in relative units) $\pm$ SD were plotted.

the kinetics of induction and decay are determined by additional factors such as posttranscriptional events and SEAP mRNA and protein stability. For example, the rate of decay in reporter gene product secretion in experiments analagous to the one shown in Fig. 6 can be increased dramatically by the incorporation of mRNA destabilizing sequences into the reporter gene message.[20] Also, similar kinetics of SEAP decay are observed when dimerizer removal is accompanied by the addition of a large excess of competing monomer, which inhibits new transcription. Thus, we believe the kinetics of dimerizer-induced gene expression are influenced heavily by posttranscriptional events and may vary substantially for different target genes.

Concluding Remarks

The components described in this article allow the tight regulation of gene expression in cell culture in response to a small molecule dimerizer. The system displays low or undetectable levels of transcription in the absence of dimerizer and high levels of dimerizer-dependent transcription. Target gene expression can be controlled accurately over a broad range and transcription can also be turned off readily by the addition of monomeric ligand to the tissue culture medium. These properties appear to be robust; in particular, we have found it relatively easy to generate stable cell lines with low basal levels of transcription and high induction ratios of target gene expression.

The incorporation of the bumped dimerizer and holed FKBP represent a significant advance in dimerizer-regulated transcription because it allows the use of small molecules that lack significant affinity for endogenous cellular FKBP. The cellular function of FKBP is unclear, although it is thought to be involved in the modulation of intracellular calcium release channels (for review, see Marks[28]) and transforming growth factor β (TGFβ) type I receptor function.[29] Whatever the precise role of FKBP turns out to be, the use of bumped dimerizers should remove the potential for unwanted effects on cellular physiology. FKBP is also an abundant protein,[10] hence an added advantage of bumped dimerizers is that they eliminate the potential for nonproductive interactions with the cellular FKBP pool, perhaps increasing the effective concentration of the dimerizer molecule. Finally, transcription factor fusions bearing holed FKBP domains display even lower levels of basal transcription in the absence of dimerizer than those observed with their unholed counterparts. This appears to be due to the elimination of unexpected low-level interactions between the unholed FKBP-activation domain and FKBP–DNA-binding domain fusions.[12]

The ability to modify the ligand protein interface raises the possibility of further improvements in dimerizer-regulated gene expression. For example, it may be possible to synthesize heterodimerizers with different bumps on each side that would eliminate the formation of nonproductive DNA-binding domain or activation domain homodimers. In addition, the generation of mutually exclusive bump/hole combinations together with DNA-binding domains bearing different DNA-binding specificities may ultimately allow the independent control of multiple target genes in the same cell.

These improvements highlight the modularity of the system that arises from the fact that dimerizer and ligand-binding domains function simply to bring the DNA-binding and activation domains into close proximity with each other to allow reconstitution of an active transcription factor. Other DNA-binding domains, activation domains, and alternative dimerizer/ligand-binding domain combinations can and have been used.[1–3] Thus, each component can be modified independently and subsequently incorporated

[28] A. R. Marks, *Physiol. Rev.* **76,** 631 (1996).

[29] T. Wang, B. Y. Li, P. D. Danielson, P. C. Shah, S. Rockwell, R. J. Lechleider, J. Martin, T. Manganaro, and P. K. Donahoe, *Cell* **86,** 435 (1996).

back into the same basic system. We have exploited this to develop modified humanized DNA-binding domains with alternative DNA-binding specificities[19] and activation domains with increased potency.[15] The flexibility of the dimerizer system and the fact that its protein components are derived exclusively from human protein sequences means that it is well placed for eventual use in regulating therapeutic gene expression in patients.

An important future development will be the development of dimerizer-regulated gene switches that can function in animals. The feasibility of this has been established by the demonstration that, in a related system, the natural heterodimerizer rapamycin can control circulating levels of human growth hormone in mice implanted with engineered human cells.[3]

Therefore, the immediate challenge is to develop synthetic dimerizers with improved pharmacological properties. The synthetic flexibility offered by molecules such as AP1510 and AP1889 provides grounds for optimism that such compounds can be developed. These would aid the creation of inducible phenotypes in mice and other organisms and perhaps lead to drugs for use in a gene therapy context.

Acknowledgments

The work described here is a result of the combined effort of the entire gene therapy group at Ariad. We thank Renate Hellmiss for art and Mike Gilman, Dennis Holt, Tim Clackson, Sridar Natesan, Dale Talbot, and Karen Zoller for comments on the manuscript.

[16] Antiprogestin Regulable Gene Switch for Induction of Gene Expression *in Vivo*

By Yaolin Wang, Sophia Y. Tsai, and Bert W. O'Malley

Introduction

In an organism, the coordinate expression of various genes is critical for development as well as growth and survival. Disregulation of the timing or the level of gene expression often results in pathological changes leading to the development of pathology. To better regulate individual genes, a "gene switch" that can control the timing and expression level of genes in a particular cell or tissue would be advantageous.

Eukaryotic genes respond to various stimuli, including metabolites, growth factors, and hormonal or environmental agents. The constitutive expression of foreign genes may result in cytotoxicity as well as undesired

0076-6879/99 $30.00

immune responses. The quest to develop a tight inducible and regulable system for controlling gene expression has been initiated over the past decade in a number of laboratories around the world.

We believe that any successful and applicable gene switch should fulfill the following criteria: (*i*) it should only induce expression of newly introduced target genes and should not affect endogenous genes; (*ii*) the target gene could be activated only by an exogenous signal, preferably a small molecule that distributes throughout all bodily tissues; (*iii*) the target induction should be reversible; (*iv*) the exogenous signal molecule should be biologically safe and easily administered, preferably orally; and (*v*) the target gene should have low basal activity and a high level of inducibility.

Several inducible systems for regulating gene expression have been established. They include the use of the metal response promoter,[1] the heat shock promoter,[2] the glucocorticoid-inducible (MMTV-LTR) promoter,[3] the *lac* repressor/operator system using IPTG as an inducer,[4,5] and the Tet repressor/operator system (tTA) with tetracycline as an inducer.[6–8] More recently, gene activator sequences and selective DNA-binding sequences have been fused to the hormone-binding domains (HBD) of steroid receptors, rendering the fusion protein responsive to steroid.[9,10] For example, a GAL4-HBD fusion protein capable of transactivating a target gene by binding to GAL4-binding sites (17-mer) located upstream of the target gene has been constructed.[11]

Construction of an RU486-Inducible Gene Expression System

An interesting discovery was made in our laboratory while studying the structural conformation of steroid hormone receptors and the role of the hormone-binding domain (HBD) in the transcriptional activation of

[1] P. F. Searle, G. W. Stuart, and R. Palmiter, *Mol. Cell. Biol.* **5,** 1480 (1985).

[2] S. A. Fuqua, M. Blum-Salingaros, and W. L. McGuire, *Cancer Res.* **49,** 4126 (1989).

[3] R. Hirt, N. Fasel, and J. P. Kraehenbuhl, *Methods Cell Biol.* **43,** 247 (1994).

[4] J. Figge, C. Wright, C. J. Collins, T. M. Roberts, and D. M. Livingston, *Cell* **52,** 713 (1988).

[5] S. B. Baim, M. A. Labow, A. J. Levine, and T. Shenk, *Proc. Natl. Acad. Sci. U.S.A.* **88,** 5072 (1991).

[6] M. Gossen and H. Bujard, *Proc. Natl. Acad. Sci. U.S.A.* **89,** 5547 (1992).

[7] S. Freudlieb, U. Baron, A. L. Bonin, M. Gossen, and H. Bujard. *Methods Enzymol.* **283,** 159 (1997).

[8] P. A. Furth, L. T. Onge, H. Boger, P. Gruss, M. Gossen, A. Kistner, H. Bujard, and L. Hennighausen, *Proc. Natl. Acad. Sci. U.S.A.* **91,** 9302 (1994).

[9] D. Picard, *Curr. Opin. Biotechnol.* **5,** 511 (1994).

[10] A. K. Walker and P. J. Enrietto, *Methods Enzymol.* **254,** 469 (1995).

[11] S. Braselmannm, P. Graninger, and M. Busslinger, *Proc. Natl. Acad. Sci. U.S.A.* **90,** 1657 (1993).

steroid-responsive target genes. We found that deletion of 42 amino acids at the C terminus of the human progesterone receptor (hPR) resulted in a loss of agonist binding and response to progesterone so that the mutated hPR could no longer activate progesterone-responsive genes.[12] Surprisingly, this mutated hPR still bound the progesterone antagonist RU486, an exogenous synthetic antiprogestin, and activated transcription of target genes containing binding sites for progesterone receptor. Further analysis of steroid hormone binding confirmed that this mutated hPR binds to no endogenous steroid hormones, such as progesterone, glucocorticoid, estrogen, androgen secosteroids, and retinoids.

Utilizing the unique features of the mutated human progesterone receptor, we set out to develop a recombinant steroid hormone receptor that could regulate target gene expression in the presence of only exogenous synthetic antiprogestin ligands. We constructed a chimeric molecule (GL) by fusing the HBD of the mutated hPR (residues 640–891), as a ligand-dependent regulatory domain, to a segment of the yeast transcriptional activator GAL4 (residues 1–94). This region of GAL4 contains a DNA-binding function (residues 1–65), a dimerization function (residues 65–94), and a nuclear localization signal (residues 1–29). By replacing the DNA-binding domain of PR with that of GAL4, we eliminated the possibility of simultaneous activation of any endogenous progesterone-responsive genes. To enhance the transcriptional activation function of the chimeric protein, we fused the C-terminal fragment of the herpes simplex virus protein VP16 (residues 411–487) to the N terminus of GAL4 in the chimera GL, creating a chimeric gene regulator (gene switch) GLVP (Fig. 1).

The expression of GLVP can be driven by any viral promoter [RSV (Rous sarcoma virus), CMV (cytomegalovirus)] or tissue-specific promoter of interest. When expressed in cells through transient transfection, we have demonstrated that this novel regulator GLVP could activate a target gene containing the GAL4 DNA-binding sites (17-mer). More importantly, the target gene expression only occurs in the presence of exogenous RU486 but not endogenous steroid hormones.[13] This tight regulation of target gene expression was confirmed in several mammalian cell lines. We also found that in addition to RU486, the GLVP can be activated by other known synthetic progesterone antagonists at low concentration.[13] Various target genes encoding intracellular proteins (chloramphenicol acetyltransferase, tyrosine hydroxylase) or secretory protein (human growth hormone) were

[12] E. Vegeto, G. F. Allan, W. T. Schrader, M.-J. Tsai, D. P. McDonnell, and B. W. O'Malley, *Cell* **69,** 703 (1992).

[13] Y. Wang, B. W. O'Malley, Jr., S. Y. Tsai, and B. W. O'Malley, *Proc. Natl. Acad. Sci. U.S.A.* **91,** 8180 (1994).

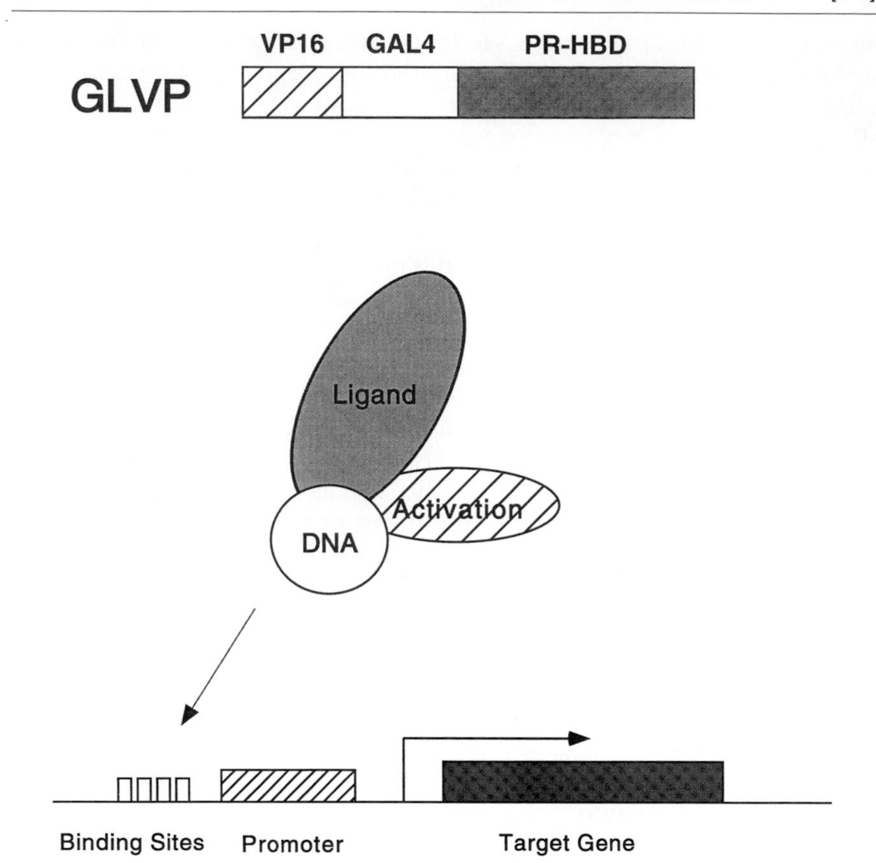

FIG. 1. Schematic diagram of the inducible system. The recombinant steroid receptor GLVP consists of the hormone-binding domain of human progesterone receptor (residues 640–891), the DNA-binding domain of GAL4, and the transcription activation domain of VP16. The reporter plasmid contains four copies of 17-mer (GAL4-binding sites) linked to the promoter and target gene of interest.

evaluated and shown to be induced effectively at a very low concentration (~1 n*M*) of RU486. In addition, this inducible system has been validated through the construction of RU486-inducible stable cell lines. Finally, we have confirmed the effectiveness of this inducible system using an *ex vivo* transplantation approach in which cells containing the stably integrated chimeric regulator GLVP and a target gene were grafted to rats. In this case, the expression of the target gene (tyrosine hydroxylase) was again tightly regulated by the administration of RU486. The dosage of RU486 used is significantly lower than that required for antagonizing *in vivo* progesterone actions.

Summary of Characteristics of RU486-Inducible System

(1) It can be regulated by an exogenous signal, in this case, RU486. (2) RU486 is a small synthetic molecule (molecular weight ~430) and has been used safely as an oral drug for other medical purposes.[14,15] (3) Because the yeast GAL4 protein has no mammalian homolog, it is unlikely that the target gene driven by the 17-mer-binding sites and promoter would be activated or repressed by endogenous proteins. (4) It does not affect endogenous gene expression as the regulator GLVP only efficiently activates the target gene bearing multiple copies of the 17-mer sequences juxtaposed to the promoter of the target gene.[16] (5) The regulator can be activated by a very low dose of RU486 (1 n*M*). At this low concentration, RU486 does not affect the normal function of endogenous progesterone and glucocorticoid receptors, as it only antagonizes the activity of these receptors at much higher doses. Thus, the high-binding affinity of antiprogestin for the GLVP gene switch ensures its specificity for regulation of only target gene expression.

Optimization of Inducible Gene Expression

Researchers working with the various inducible systems currently available [RU486, tetracycline, IPTG(isopropylthiogalactoside)] sometimes encounter a problem of leaky expression of the target gene. In certain instances, this could significantly affect the outcome of an experiment. To address this question, we can choose to use a minimal promoter (E1B TATA box) instead of a stronger thymidine kinase (*tk*) promoter, thereby reducing the basal level of target gene expression significantly (Fig. 2). In cases where a strong promoter is required to maximally induce the expression of a target gene, we found it necessary to optimize the ratio of the GLVP plasmid vs the reporter plasmid in transient transfection assays. In general, we use a ratio of 1 : 10 to 1 : 5 of GLVP vs reporter. To ensure the successful application of the inducible system, it is important that the individual researcher empirically determines the amount and the ratio of the regulator and the reporter to be used in their investigation. When constructing a stable cell line harboring the inducible system, it is useful to screen for multiple clones until a desired one has been identified. In most cases, some residual basal expression would be detected (depending on the sensitivity of assay), as in the case of many endogenous genes, and

[14] R. N. Brogden, K. L. Goa, and D. Faulds, *Drugs* **45,** 384 (1993).

[15] S. M. Grunberg, M. H. Weiss, I. M. Spitz, J. Ahmadi, A. Sadun, C. A. Russell, L. Lucci, and L. L. Stevenson, *J. Neurosurg.* **74,** 861 (1991).

[16] Y.-S. Lin, M. F. Carey, M. Ptashne, and M. R. Green, *Cell* **52,** 713 (1988).

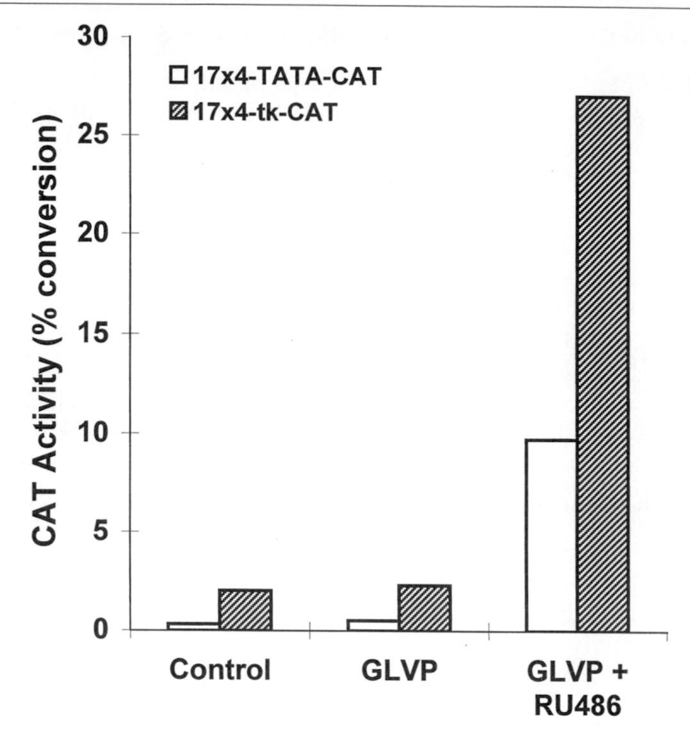

FIG. 2. Optimization of target gene expression. CAT assay showing cotransfection of 5 μg of reporter plasmid (17 × 4-TATA-CAT or 17 × 4-tk-CAT) with 2 μg of GLVP expression plasmid (driven by RSV promoter). CV-1 cells were then incubated with RU486 (10-7 M) as indicated.

should not be of general concern. We found that clones with more than 10-fold induction will likely be sufficient to present the researcher with the desired biological function or phenotype.

Protocol for Transient Transfection of Cell Cultures

The RU486-inducible system has been used successfully in various cell cultures such as HeLa, CV1, and HepG2. There may be some background variation in other cell types, and we suggest that one of the previous cell lines be employed as a control. The following protocol describes the Polybrene-mediated transfection of HeLa cells. When using different cell lines, appropriate media should be employed according to the supplier's specification. Other transfection methods, such as calcium precipitation, and liposome-mediated transfection also can be used.

1. Cells are grown in Dulbecco's modified Eagle's medium (DMEM) with 10% fetal bovine serum (FBS), 100 U/ml penicillin G, and 100 mg/ml streptomycin (GIBCO-BRL, Gaithersburg, MD).
2. Split exponentially growing cells into 10-cm tissue culture dishes the day before transfection at 1×10^6 per dish. The optimal cell density before transfection should be around 30–50% confluency.
3. To a 15-ml sterile Falcon tube, add 1 ml of sterile transfection buffer (prepared by mixing 0.2 g $MgCl_2 \cdot 6H_2O$, 0.132 g $CaCl_2 \cdot 2H_2O$, 8 g NaCl, 0.38 g KCl, 0.19 g $Na_2HPO_4 \cdot 7H_2O$, and 23.5 ml 1 *M* HEPES to 1 liter and titrate to pH 7.05).
4. Add 10–50 mg DNA of choice to each tube followed by gentle vortexing. We normally use 0.5–2 mg of transactivator plasmid (pGLVP) and 10 mg of reporter plasmid ($17 \times$ 4-tk-CAT) for each transfection (dish) and add calf thymus DNA to balance the amount of DNA in each tube to 20 mg. To achieve optimal induction and low basal activity of reporter gene expression, it is critical to titrate the amount of transactivator and reporter plasmid used in the initial experiment by varying the ratio between transactivator and reporter.
5. To each tube, add 5 ml of Polybrene solution (10 mg/ml, prepared in transfection buffer) while vortexing the tube gently. Polybrene solution can be stored as frozen aliquots in the freezer and fresh aliquot should be used each time.
6. Let the mixture sit at room temperature for 30 min.
7. While incubating the DNA mixture, wash the cells twice with 5 ml of Hanks' balanced salt solutions (HBSS) (GIBCO-BRL) to remove serum from the media. Then add 10 ml of serum-free DMEM to each plate.
8. Vortex the tube containing the DNA mixture gently and add the mixture drop by drop to the cell culture plate. Swirl the plate so that the media and the DNA mixture are well mixed.
9. Incubate the cells for 4–5 hr in a cell culture incubator.
10. Remove the plates, aspirate the media, and immediately add 3 ml of 25% glycerol (prepared in DMEM).
11. Remove the glycerol after 30 sec. Do not incubate longer than 1 min. Wash the cells twice with 5 ml DMEM to remove glycerol.
12. Feed the cells with 10 ml DMEM [with 10% (w/v) FBS].
13. At this point, 10 ml of RU486 (10^{-6} or 10^{-5} *M*) can be added to each plate to make a final concentration of 10^{-9} *M* (or 10^{-8} *M*). A stock solution of RU486 can be prepared by dissolving RU486 in 80% (v/v) ethanol and can be stored in a freezer for 12 months.

14. Incubate the cells for 36 hr and harvest the cells for reporter assay using the specified lysis buffer for the designated assay method.

Combining Inducible Transactivator and Reporter into One Vector

One way to increase the efficiency of simultaneous uptake of the inducible chimeric regulator GLVP and reporter DNA by cells is to construct a plasmid vector containing both gene products. For this purpose, we first construct a plasmid vector of the desired target gene and GLVP driven by a tissue-specific promoter. In the reporter plasmid, we then fuse the human growth hormone gene (genomic fragment) with the adenoviral E1B minimal TATA promoter linked to GAL4-binding sites (four copies of a 17-mer).[17] For the expression of GLVP, we choose the transthyretin enhancer/promoter shown previously to confer liver-specific expression.[18] These two plasmids are constructed and transiently transfected into a liver-specific cell line, HepG2. In the presence of RU486, approximately 30- to 90-fold induction of human growth hormone expression is observed.[17,19] We then fuse the two genes by the standard cloning procedure, generating a plasmid containing both the regulator and target genes as illustrated in Fig. 3A. When this plasmid is introduced into HepG2 cells, approximately 120-fold induction of hGH expression is observed. This study demonstrates that a single vector incorporating both the regulator and the reporter gene works as efficiently as two individual plasmid vectors. This single vector-inducible system should be particularly suitable for application to gene therapy where viral vectors are utilized.

Generation of a More Potent Gene Switch

We employ mutagenesis to generate a gene switch regulator that is more potent in the presence of RU486 while maintaining the same low basal level of target gene expression in the absence of the ligand.[20] We find that deletion of fewer amino acids from the C terminus of the human progesterone receptor hormone-binding domain, 19AAs instead of 42AAs, results in a more potent RU486-dependent regulator ($GL_{914}VP$) that activates target gene expression more strongly because the expanded sequence allows greater dimerization of the regulator (data not shown). Interestingly, when we fuse the VP16 activation domain to the C terminus of the hPR

[17] Y. Wang, B. W. O'Malley, and S. Y. Tsai, *Methods Mol. Biol.* **63,** 401 (1997).
[18] C. Yan, R. H. Costa, J. E. Darnell, Jr., J. Chen, and T. A. Van Dyke, *EMBO J.* **9,** 869 (1990).
[19] Y. Wang, F. J. DeMayo, S. Y. Tsai, and B. W. O'Malley, *Nature Biotechnol.* **15,** 239 (1997).
[20] Y. Wang, J. Xu, T. Pierson, B. W. O'Malley, and S. Y. Tsai, *Gene Ther.* **4,** 432 (1997).

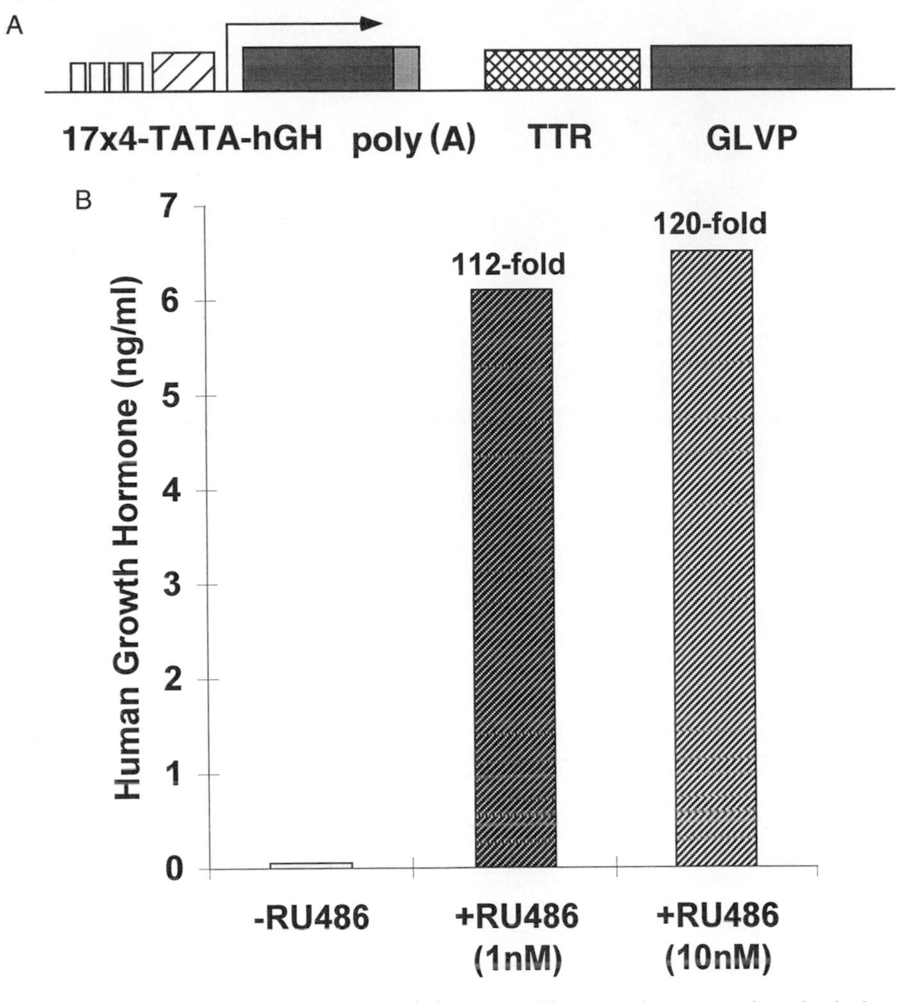

FIG. 3. Inducible system in one vector. (A) Diagram illustrates the construction of a single plasmid vector carrying both the regulator GLVP, driven by the liver-specific transthyretin enhancer/promoter (TTR) and reporter gene, human growth hormone (hGH). (B) Two micrograms (0.3 pmol) of the plasmid DNA was transfected into HepG2 cells and hGH expression in response to various concentration of RU486 was measured using RIA.[17] Numbers above the bar indicate fold of induction in the presence of RU486.

hormone-binding domain (640–914), the new chimeric regulator ($GL_{914}VP_C$) is even more potent than the $GL_{914}VP$ in which the VP16 activation domain is located at the N terminus of the chimeric molecule. More importantly, the $GL_{914}VP_C$ activates target gene expression at a lower

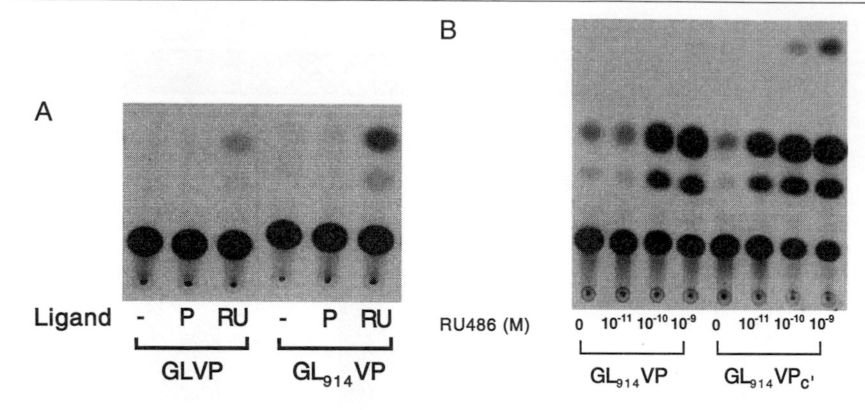

FIG. 4. Comparison of transactivation activity of the original gene switch GLVP with the new and potent version $GL_{914}VP$ and $GL_{914}VP_C$ in response to RU486. HeLa cells were transfected with 4 μg of chimeric regulator (driven by RSV promoter) as indicated and 10 μg of reporter 17 × 4-TATA-CAT. After transfection, cells were incubated with either solvent control (−) (85% ethanol, final concentration in the media is diluted 1000-fold or 0.085%) or RU486 (RU) at various concentrations. (A) $GL_{914}VP$ activation on reporter gene (CAT) expression is stronger than that of the original GLVP. (B) Dose–response analysis of chimeric regulators demonstrates that $GL_{914}VP_C$ responds to a very low concentration of RU486 (0.01 n*M*) and exhibit maximal induction of target gene expression at 1 n*M* RU486.

concentration of RU486 (0.01 n*M*) (Fig. 4). It is likely that this new fusion protein adopts a slightly different conformation, allowing better binding to RU486 as well as better interaction with the general transcription machinery.

Construction of Inducible Repressor

While most current inducible systems involve activation of target genes, it is obvious that development of an inducible transcriptional repression system would be advantageous in certain instances. To explore the possibility of converting the chimeric transactivator GLVP into a regulatable repressor, we replaced the VP16 activation domain with a transcriptional repression domain, the Krüppel-associated box-A (Krab), from the zinc finger protein *Kid-1. Kid-1* is identified as a kidney-specific transcription factor that is regulated during renal ontogeny and injury.[21] The Krab domain (residues 1–70) is inserted at the C terminus of GL_{914} generating GL_{914}Krab

[21] R. Witzgall, E. O'Leary, A. Leaf, D. Onaldi, and J. V. Bonventre, *Proc. Natl. Acad. Sci. U.S.A.* **91,** 4514 (1994).

(Fig. 5). The reporter plasmid 17 × 4-tk-CAT is cotransfected with GL_{914}Krab into HeLa cells and repression by RU486 is analyzed. Because we are looking at repression of the basal promoter and enhancer (tk) activity from the reporter construct, we use a higher amount of reporter plasmid (10 μg) and 4 μg of regulator plasmid (driven by CMV promoter). As shown in Fig. 5, the chimeric regulator GL_{914}Krab strongly represses (88% reduction) expression of the target gene (CAT) in an RU486-dependent manner. It should be noted that achieving maximal repression of a target promoter is more difficult than activation of a target gene, as it usually requires almost maximal occupation (competitive binding) of the promoter site by the regulator. A 10-fold repression would correspond to ~90% reduction of the basal activity. These results suggest that the Krab domain contains a potent transcriptional repression function and that the

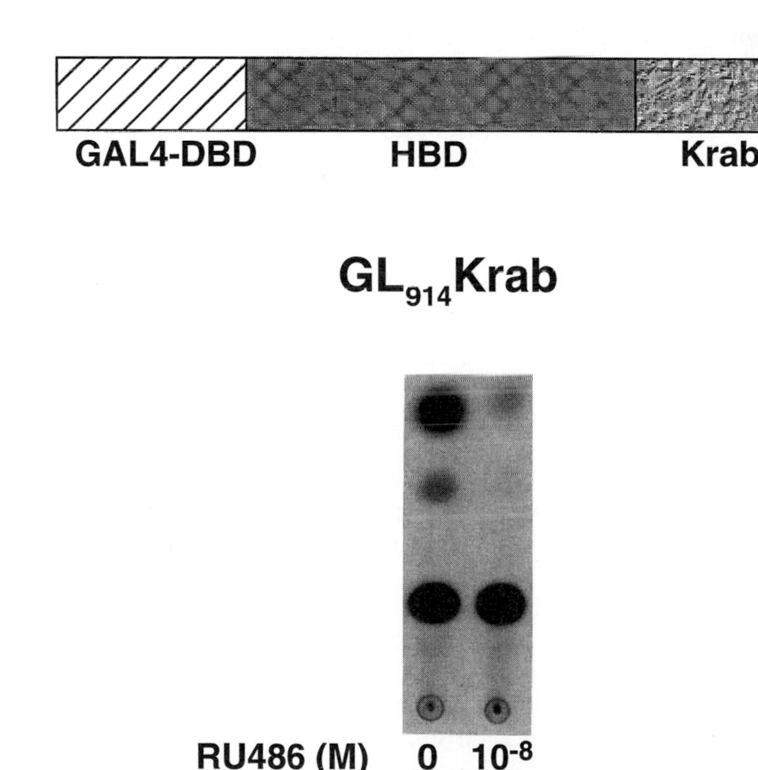

FIG. 5. Inducible repression of target gene expression. HeLa cells were transfected with 4 μg of expression plasmid pGL_{914}Krab (driven by CMV promoter) and 10 μg of reporter plasmid 17 × 4-tk-CAT. CAT activity was measured in the absence or presence of RU486 (10 n*M*).

chimeric repressor molecule GL_{914}Krab binds to DNA efficiently in the presence of RU486 to repress gene expression.

Regulable and Tissue-Specific Gene Expression in Transgenic Mice

One of the most challenging aspects of research in gene regulation is to achieve regulable gene expression *in vivo* in an intact animal. The major problem for biologists is the fact that integration of a foreign gene into the chromosome occurs at random sites. For example, integration into a heterochromatin region may result in complete silence of the foreign gene. Not surprisingly, multiple transgenic mouse lines need to be screened before one with satisfactory transgene expression can be obtained. To express the chimeric transactivator GLVP in a tissue-specific fashion we again chose the enhancer/promoter of the (TTR) gene, known to target transgene expression specifically to the liver in transgenic mouse.[18]

Our initial attempts to express the TTR-GLVP transgene in the liver of transgenic mice failed to produce a GLVP-expressing line in 10 founders. This could have resulted from the integration of the GLVP transgene into a heterochromatin region or a site influenced by a neighboring negative regulatory element. To minimize the effect associated with the site of integration, a tandem repeat of a chromosomal insulator sequence obtained from the 5′ region of the chicken β-globin gene is placed upstream of the TTR-GLVP construct.[22] This insulator sequence possesses no intrinsic enhancing or silencing activity.[22] Four founder lines bearing the insulator–TTR–GLVP transgene were established and three showed liver-specific expression of GLVP.

A target mouse line (17 × 4-tk-hGH) is generated and its serum is examined for hGH expression. No hGH is detected above the assay background in the serum of these mice, indicating that endogenous proteins do not activate this target gene, supporting the notion that no mammalian homolog of the yeast transcriptional activator GAL4 exists in higher eukaryotes[23] and vertebrates.[24]

By crossing individual lines of mice containing the regulator GLVP to mice containing the target gene, bigenic mice harboring both transgenes (TTR-GLVP/hGH) are selected for further analysis. No, or very little, expression of hGH is detected in the serum of bigenic mice, demonstrating that the inducible system is not activated by endogenous ligand. In contrast, intraperitoneal administration of a single dose of RU486 (250 μg/kg) results

[22] J. Chung, M. Whiteley, and G. Felsenfeld, *Cell* **74,** 505 (1993).
[23] A. H. Brand and N. Perrimon, *Development* **118,** 401 (1993)
[24] I. M. Spitz and C. W. Bardin, *N. Engl. J. Med.* **329,** 404 (1993)

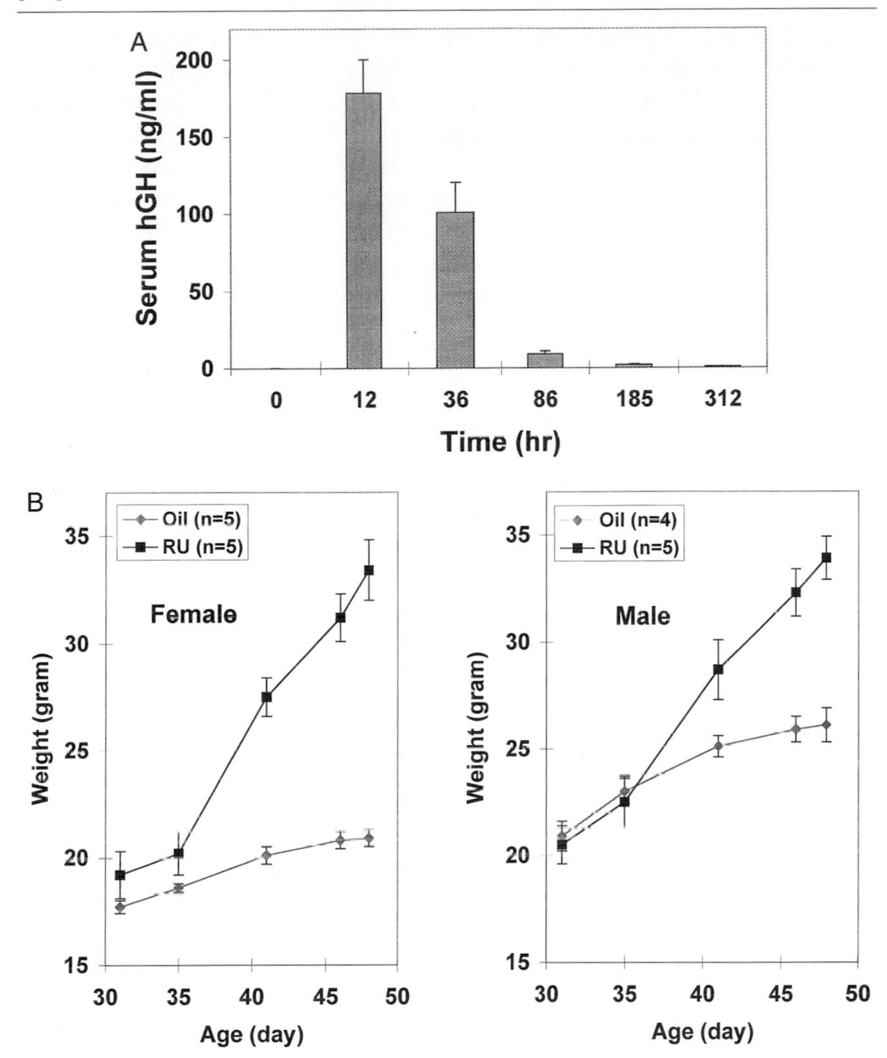

FIG. 6. RU486 inducible gene switch system in transgenic mice. (A) Kinetics of RU486 inducible hGH expression in transgenic mice. Bigenic mice ($n = 14$) were given a single injection (*ip*) of 250 μg/kg dose of RU486 (dissolved in sesame oil) and serum hGH was measured[19] at different time points as indicated. (B) Phenotype of bigenic mice in response to RU486 treatment. Bigenic mice (grouped in female and male, respectively) were given either sesame oil or RU486 (250 μg/kg) at day 31 after birth and continued to receive the same dose every 2 days. The mouse weight was recorded at the indicated points. The number of mice used in each study group is shown in the legend box.

in a 1500-fold induction of hGH expression in serum (Fig. 6A). High hGH expression (~200 ng/ml) is achieved by 12 hr. For comparison, the mouse endogenous GH level peaks at ~10 ng/ml, depending on the time of the day, age, and sex of the mouse. Transgene hGH expression peaks at ~12 hr after RU486 administration and returns to background levels ~100 hr postinjection. The RU486 inducible expression of hGH transgene is dose dependent and liver specific.[19] A repeated injection of RU486 (one per every 2 days) to bigenic mice results in a significant increase (30–37%) in weight (within 10 days) compared to the control group (Fig. 6B).

These studies represent the first kinetic study of inducible and dose-dependent transgene activation *in vivo* accompanied by a gross phenotypic change in response to a synthetic inducer. The use of such bigenic mice should allow the creation of an array of novel transgenic mouse models mimicking human diseases that might otherwise be impossible due to the deleterious effects of constitutive expression, such as genes involved in apoptosis and tumorigenesis. Because RU486 (Mifeprestone) can be administered orally and has an established safety record,[24] we envision that this inducible system should have broad application both in the regulation of target gene expression in transgenic mice and in the future development of methodology for human gene therapy.

Acknowledgments

We thank Dr. Ming-Jer Tsai and members of our laboratory for helpful discussions throughout the development of the RU486-inducible gene expression system.

Scction VI

Viral Systems for Recombinant Gene Expression

[17] Expression of SV40 Large T Antigen in Baculovirus Systems and Purification by Immunoaffinity Chromatography

By PAUL CANTALUPO, M. TERESA SÁENZ-ROBLES, and JAMES M. PIPAS

Introduction

The DNA tumor virus SV40 (simian virus 40) has been studied extensively as a model system for eukaryotic gene replication and expression, as well as for tumorigenesis. The large tumor antigen (T antigen) encoded by SV40 is a 708 amino acid regulatory protein that is required for viral productive infection and is necessary and sufficient for SV40-mediated tumorigenesis. T antigen orchestrates many aspects of the SV40 infectious cycle, including (1) DNA replication, (2) transcriptional control, and (3) virion assembly. T antigen is a multifunctional protein, with some of its biochemical activities mapping to discrete domains capable of functioning independent of the rest of the polypeptide.[1] For example, the sequence-specific DNA-binding activity of T antigen maps between amino acids 135 and 249, while the ATPase activity requires only amino acids 371–625. However, T antigen action often requires the coordinate action of multiple domains.[2] Thus, the DNA helicase activity of T antigen requires amino acids 135–625, which include the DNA-binding domain, the ATPase domain, and the zinc-finger regions of T antigen (Fig. 1). The mechanisms by which interdomain communication is effected and how the actions of independent domains are coordinated is a current focus of T antigen research.

Aside from possessing intrinsic biochemical activities (DNA binding, ATPase), T antigen also exerts its biological effects by influencing the biochemical activities of key cellular regulatory proteins. For example, in inducing tumorigenesis, T antigen acts on both the retinoblastoma family (pRb, p107, and p130) and p53 proteins, all of which are tumor suppressors. In addition, T antigen is a DnaJ-like molecular chaperone, and this chaperone function is required for T antigen action on the Rb family.[3,4] Thus, T

[1] C. N. Cole, *in* "Fields Virology" (B. N. Fields, D. M. Knipe, P. M. Howley *et al.*), 3rd ed., p. 1997. Lippincott-Raven, Philadelphia, PA, 1996.

[2] A. M. Castellino, P. Cantalupo, I. M. Marks, J. V. Vartikar, K. W. C. Peden, and J. M. Pipas, *J. Virol.* **71,** 7549 (1997).

[3] A. Srinivasan, A. J. McClellan, J. Vartikar, I. Marks, P. Cantalupo, Y. Li, P. Whyte, K. Rundell, J. L. Brodsky, and J. M. Pipas, *Mol. Cell. Biol.* **17,** 4761 (1997).

[4] J. Zalvide, H. Stubdal, and J. A. DeCaprio, *Mol. Cell. Biol.* **18,** 1408 (1998).

0076-6879/99 $30.00

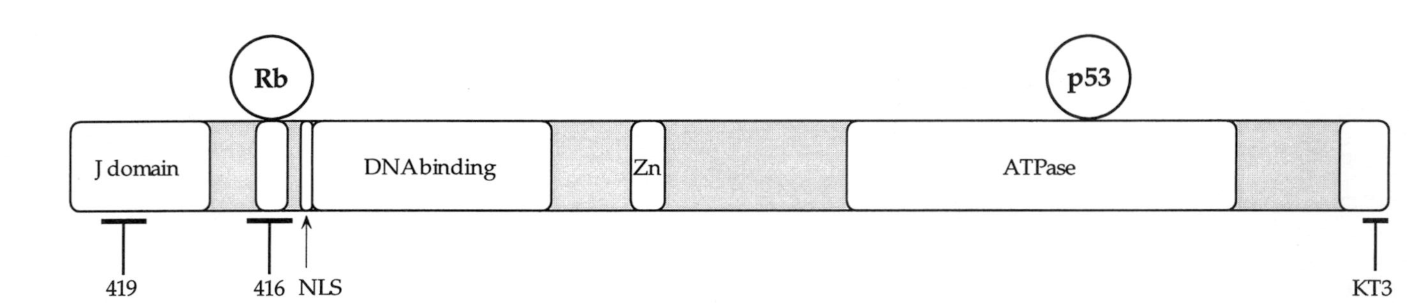

FIG. 1. Schematic representation of the SV40 large T antigen. The T antigen comprises multiple domains of which several are represented here as open boxes. The first 82 amino acids of T antigen is a J domain and T antigen functions as a DnaJ chaperone. T antigen binds to the tumor suppressors Rb and p53 in the N terminus and C terminus, respectively. The approximate location of the epitopes for the T antigen-specific monoclonal antibodies, PAb419 and PAb416, is shown in the N terminus. The antibody PAb419 recognizes an epitope in the first 82 amino acids of T antigen and PAb416 recognizes an epitope between amino acids 83 and 125. The monoclonal antibody KT3[9] recognizes a defined epitope at the far C terminus of T antigen, amino acids 698–708. Zn, zinc finger; NLS, nuclear localization signal; Rb, retinoblastoma protein.

antigen possesses at least three activities important for neoplastic transformation: (1) chaperone function, (2) Rb family binding, and (3) p53 binding. The precise biochemical nature of the T antigen action on the Rb family and p53 is not known. Such an understanding requires the purification of each of these proteins and the characterization of their interactions *in vitro*.

An excellent review of the SV40 T antigen has already been published[1] so we will not describe the methods used to map functional domains or assess the biological activity of T antigen mutants using cell culture systems. Rather we will provide methods for expression and purification of T antigen. These methods provide milligram quantities of pure T antigen suitable for use in biochemical studies and are suitable for application to other proteins of interest.

Purification of SV40 T Antigen by Immunoaffinity Chromatography

We routinely use baculovirus vectors to produce T antigen for purification. Originally, T antigen was cloned into baculovirus with the pVL941 vector and this system produces milligram quantities of T antigen.[5] Previously, our laboratory used the BaculoGold transfection kit with the pVL1392 and pVL1393 transfer vectors (PharMingen, San Diego, CA). Currently, we are using the "Bac to Bac" baculovirus expression system (GIBCO-BRL, Gaithersburg, MD) to make recombinant viruses. One advantage of this system is that the procedure utilizes a quick and simple color selection for the identification of recombinant baculoviruses.

After infecting insect cells with recombinant baculovirus, T antigen is purified by immunoaffinity chromatography using monoclonal antibodies conjugated to protein G Sepharose 4 Fast Flow (Pharmacia, Piscataway, NJ). This same method can be used to purify T antigen from other sources such as virus-infected or transformed mammalian cells or bacteria.

Materials

Cell Lines

The High Five cell line BTI-TN-5B1-4 was purchased from Invitrogen (Carlsbad, CA) and maintained in suspension culture as described by the manufacturer. This line was derived from the ovarian cells of the cabbage looper, *Trichoplusia ni.* The Sf9 cell line was derived from the pupal ovarian tissue of the fall army worm, *Spodoptera frugiperda.* Sf9 cells were provided

[5] R. E. Lanford, *Virology* **167,** 72 (1988).

by Dr. Max Summers and maintained in suspension culture as described.[6] The insect cell lines were maintained in 100-ml spinner culture flasks (Kontes, Vineland, NJ).

Hybridoma cell lines producing the T antigen-specific monoclonal antibodies PAb419, PAb416, and KT3 have been described.[7–9]

Growth Media and Supplements

High Five cells are grown at 27° in EXCELL-405 medium (Invitrogen). The Sf9 cells were originally maintained at 27° in TNM-FH medium, which is prepared from Grace's medium (GIBCO-BRL) by the addition of 3.33 g/liter of TC yeastolate and 3.3 g/liter lactalbumin hydrolyzate (Difco) and supplemented with 10% fetal bovine serum (Hyclone). They were then adapted to serum-free medium Sf-900 II as instructed by the manufacturer (GIBCO-BRL).

PAb419, PAb416, and KT3 are maintained in DMEM (GIBCO-BRL) + 10% fetal bovine serum in 5% (v/v) CO_2 at 37°.

All media are supplemented with penicillin/streptomycin (GIBCO-BRL) at 50 U/ml and 50 μg/ml, respectively.

Methods

The methods described here are separated into three steps: (1) purification of antibody, (2) preparation of the immunoaffinity column, and (3) immunoaffinity purification of T antigen.

Purification of Antibody

The purification of antibody[10] relies on the observation that the bacterial surface protein, protein G, can bind to the Fc domain of antibodies. Protein G that is covalently bound to a matrix, Sepharose 4 Fast Flow, was purchased from Pharmacia. Hybridoma supernatant or ascites is then passed over the column to bind the antibody. Following washing steps, the antibody is eluted from the column. This procedure can be used for purifying antibody

[6] M. D. Summers and G. E. Smith, "A Manual of Methods for Baculovirus Vectors and Insect Cell Culture Procedures." Texas Agricultural Experiment Station, College Station, TX, 1987.

[7] E. Harlow, L. V. Crawford, D. C. Pim, and N. M. Williamson, *J. Virol.* **39,** 861 (1981).

[8] L. Crawford and E. Harlow, *J. Virol.* **41,** 709 (1982).

[9] H. MacArthur and G. Walter, *J. Virol.* **52,** 483 (1984).

[10] E. Harlow and D. Lane, "Antibodies: A Laboratory Manual." Cold Spring Harbor Laboratory, Cold Spring Harbor, NY, 1988.

either from a hybridoma supernatant or from ascites. All solutions that come in contact with the column should be filter sterilized through a 0.2-μm filter.

1. Clarify starting material (hybridoma supernatant or ascites) by centrifuging in a Sorvall SS-34 rotor at 15,000 rpm (27,000*g*) for 30 min at 4°.
2. Transfer supernatant to a Costar 0.2-μm bottle filter system (CA).
3. If ascites is used, dilute 1:25 with antibody column wash buffer (ACW) [1× phosphate-buffered saline (PBS), pH 7.2 (137 m*M* NaCl, 2.7 m*M* KCl, 4.3 m*M* Na_2HPO_4, 1.4 m*M* KH_2PO_4) and 0.05% (v/v) Tween 20] hybridoma supernatant does not need further dilution. Store at 4° until needed.
4. The antibody purification column should be prepared at 4°. Take a 4-ml (0.7 × 10 cm) flex column (Kontes) and attach it to a stand. Attach an 18-inch piece of tubing to the bottom of the column. Fill the column to a few millimeters past the top of the glass body with protein G Sepharose 4 Fast Flow and let the column pack by gravity. The tubing will increase the flow rate of the column and allow more efficient packing. After all the beads have settled, add 5 ml of ACW and allow the buffer to flow through. Connect the column to a peristaltic pump and equilibrate it with 50 ml of ACW at a flow rate of 1 ml/min. It is good practice to use a different column for each antibody to eliminate the possibility of cross-contamination.
5. Load the sample at a rate of 0.25 ml/min. This is generally done overnight. Be careful not to let the column dry out. Make sure there is about 1 cm of liquid on top of the packed bed. Columns can be saved if let dry for several minutes, but ones that have dried overnight have not been satisfactory. The flow through should be saved and analyzed for antibody that fails to bind to the column.
6. Wash the column with 50 ml of ACW at a flow rate of 1 ml/min.
7. Elute the antibody with 15 ml of 0.1 *M* glycine, pH 2.5, at a flow rate of 1 ml/min. Collect 1-ml fractions into 100 μl of 1 *M* Tris, pH 9.6, to neutralize the pH of the solution.
8. After the fractions have been collected, wash the column with 50 ml of antibody column storage buffer [ACW + 0.02% (v/v) NaN_3] and store at 4°.
9. Analyze the fractions for the presence of antibody using the protein assay dye (Bio-Rad, Richmond, CA) based on the method of Bradford.[11] Pool the protein containing fractions.

[11] M. Bradford, *Anal. Biochem.* **72,** 248 (1976).

10. Dialyze against 0.1 *M* borate, pH 8.2, overnight at 4° and determine the antibody concentration using the protein assay dye.

Preparation of Immunoaffinity Column

The following method is based on a previously published protocol[12] that utilizes a cross-linker to attach the antibody of choice to an immobilized support, namely protein G Sepharose 4 Fast Flow. This method describes how to make one immunoaffinity column for T antigen purification. To avoid cross-contamination, we use a different column for each different mutant of T antigen. All steps for this protocol are carried out at room temperature unless otherwise noted. In order to estimate the efficiency of attachment, it is imperative to save and analyze each supernatant for the presence of unbound antibody after the protocol is completed.

1. Wash 1.5 ml (packed bead volume) of protein G Sepharose 4 Fast Flow with 0.1 *M* borate, pH 8.2, in two 1.5-ml Eppendorf tubes and centrifuge for 30 sec. Repeat this step three times. Transfer the beads to a 15-ml conical tube and add 4.5 mg of purified antibody to the washed beads. Mix and allow to rock for 2 hr. If the volume is low, add some 0.1 *M* borate, pH 8.2, to allow smooth rocking of the beads and antibody.
2. Centrifuge at 3000 rpm (1800*g*) in a Sorvall RT6000B for 10 min. Decant the supernatant.
3. Wash the beads with 10 ml of 0.1 *M* borate, pH 8.2. Centrifuge at 3000 rpm (1800*g*) for 5 min and decant supernatant. Repeat this step twice. Wash the beads with 10 ml of 0.2 *M* triethanolamine, pH 8.2. Centrifuge at 3000 rpm (1800*g*) for 5 min and decant the supernatant.
4. During the 0.2 *M* triethanolamine wash, prepare 50 ml of cross-linking solution (20 m*M* dimethyl pimelimidate in 50 ml 0.2 *M* triethanolamine, pH 8.2, adjust to pH 8.2 with NaOH). Resuspend beads in 20 volumes (30 ml) of cross-linking solution and transfer to a 50-ml conical tube. Allow the cross-linking reaction to occur with rocking for 45 min. Inspect periodically and manually mix beads back into solution to avoid bead settling.
5. Stop the reaction by centrifuging the beads at 3000 rpm (1800*g*) for 10 min and resuspending them in 10 ml of stop solution (20 m*M* ethanolamine in 0.2 *M* triethanolamine, pH 8.2). Transfer the suspension to a 15-ml conical tube and rock for 5 min. Centrifuge at 3000 rpm (1800*g*) for 5 min. Decant the supernatant.

[12] C. Schneider, R. A. Newman, D. R. Sutherland, U. Asser, and M. F. Greaves, *J. Biol. Chem.* **257,** 10766 (1982).

6. Wash the beads (now cross-linked to antibody) with 10 ml of 0.1 *M* borate, pH 8.2, at 4°. Centrifuge at 3000 rpm (1800*g*) for 5 min and decant supernatant. Repeat this step twice. Resuspend beads in 0.1 *M* borate, pH 8.2, + 0.02% (w/v) NaN_3 at 4°.
7. The T antigen immunoaffinity column should be prepared at 4°. Attach a 2-ml (0.7 × 4 cm) flex column (Kontes) to a stand. Connect an 18-inch piece of tubing to the bottom of the column. Pour the antibody beads into the column and let it pack by gravity. The tubing will increase the flow rate of the column and allow more efficient packing. After all the beads have settled, add 15 ml, in 5-ml increments, of column storage buffer [1 m*M* Tris, pH 7.4, 0.15 *M* NaCl, 1 m*M* EDTA, 0.02% (w/v) NaN_3] and allow the buffer to flow through. Top off the column with 2 ml of the same buffer. Store the column at 4°.
8. Calculate the efficiency of antibody attachment onto the column. Generally, we obtain an attachment efficiency of 85–95%.

Immunoaffinity Purification of T Antigen

The following method consists of three parts: (1) infection of insect cells with baculovirus, (2) preparation of the lysate, and (3) purification of T antigen from the lysate and is based on protocols described previously.[5,13–15]

Infection of Insect Cells with Baculovirus

1. Centrifuge 4×10^8 Sf9 cells (or 3×10^8 High Five) at 1000 rpm (150 g) for 10 min in an IEC clinical centrifuge and resuspend the cell pellet in 7.5 ml of media (i.e., TNM-FH for Sf9 cells and EXCELL-405 for High Five cells).

2. Add 30 ml of viral stock[6] to cell suspension, mix well, and incubate at room temperature for 45 min to 1 hr with periodic agitation every 15 min to avoid cell settling.

3. Aliquot 2.5 ml of cell–virus suspension to 15 T75 tissue culture flasks, each with 8 ml of appropriate media in them. Rock flasks to distribute cells evenly.

4. Incubate flasks at 27° for 43 to 45 hr.

[13] J. J. Li and T. J. Kelly, *Mol. Cell. Biol.* **5,** 1238 (1985).
[14] V. Simanis and D. P. Lane, *Virology* **144,** 88 (1985).
[15] R. A. F. Dixon and D. Nathans, *J. Virol.* **53,** 1001 (1985).

Preparation of Lysate

1. On the previous day, prepare the Sephadex G-25 medium (SG25) (Pharmacia, Piscataway, NJ) by swelling 4 g of beads in 50 ml lysis buffer [0.2 *M* LiCl, 20 m*M* Tris, pH 8.0, 1 m*M* EDTA, 0.5% (v/v) Nonidet P-40 (NP-40)] O/N at room temperature and then place at 4° until use. The beads will hydrate and swell to 20 ml of bead volume. The exclusion limit of SG25 is 5×10^3 Da.
2. Scrape cells gently from the flasks with cell scrapers (Becton Dickinson Labware, Franklin Lakes, NJ) and pipette them into 50-ml conical tubes. Centrifuge the cells at 4° for 10 min at 1000 rpm (150*g*).
3. While cells are spinning, wash the scraped flasks with 20 ml cold PBS–EDTA [1× PBS, pH 7.2 (137 m*M* NaCl, 2.7 m*M* KCl, 4.3 m*M* Na_2HPO_4, 1.4 m*M* KH_2PO_4) and 0.5 m*M* EDTA, pH 8.0] by rinsing out the first flask and then transferring the PBS–EDTA to the next flask. Continue this with all the flasks.
4. Decant the supernatant carefully and resuspend the cells in PBS–EDTA from step 3. This is the first wash. Centrifuge cells at 4° for 10 min at 1000 rpm (150*g*).
5. Wash cell pellet twice with fresh cold PBS–EDTA.
6. Prepare 25 ml of lysis buffer with 1 m*M* dithiothreitol (DTT) and protease inhibitor cocktail (Boehringer Mannheim, Indianapolis, IN) [listed as final concentration: leupeptin, 2 μg/ml; pepstatin, 1 μg/ml; *N*-[*N*-(L-3-*trans*-carboxirane-2-carbonyl)-L-leucyl]agmatine (E-64), 5 μg/ml; Pefabloc SC, 1 m*M*; aprotinin, 2 μg/ml; trypsin inhibitor from soybean, 10 μg/ml; and tosyl-L-phenylalanine chloromethyl ketone (TPCK), 10 μg/ml] on ice during the third and final wash. Add DTT and the protease inhibitor cocktail right before use. Remove the final wash and lyse the cells with lysis buffer. Transfer the lysate to an Oakridge tube (Nalgene, Rochester, NY) and incubate on ice for 25 min.
7. Centrifuge the crude lysate at 4° in a Sorvall SS-34 rotor at 15,000 rpm (27,000*g*) for 30 min.
8. Decant supernatant (crude lysate) to a 50-ml conical tube and record the volume. Save 50 μl of crude lysate for later analysis (Fig. 2A, lane 4; Fig. 2B, lane 3). Add the remaining crude lysate to 4 g of swelled SG25 (step 1) and rotate at 4° for 20 min. To recover T antigen efficiently, pack the Sephadex mixture into a 27-ml (1.5 × 30 cm) flex column (Kontes) and recover the T antigen by saving the eluate (Sephadex lysate). Record the volume and save 50 μl for later analysis (Fig. 2B, lane 4). Alternatively, centrifuge the Sephadex mixture for 10 min at 1000 rpm (200 g) in a Sorvall RT6000B at 4°

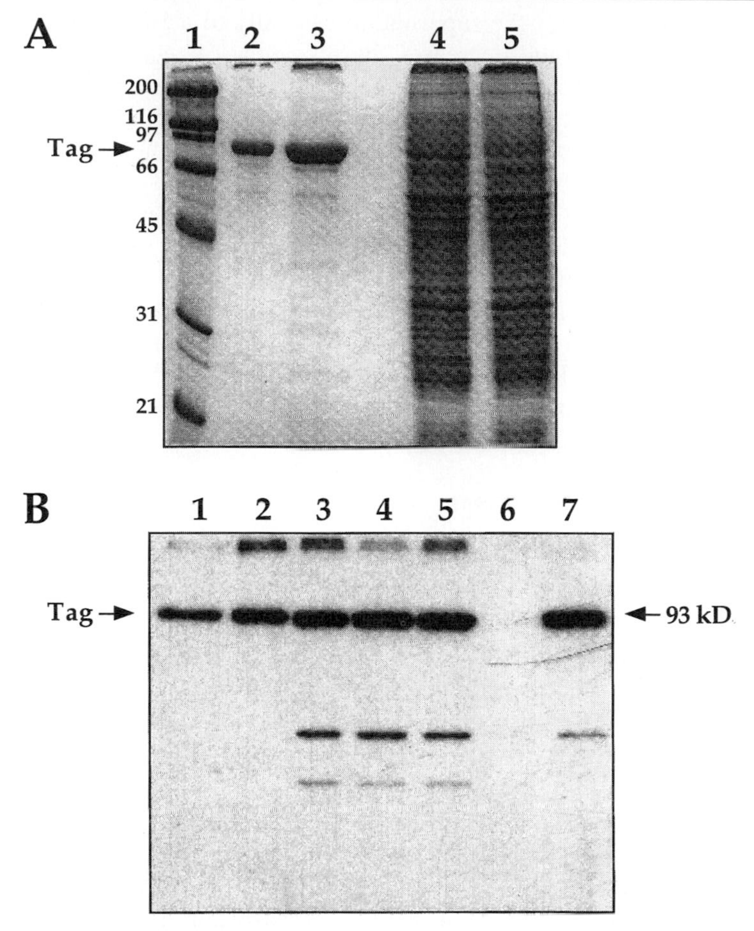

FIG. 2. Analysis of the immunoaffinity purification of T antigen. All samples were obtained as explained in the text. (A) Coomassie blue staining of purified T antigen. Samples were resolved on an 11% SDS–PAGE gel, stained with 0.2% (w/v) Coomassie Brilliant Blue R-250, 9% (v/v) acetic acid, and 30% (v/v) methanol for 20 min, and destained in 9% acetic acid and 30% methanol overnight. Lane 1, molecular weight standards (Bio-Rad); lane 2, purified T antigen (2 μg); lane 3, purified T antigen (5 μg); lane 4, crude lysate (10 μl); and lane 5, flow through (10 μl). The migration of T antigen (Tag) is indicated. Notice that the band corresponding to T antigen in the crude lysate is reduced greatly in the flow through. (B) Western blot analysis of the purification of T antigen. Samples were resolved on an 11% SDS–PAGE gel and were transferred to a polyvinyl difluoride (PVDF) membrane (Millipore, Bedford, MA) overnight. The membrane was immunoblotted with PAb419 and then incubated with an HRP-conjugated goat antimouse antibody. T antigen was visualized by the ECL system (Amersham, Buckinghamshire, England). Lane 1, pure T antigen (50 ng); lane 2, pure T antigen (100 ng); lane 3, crude lysate (5 μl); lane 4, Sephadex lysate (5 μl); lane 5, filtered lysate (5 μl); lane 6, flow through (5 μl); and lane 7, Sephadex beads (20 μl). The position of the T antigen (Tag) and the 93-kDa protein marker is indicated.

and then transfer the supernatant carefully to a 50-ml conical tube. The latter method decreases the efficiency of recovery about 25% due to proteins remaining in the void volume (approximately 30%) of the Sephadex pellet.

9. Filter Sephadex lysate through a 0.45-μm sterile acrodisc (Gelman Sciences, Ann Arbor, MI) connected to a 30-ml syringe over a 50-ml conical tube. After filtering 5–10 ml, the acrodisc will clog and a new filter needs to be attached. In order to avoid sample loss, draw up on the syringe before attaching a new filter. This will pull the lysate in the filter back into the syringe. After filtering, record the volume of filtered lysate and save 50 μl for later analysis (Fig. 2B, lane 5). The filtered lysate is ready to be passed over the immunoaffinity column. We do not freeze or store the lysate; it is used immediately for immunoaffinity purification.

Purification of Tag from Lysate

Filter sterilize (0.2 μm) all solutions that come in contact with the column and add DTT and the protease inhibitor cocktail to the appropriate solutions right before use.

1. During the "Preparation of the Lysate," connect the T antigen immunoaffinity column (see step 7 of "Preparation of Immunoaffinity Column") to a peristaltic pump and equilibrate the column with 30 ml of lysis buffer (without DTT and protease inhibitor cocktail) at a flow rate of 0.5 ml/min.
2. Load the lysate onto the column at a flow rate of 0.2 ml/min. Collect and save the flow through and record the volume (Fig. 2A, lane 5; Fig. 2B, lane 6).
3. Wash column with 25 ml of the first wash buffer [20 m*M* Tris, pH 8.0, 0.5 *M* NaCl, 1 m*M* EDTA, 1% (v/v) NP-40, 10% (v/v) glycerol, 1 m*M* DTT, and protease inhibitor cocktail] and slowly, over the course of 10 min, increase the flow rate of the first wash buffer to 0.66 ml/min. Next, wash the column with 25 ml of the second wash buffer (same as first wash buffer but without NP-40) and then 25 ml of third wash buffer (same as second wash buffer except use Tris, pH 9.0). During the end of the third wash buffer, increase speed slowly to 1 ml/min over a period of 10 min.
4. While the third wash buffer is passing over the column, prepare 10 ml of elution buffer [20 m*M* CAPS, pH 11.0, 0.5 *M* NaCl, 1 m*M* EDTA, 10% (v/v) glycerol, 1 m*M* DTT, and protease inhibitor cocktail]. Test the pH and adjust it to pH 11 accordingly. Also, prepare 5 ml of pH adjustment buffer (PAB) [0.5 *M* Tris, pH 7.0, 10% (v/v)

glycerol, 1 mM EDTA, 1 mM DTT, and protease inhibitor cocktail] and prepare fraction tubes by aliquoting 250 μl of PAB to 10 1.5-ml Eppendorf tubes.

5. Proceed to pass the elution buffer onto the column but stop the pump and the column flow when it is about to enter the column. At this time, remove the top cap of the column and pipette out most of the buffer on top of the packed bed. Align the column over the top of the opened fraction tubes that contain PAB. Add 1.5 ml of elution buffer onto the top of the packed bed. Connect the top cap onto the column, open the column flow, and turn on the pump. Collect 750 μl into each fraction tube (total 1 ml per fraction). After each fraction is collected, mix gently to equilibrate the pH of the solution.
6. After all the fractions have been collected, remove the top cap of the column and pipette out most the buffer on top of the packed bed. Add 2 ml of column storage buffer [1 mM Tris, pH 7.4, 0.15 M NaCl, 1 mM EDTA, 0.02% (w/v) NaN_3]. Flush the column with 40 ml of the same buffer and store the column at 4°.
7. Analyze the fractions for the presence of T antigen with the protein assay dye (Bio-Rad) and pool the fractions that show protein. Dialyze against dialysis buffer [10 mM Tris, pH 8.0, 1 mM EDTA, 0.1 M NaCl, 10% (v/v) glycerol, 1 mM DTT] overnight at 4°.
8. After collecting pure T antigen, the concentration of T antigen is determined by using the protein assay dye (Bio-Rad). Store pure T antigen in small aliquots at −80°.
9. Analyze purification by Coomassie blue staining and Western blot analysis (Fig. 2). The purity of the T antigen is approximately 95%.

Conclusion

We normally recover about 75% of the total T antigen in the lysate yielding 0.5–2 mg of pure T antigen. The flow through of the column can be repassaged over the column if desired to increase the recovery of T antigen.

We assess the biological activity of each T antigen preparation by utilizing a number of biochemical assays,[2] most commonly an ATPase assay.

[18] Applications of *ori*P Plasmids and Their Mode of Replication

By DAVID MACKEY and BILL SUGDEN

Introduction

The function and regulation of mammalian genes can be studied effectively when these genes are contained in vectors maintained as plasmids within mammalian cells. Nonreplicating vectors introduced into cells are present for several days, a time sufficient for short-term assays, but inadequate for long-term study. The origin of plasmid replication (*ori*P) of the human herpesvirus, Epstein–Barr virus (EBV), supports stable replication of plasmids, and the cloned DNAs they contain, in proliferating cells from many species. This article discusses various applications of *ori*P plasmids, including the generation of cell lines stably or inducibly expressing a cloned gene, study of the regulation of transcription from a cloned promoter, study of mutagenesis of a particular target sequence, and isolation of plasmids containing desired cDNAs for which positive selections are available.

Replication of *ori*P requires a single viral *trans*-acting factor, Epstein–Barr nuclear antigen 1 (EBNA1). In cells expressing EBNA1, plasmids containing *ori*P replicate during S phase and segregate to daughter cells efficiently. EBNA1 and *ori*P, therefore, provide a model system for the study of replication and segregation of DNAs in mammalian cells. In order to provide a framework for the informed application of *ori*P plasmids, we also describe our current understanding of their function.

Applications of *ori*P Plasmids

Introduction of a specific sequence of DNA into mammalian cells is frequently useful for its analysis. DNA can be introduced permanently into a cell through its integration into a chromosome. However, the utility of this approach is compromised by position effect variegation (PEV); the site of integration variably affects the activity of the integrated DNA. Integrants must be screened for function of the inserted DNA, and the chromosomal site of insertion is usually unknown. Also, retrieval and isolation of integrated DNA are cumbersome. Plasmids containing *ori*P are maintained extrachromosomally in many cells, and thus the DNA is in a known context, not subject to PEV, and is retrieved easily from cells. Plasmids containing *ori*P and a drug resistance marker are maintained in

0076-6879/99 $30.00

cells propagated under appropriate selection, and the number of plasmids per cell (copy number) is stable. Cell lines selected for drug resistance conferred by *ori*P plasmids maintain various numbers of plasmids, ranging from 1 to approximately 100 per cell,[1–5] copy numbers similar to those for EBV in immortalized cells.[6] However, when propagated in the presence of selection, individual clones maintain their plasmids at a constant copy number.[3,5] The copy number of *ori*P plasmids appears to be a function of the number of DNAs initially introduced into the recipient cell.[4] Introduction of DNA into cells on *ori*P plasmids has been applied adeptly to varied types of research.

Analyses of Gene Expression

A common application of *ori*P plasmids is to express stably a cloned gene in proliferating cells. High levels of reporter genes,[7–10] receptors,[10–13] enzymes,[14–17] glycoproteins,[18,19] secreted proteins,[14,16,19] and viral proteins[18–21] have been expressed stably from *ori*P plasmids and studied. Clones that stably express varying amounts of the gene product can be generated

[1] D. Reisman, J. Yates, and B. Sugden, *Mol. Cell. Biol.* **5,** 1822 (1985).
[2] J. L. Yates, N. Warren, and B. Sugden, *Nature* **313,** 812 (1985).
[3] B. Sugden and N. Warren, *Mol. Biol. Med.* **5,** 84 (1988).
[4] J. L. Yates and N. Guan, *J. Virol.* **65,** 483 (1991).
[5] A. L. Kirchmaier and B. Sugden, *J. Virol.* **69,** 1280 (1995).
[6] L. Sternås, T. Middleton, and B. Sugden, *J. Virol.* **64,** 2407 (1990).
[7] G. Cachianes, C. Ho, R. F. Weber, S. R. Williams, D. V. Goeddel, and D. W. Leung, *BioTechniques* **15,** 255 (1993).
[8] J. Raimond, C. Plessis, F. Fargette, A. Legrand, and F. Herbert, *Plasmid* **32,** 70 (1994).
[9] E. S. Shen, G. M. Cooke, and R. A. Horlick, *Gene* **156,** 235 (1995).
[10] S. Mucke, A. Polack, M. Pawlita, D. Zehnpfennig, N. Massoudi, H. Bohlen, W. Doerfler, G. Bornkamm, V. Diehl, and J. Wolf, *Gene Ther.* **4,** 82 (1997).
[11] Y. Shimizu, B. Koller, D. Geraghty, H. Orr, S. Shaw, P. Kavathas, and R. DeMars, *Mol. Cell. Biol.* **6,** 1074 (1986).
[12] D. Kioussis, F. Wilson, C. Daniels, C. Leveton, J. Taverne, and J. H. Playfair, *EMBO J.* **6,** 355 (1987).
[13] R. A. Horlick, K. Sperle, L. A. Breth, C. C. Reid, E. S. Shen, A. K. Robbins, G. M. Cooke, and B. L. Largent, *Protein Expr. Purif.* **9,** 301 (1997).
[14] J. M. Young, C. Cheadle, J. J. Foulke, W. N. Drohan, and N. Sarver, *Gene* **62,** 171 (1988).
[15] A. Jalanko, J. Pirhonen, G. Pohl, and L. Hansson, *J. Biotechnol.* **15,** 155 (1990).
[16] M. Kolmer, T. Ord, and I. Ulmanen, *J. Biotechnol.* **20,** 131 (1991).
[17] R. S. McIntosh, A. F. Mulcahy, J. M. Hales, and A. G. Diamond, *J. Autoimmun.* **6,** 353 (1993).
[18] A. Jalanko, A. Kallio, M. Salminen, and I. Ulmanen, *Gene* **78,** 287 (1989).
[19] S. A. Jorge, C. Hera, A. M. Spina, R. C. Moreira, J. R. Pinho, and C. F. Menck, *Appl. Microbiol. Biotechnol.* **46,** 533 (1996).
[20] C. Caravokyri and K. N. Leppard, *J. Virol.* **69,** 6627 (1995).
[21] B. Kempkes, D. Spitkovsky, D. P. Jansen, J. W. Ellwart, E. Kremmer, H. J. Delecluse, C. Rottenberger, G. W. Bornkamm, and W. Hammerschmidt, *EMBO J.* **14,** 88 (1995).

and studied. Retroviral RNA expressed from an *ori*P plasmid led to the production of high titers of helper-free virus.[22] Antisense RNAs transcribed from *ori*P plasmids have inhibited protein production from genes to which they are complementary.[23,24] *ori*P plasmids can also be used to establish assays for the function of encoded enzymes. For example, derivatives of an L1 retrotransposon maintained stably on *ori*P plasmids have been used to measure the frequency of retrotransposition into cellular locations mediated by the encoded transposase.[25,26] These examples underscore the utility of *ori*P plasmids as a tool for expressing genes and analyzing their function.

Regulatable expression of a cloned gene is often desirable. Numerous protocols to regulate gene expression in mammalian cells have been described. "Position effects" on integrated, regulatable promoters have hindered their use.[27] The activity of various regulatable promoters on *ori*P plasmids has been appropriately controlled by hormones, metals, and tetracycline.[19,27–29] An estrogen-dependent fusion between a nuclear factor and the ligand-binding domain of the estrogen receptor has been expressed successfully from an *ori*P plasmid.[21] Because *ori*P plasmids containing cassettes for regulatable gene expression are maintained stably in cells, individual clones that have low basal expression and high induced expression can be identified and studied.

Frequently, expression of high levels of a cloned gene product is desirable. High level expression can facilitate detection or purification of the protein, as well as enhance phenotypes induced by it. One portion of *ori*P acts as a transcriptional enhancer to which numerous, but not all, promoters are responsive.[30–33] Those promoters activated by *ori*P are not useful for regulated expression, but will frequently mediate expression of high levels of cloned genes.[7–10,13,14,16,18]

*ori*P plasmids encoding resistance to a drug (Fig. 1) can be used to generate resistant clonal lines in a variety of cells following a standard

[22] T. M. Kinsella and G. P. Nolan, *Hum. Gene. Ther.* **7,** 1405 (1996).

[23] J. E. Hambor, C. A. Hauer, H. K. Shu, R. K. Groger, D. R. Kaplan, and M. L. Tykocinski, *Proc. Natl. Acad. Sci. U.S.A.* **85,** 4010 (1988).

[24] L. Lindenboim, D. Anderson, and R. Stein, *Cell. Mol. Neurobiol.* **17,** 119 (1997).

[25] Q. Feng, J. V. Moran, H. H. J. Kazazian, and J. D. Boeke, *Cell* **87,** 905 (1996).

[26] J. V. Moran, S. E. Holmes, T. P. Naas, R. J. DeBerardinis, J. D. Boeke, and H. H. J. Kazazian, *Cell* **87,** 917 (1996).

[27] M. Jost, C. Kari, and U. Rodeck, *Nucleic Acid. Res.* **25,** 3131 (1997).

[28] C. A. Hauer, R. R. Getty, and M. L. Tykocinski, *Nucleic Acid. Res.* **17,** 1989 (1989).

[29] M. R. James, A. Stary, G. L. Daya, C. Drougard, and A. Sarasin, *Mutat. Res.* **220,** 169 (1989).

[30] B. Sugden and N. Warren, *J. Virol.* **63,** 2644 (1989).

[31] D. Wysokenski and J. L. Yates, *J. Virol.* **63,** 2657 (1989).

[32] T. A. Gahn and B. Sugden, *J. Virol.* **69,** 2633 (1995).

[33] M. T. Puglielli, N. Desai, and S. H. Speck, *J. Virol.* **71,** 120 (1997).

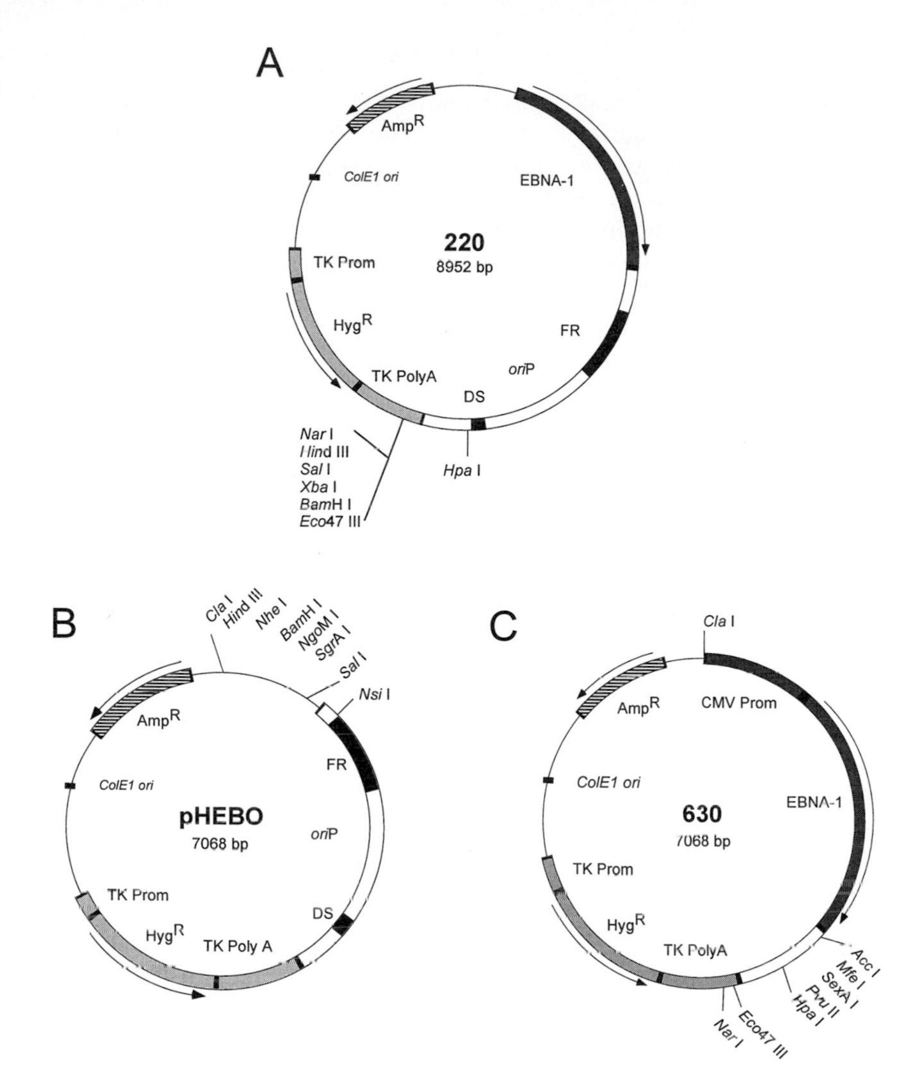

FIG. 1. Plasmids containing *ori*P and/or EBNA1. (A) Plasmid 220 contains sequences from pBR322 for propagation in *Escherichia coli,* the ColE1 origin (white box), and selection with ampicillin, β-lactamase gene (*bla,* hatched box). p220 also contains the *trans*-acting factor, EBNA1 (the expression of which is driven by a fortuitous promoter within pBR322), and the *cis*-acting element, *ori*P (family of repeats, FR, and dyad symmetry element, DS),[1] necessary for its stable replication in proliferating mammalian cells of several species. Finally, p220 contains the hygromycin B phosphotransferase gene that confers resistance to hygromycin (Hyg^R, hatched box),[120] which is flanked by the promoter and polyadenylation sequences of the thymidine kinase gene (TK) of herpes simplex virus type 1. The poly-cloning region inserted into the TK sequences can be used to clone a gene of interest into 220. (B) pHEBO is similar to p220, but lacks EBNA1. In place of EBNA1 are several indicated restriction sites derived from pBR322. (C) Plasmid 630 contains the gene encoding EBNA1 under control of the immediate early promoter of cytomegalovirus (CMV) and lacks *ori*P. p630 also contains sequences conferring resistance to hygromycin and for propagation in *E. coli.* The arrows next to genes indicate the direction of their transcription. Derivatives of these plasmids conferring resistance to G418 are also available.

protocol.[34] First, the lowest concentration of drug to which the recipient cells are completely sensitive needs to be determined for each recipient cell used. This concentration varies (50 to 1000 μg/ml for hygromycin B and 50 to 2500 μg/ml for G418), and cell lines propagated for long times may evolve distinct sensitivities to the drugs. Second, after introduction of the plasmid, the cells are passaged in the presence of the drug. The limiting factor for toxicity of these drugs appears to be the degree of permeability of cells for them. It is necessary to allow the cells time to recover after introduction of DNA, prior to exposing them to drug, particularly if the process of introducing the DNA affects the permeability of the cells. Generation of cell lines in this manner is extremely efficient. When using adherent cells, the number of drug-resistant colonies approaches the number of cells that take up DNA and express a gene from it. Cells grown in the presence of drug will maintain a stable copy number of plasmid for months or longer,[3,5] ample time to characterize and analyze them.

Analyses of Promoter Activity

The activity of a promoter can also be studied on *ori*P plasmids.[35,36] When maintained in cells, promoters on *ori*P plasmids can be studied in the context of chromatin. During "transient" assays with nonreplicating vectors, the chromatin structure of transcriptionally active templates is unknown. Promoters integrated into cellular chromosomes can be studied in the context of chromatin, but the comparison of derivatives of the promoter is difficult or impractical because random integration will result in PEV.[37] Use of cre/lox sites to target integration may homogenize the "position effects" on integrated DNA, but verifying the status of each integrant is labor intensive. Derivatives of a promoter within *ori*P plasmids are maintained in a known context, are subject only to the "position effects" intrinsic to the plasmid, and cells maintaining them can be generated simply.

The chromatin structure on a particular DNA, within an *ori*P plasmid, likely is a function of the sequence of that DNA. The placement of histones on a human immunodeficiency virus (HIV) provirus was determined by its sensitivity to a DNA-cleaving reagent. The pattern of sensitivity for the same HIV–DNA on an *ori*P plasmid was very similar to that of the provirus indicating that the chromatin structure is dictated by the sequence of the

[34] B. Sugden, K. Marsh, and J. Yates, *Mol. Cell. Biol.* **5,** 410 (1985).
[35] D. L. Clemens and J. O. Carlson, *J. Virol.* **63,** 2737 (1989).
[36] O. Mazda, K. Teshigawara, S. Fujimoto, N. Hattori, Y. M. Dou, and Y. Katsura, *J. Immunol. Methods* **169,** 53 (1994).
[37] M. Emerman and H. M. Temin, *Cell* **39,** 449 (1984).

DNA.[38] If this observation proves to be general, *ori*P plasmids should be useful vectors to study both the *cis*-acting elements that define chromatin structure and the effects of chromatin structure on promoter activity.

Additional Applications of oriP Plasmids

*ori*P plasmids can be used to isolate cDNAs that confer a selective advantage to cells (Fig. 2A). The selective advantage can be survival,[39–41] increased proliferation under defined growth conditions, increased adherence to a given substratum, or other positive selections. Alternatively, the cDNA can allow cells to be isolated by sorting based on phenotypic changes at the cell surface, e.g., expression of a protein[39,42] or binding of a ligand.[43] This approach uses *ori*P plasmids containing a cDNA library in an expression cassette.[44,45] Plasmids containing the library are introduced into cells, and the selection is applied. As the selected cells are propagated to increase their number, the *ori*P plasmids they contain are similarly increased in number. Plasmid DNA is then isolated from the cells,[46] introduced into *Escherichia coli,* further expanded, and characterized. If necessary, the rescued plasmids are reintroduced into mammalian cells and the process is repeated. In each cycle, plasmids with "positive" cDNAs are enriched through selection and are expanded through replication in cell culture. By repeating the procedure multiple times, rare cDNAs that provide only a moderate selective advantage can be exponentially enriched and identified (Fig. 2B).

The action of a protein that kills cells or limits their proliferation can be studied quantitatively using *ori*P plasmids.[47,48] An *ori*P plasmid, expressing a selectable marker and the gene of interest, is introduced into cells that are plated at limiting dilutions in multiwell dishes. After a recovery period, the cells are maintained in growth conditions under selection. After a defined time, wells are scored as positive or negative based on whether they contain

[38] S. A. Stanfield-Oakley and J. D. Griffith, *J. Mol. Biol.* **256,** 503 (1996).

[39] R. F. Margolskee, P. Kavathas, and P. Berg. *Mol. Cell. Biol.* **8,** 2837 (1988).

[40] P. B. Belt, W. Jongmans, J. de Wit, J. H. Hoeijmakers, P. van de Putte, and C. Backendorf, *Nucleic Acid. Res.* **19,** 4861 (1991).

[41] R. Legerski and C. Peterson, *Nature* **359,** 70 (1992).

[42] L. C. Pan, R. F. Margolskee, and H. M. Blau, *Somat. Cell Mol. Genet.* **18,** 163 (1992).

[43] R. A. Heller, K. Song, D. Villaret, R. Margolskee, J. Dunne, H. Hayakawa, and G. M. Ringold, *J. Biol. Chem.* **265,** 5708 (1990).

[44] P. B. Belt, H. Groeneveld, W. J. Teubel, P. van de Putte, and C. Backendorf, *Gene* **84,** 407 (1989).

[45] C. Peterson and R. Legerski, *Gene* **107,** 279 (1991).

[46] B. Hirt, *J. Mol. Biol.* **26,** 365 (1967).

[47] W. Hammerschmidt, B. Sugden, and V. R. Baichwal, *J. Virol.* **63,** 2469 (1989).

[48] A. Kaykas and B. Sugden, manuscript in preparation.

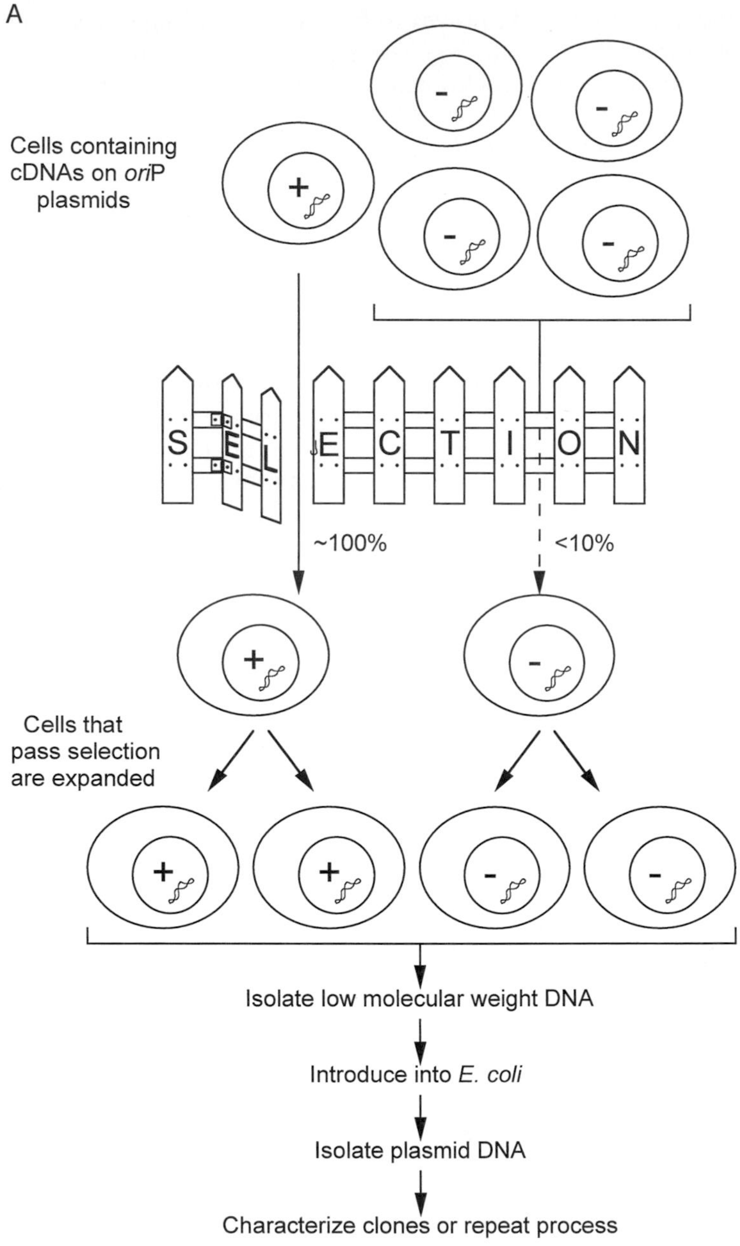

A

Cells containing cDNAs on *ori*P plasmids

+

-

-

-

-

S E L E C T I O N

~100%

<10%

+

-

Cells that pass selection are expanded

+

+

-

-

Isolate low molecular weight DNA

Introduce into *E. coli*

Isolate plasmid DNA

Characterize clones or repeat process

B

$$\text{Frequency (f)} = \frac{\text{positive cDNAs}}{\text{total cDNAs}}$$

		Percentage of cells containing negative cDNAs that pass selection (false positives)		
		5	0.5	0.05
Number of cycles	0	1×10^{-6}	1×10^{-6}	1×10^{-6}
	1	2×10^{-5}	2×10^{-4}	2×10^{-3}
	2	4×10^{-4}	4×10^{-2}	~1
	3	8×10^{-3}	~1	
	4	0.16		

FIG. 2. Isolation of cDNAs whose expression allows positive selection in mammalian cells. (A) Schematic representation of the process by which selected cDNAs are isolated. *ori*P plasmids from which cDNAs and drug resistance are expressed are introduced into cells, and those cells harboring plasmids are selected with drug. A positive selection is applied to that population of cells, and the cells that pass selection are expanded. The number of *ori*P plasmids increases within the expanded pool of cells, and those plasmids are isolated and reintroduced into *E. coli.* The identity of cDNAs can then be determined or the process can be repeated to decrease the frequency of "false positives." (B) Relationship between the stringency of selection, number of cycles of selection, and the frequency of positive cDNAs among the pool of cDNAs. Frequency (f) is defined as the fraction of positive cDNAs among the total pool of cDNAs. The theoretical value of f is shown for zero to four rounds of selection in which 0.05–5% of cells will nonspecifically pass selection. For all calculations, f of the initial pool of cDNA-containing plasmids is 1 in one million.

more or less viable cells than a defined background number. By determining the frequency with which "positive" growth occurs in wells that received various numbers of cells, this assay can been used to quantify the cytostatic or toxic affect of a protein. The assay can similarly be used to study the inhibition of cellular proliferation by derivatives of the protein or the affects of other variables, such as the presence of soluble factors in the growth medium, on the activity of the protein.

Specific alterations to DNA, resulting from various mutagenic treatments, have been identified and enumerated with *ori*P plasmids. Mutagenic treatments have been performed *in vitro,* and the treated plasmids introduced subsequently into cells,[49] or *in vivo,* for example, by exposing cells that stably maintain the plasmid to UV light.[50] The plasmids can then be isolated from cells,[46] rescued in *E. coli,* and the sequence of individual plasmids determined. Because *ori*P plasmids are replicated by cellular machinery synchronously with cellular DNA, they are subject to normal cellular proofreading and repair. It has been shown that the mutational rates and specificities within *ori*P plasmids are identical to those on cellular chromosomes.[51] Thus, mutational rates and specificities of various treatments on DNA can be measured with *ori*P plasmids.

Yeast artificial chromosomes (YACs) replicate as plasmids in yeast. YACs that contain *ori*P (*ori*PYACs) have been characterized. *ori*PYACs ranging from 90 to 660 kbp were maintained intact after introduction into human cells.[52,53] These plasmids can be shuttled between yeast and human cells. Powerful cloning techniques available in yeast can thus be used to manipulate large plasmids studied in mammalian cells.

*ori*P Plasmids also mediate long-term expression of cloned genes in the absence of drug selection.[54] *ori*P plasmids establish themselves efficiently when introduced into cells that express EBNA1.[45] This characteristic is demonstrated by our observation that greater than 50% of 143/EBNA1[55] cells, which, by electroporation, received an *ori*P plasmid conferring resistance to G418 grow out to yield resistant clones.[56] The rate of loss of *ori*P plasmids in the absence of selection is low (2–4% per cell per generation[3,5]). From cells in which *ori*P plasmids are established, approximately 50% of their descendants still contain *ori*P plasmids after 20 cell generations. Therefore, "transient" assays can be conducted over much longer time scales than is typically possible with nonreplicating vectors.

[49] C. A. Ingle and N. R. Drinkwater, *Mutat. Res.* **220,** 133 (1989).

[50] A. Calcagnile, Z. T. Basic, F. Palombo, and E. Dogliotti, *Nucleic Acid. Res.* **24,** 3005 (1996).

[51] N. R. Drinkwater and D. K. Klinedinst, *Proc. Natl. Acad. Sci. U.S.A.* **83,** 3402 (1986).

[52] K. Simpson and C. Huxley, *Nucleic Acid. Res.* **24,** 4693 (1996).

[53] K. Simpson, A. McGuigan, and C. Huxley, *Mol. Cell. Biol.* **16,** 5117 (1996).

[54] J. G. Wohlgemuth, S. H. Kang, G. H. Bulboaca, K. A. Nawotka, and M. P. Calos, *Gene Ther.* **3,** 503 (1996).

[55] T. Middleton and B. Sugden, *J. Virol.* **68,** 4067 (1994).

[56] D. Mackey and B. Sugden, unpublished observation, 1998.

Function of oriP Plasmids Requires EBNA1

All the applications just described require the long-term function of *ori*P and therefore require expression of EBNA1.[2] The ability of *ori*P to act as a transcriptional enhancer also requires EBNA1.[57] EBNA1 can be expressed from a variety of sources. Cells immortalized by EBV express EBNA1 from the viral genome and can support replication of additional *ori*P plasmids.[1] A variety of cell lines that express EBNA1 from integrated DNA (e.g., plasmid 630, Fig. 1) are also available. Generation of such cell lines is straightforward and has been facilitated by the development of a plasmid from which EBNA1 and a selectable marker are expressed from a dicistronic mRNA.[58] When appropriate cell lines expressing EBNA1 are not available, EBNA1 can be expressed from the plasmid containing *ori*P[2] (e.g., plasmid 220, Fig. 1). Plasmids containing *ori*P function similarly in cells in which EBNA1 is provided in each of these ways. However, *ori*P-positive transformants arise with the highest efficiency when the recipient cells stably express EBNA1.[45]

A Functional Overview of *ori*P

Plasmids containing *ori*P can replicate in human, primate, feline, and canine, but not in rodent cells.[2,35] EBNA1 is the only Epstein–Barr viral protein necessary for the replication and segregation of *ori*P plasmids; all other factors are provided by the cell.[2,59] When EBNA1 is provided *in trans*, *ori*P plasmids replicate during S phase[4,60] and segregate to daughter cells efficiently.[3,5] *ori*P is composed of two clusters of binding sites for EBNA1[61] (Fig. 3). The family of repeats (FR) contains 20 imperfect copies of a 30-bp repeat. Within each of these 20 copies is an 18-bp palindrome to which dimers of EBNA1 bind with high affinity ($K_D = 10^{-9}$).[62–64] Located approximately 1 kbp away from FR are four sites that EBNA1 binds with lower affinity than the sites in FR.[65] Two of these four sites are part of an

[57] D. Reisman and B. Sugden, *Mol. Cell. Biol.* **6,** 3838 (1986).

[58] A. D. Ramage, A. J. Clark, A. G. Smith, P. S. Mountford, and D. W. Burt, *Gene* **197,** 83 (1997).

[59] S. Lupton and A. J. Levine, *Mol. Cell. Biol.* **5,** 2533 (1985).

[60] A. Adams, *J. Virol.* **61,** 1743 (1987).

[61] D. R. Rawlins, G. Milman, S. D. Hayward, and G. S. Hayward, *Cell* **42,** 859 (1985).

[62] R. F. Ambinder, W. A. Shah, D. R. Rawlins, G. S. Hayward, and S. D. Hayward, *J. Virol.* **64,** 2369 (1990).

[63] L. Frappier and M. O'Donnell, *J. Biol. Chem.* **266,** 7819 (1991).

[64] T. Middleton and B. Sugden, *J. Virol.* **66,** 489 (1992).

[65] C. H. Jones, D. Hayward, and D. R. Rawlins, *J. Virol.* **63,** 101 (1989).

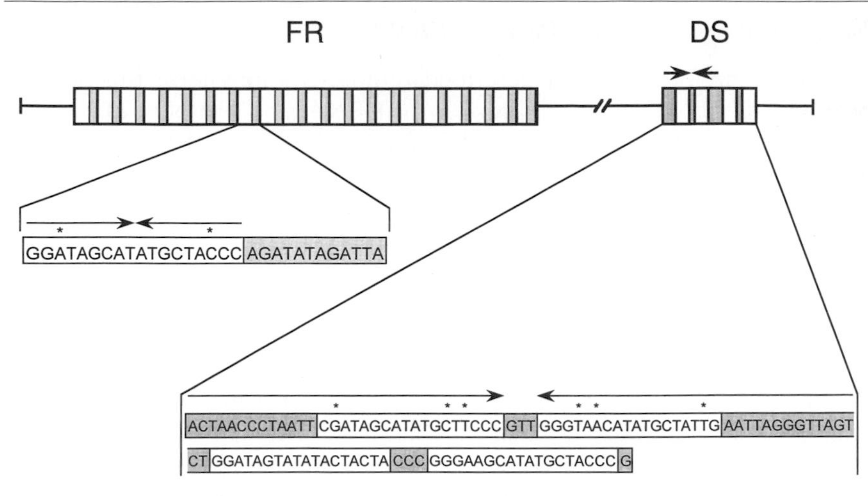

FIG. 3. Schematic representation of the Epstein–Barr virus origin of plasmid replication, *ori*P. *ori*P is composed of two distinct *cis*-acting elements: the family of repeats (FR) and the dyad symmetry element (DS). DS is an EBNA1-independent origin of DNA synthesis and FR is an EBNA1-dependent enhancer of transcription. Both FR and DS are required for stable replication of plasmids containing *ori*P in cells expressing EBNA1. FR contains 20 imperfect copies of a 30-bp repeat. Within the repeated unit is a sequence of 18 bp to which EBNA1 binds with high affinity (white boxes). The nucleotide sequence of one of the repeats is shown below, the palindromic nature of the EBNA1-binding site is indicated by the arrows above, and the mismatches within that palindrome are marked with asterisks. DS, located approximately 1 kbp from FR, contains four sites to which EBNA1 binds with a lower affinity than the sites in FR. Two of the binding sites within DS are part of an imperfect 65 base dyad, indicated by the arrows. The sequence of DS is shown below with the EBNA1-binding sites boxed in white, the 65-bp dyad indicated by arrows, and the mismatches within that dyad marked with asterisks.

extended palindrome, or dyad, for which the dyad symmetry element (DS) was named.[1] This section discusses current understanding of long-term replication of *ori*P, short-term replication of *ori*P, and the contribution(s) of EBNA1 to the replication of *ori*P.

Long-Term Replication of oriP

The long-term replication of *ori*P requires both FR and DS *in cis;* deletion of either of these elements renders *ori*P nonfunctional.[1,59] The functions of FR and DS appear distinct. It was determined by two-dimensional gel electrophoresis that the site of initiation of DNA synthesis in an *ori*P plasmid is within or near DS.[66] This observation, and the presence of

[66] T. A. Gahn and C. L. Schildkraut, *Cell* **58,** 527 (1989).

an extended dyad in DS, led to the hypothesis that DS is the origin of DNA synthesis. FR is an auxiliary element required *in cis* for the long-term replication of DS[1,59]; FR also functions independently as an EBNA1-dependent transcriptional enhancer.[57] Together FR and DS compose a *cis*-acting element, *ori*P, which allows efficient replication and segregation of plasmids in mammalian cells.

A single, viral *trans*-acting factor, EBNA1, is required for the long-term function of *ori*P.[2,59] All other factors contributing *in trans* to the activity of *ori*P are provided by the cell. The inability of *ori*P plasmids to replicate in rodent cells presumably stems from an incompatibility between a rodent replication factor (EBNA1 can activate transcription in rodent cells[31]) and *ori*P, EBNA1, or both. An EBNA1 gene within latently replicating EBV, integrated into a cellular chromosome or on an *ori*P plasmid itself, can support stable replication of an *ori*P plasmid. Possible contributions of EBNA1 at *ori*P are discussed later.

Replication and segregation of *ori*P plasmids in cells are efficient. Cells harboring an *ori*P plasmid containing a selectable marker will, when grown under selective pressure, maintain the plasmid at a stable copy number for extended periods of time.[5] When such cells are grown in the absence of selection the rate of loss of the plasmid is low; 2–4% of cells per generation lose the plasmid.[3,5] This rate of loss is the same for cells with copy numbers ranging from 1 to 35.[5] If loss of plasmid from cells resulted in a gradually diminishing copy number, the appearance of *ori*P-negative cells would decrease with increasing copy number; it does not. Therefore, it is likely that the loss of plasmid(s) from a cell is an all-or-nothing event. The mechanism by which daughter cells either receive the appropriate complement of *ori*P plasmid(s) or do not receive plasmid is enigmatic.

The rate of loss of *ori*P plasmids is similar to that of ARS–CEN plasmids in *Saccharomyces cerevisiae*.[67,68] ARS–CEN plasmids contain two *cis*-acting elements: an origin of replication (ARS) that mediates the initiation of DNA synthesis during each S phase and a centromere (CEN) that mediates the segregation of replicated plasmids by the mitotic spindle during M phase. The function of each of these *cis*-acting elements requires *trans*-acting factors provided by the yeast. *ori*P contains an origin of replication that functions analogously to an ARS. This origin of replication, DS, functions independently of EBNA1.[69] However, *ori*P does not behave as if it contains a CEN-like element. Plasmids that contain two CEN elements are unstable, presumably because they are pulled apart by the mitotic

[67] P. Hieter, C. Mann, M. Snyder, and R. W. Davis, *Cell* **40,** 381 (1985).

[68] D. Koshland, J. C. Kent, and L. H. Hartwell, *Cell* **40,** 393 (1985).

[69] A. Aiyar, C. Tyree, and B. Sugden, *EMBO J.* **17,** 6394 (1998).

spindle.[70,71] In contrast, plasmids that contain one or two copies of *ori*P are equally stable in mammalian cells.[3] The mechanism by which *ori*P and EBNA1 act together to support the maintenance of *ori*P plasmids in proliferating cells is unknown, but known biochemical activities of EBNA1 provide some insights into this process.

Short-Term Replication of oriP

The initiation of DNA synthesis mediated by *ori*P does not require EBNA1. Plasmids containing *ori*P or DS alone can undergo at least two rounds of DNA synthesis following their introduction into human cells.[69] Two independent types of experiments demonstrate DNA synthesis of *ori*P in the absence of EBNA1. Plasmid DNA isolated from *dam*$^+$ *E. coli* is sensitive to the restriction endonuclease *Dpn*I. In the absence of EBNA1, progeny plasmids containing *ori*P or DS, which are *Dpn*I resistant (and therefore are the product of two rounds of semiconservative replication), can be detected 48 hr after the introduction into human cells. At this time point, *Dpn*I-resistant plasmids containing *ori*P or DS synthesized in the absence of EBNA1 accumulate to levels equivalent to those found in the presence of EBNA1, levels 10- to 30-fold greater than those measured for plasmids lacking *ori*P.[69] Similar short-term studies using density shift experiments with BrdU also demonstrate that *ori*P can direct at least two rounds of semiconservative replication in the absence of EBNA1 within 60 hr posttransfection.[69] In these studies, approximately 40% as much plasmid containing *ori*P was found in the "heavy/heavy" peak in the absence of EBNA1 as in its presence. Significantly less "heavy/heavy" DNA was detected for plasmids lacking *ori*P than for those containing it. These experiments demonstrate independently that, in the absence of EBNA1, cellular factors recognize *ori*P and mediate replication of plasmids containing it.

In the absence of EBNA1, plasmids containing *ori*P do not persist in cells. By 96 hr after their introduction, plasmids that contain *ori*P are being lost from cells in the absence of EBNA1; the level of *Dpn*I-resistant DNA present in the absence of EBNA1 can be $\leq$5% of that in the presence of EBNA1.[1,69,72] In EBNA1-positive cells, *ori*P plasmids are maintained at the same number of copies per transfected cell at 96 hr as they were at 48 hr.[69] Therefore, EBNA1 is not required for the first two rounds of DNA synthesis of *ori*P plasmids in proliferating cells, but is required for the persistence and continued replication of those plasmids during the next few cell cycles and thereafter.

[70] C. Mann and R. W. Davis, *Proc. Natl. Acad. Sci. U.S.A.* **80,** 228 (1983).

[71] D. Koshland, L. Rutledge, M. Fitzgerald-Hayes, and L. H. Hartwell, *Cell* **48,** 801 (1987).

[72] A. L. Kirchmaier and B. Sugden, *J. Virol.* **71,** 1766 (1997).

EBNA1 contributes to persistence and continued replication of *ori*P plasmids through FR. In the first 48 hr after introduction into cells containing EBNA1, plasmids that contain only DS replicate approximately as well as those that contain all of *ori*P.[69] By 96 hr after introduction in the presence of EBNA1, plasmids that contain only DS are undetectable, whereas plasmids containing all of *ori*P are present at levels equivalent to those seen at 48 hr.[69] In the absence of EBNA1, plasmids containing either DS or *ori*P are undetectable by 96 hr. These observations indicate that FR is a *cis*-acting element, which, in the presence of EBNA1, mediates the persistence and continued replication of plasmids containing *ori*P. This hypothesis is supported by experiments demonstrating that FR *in cis* and EBNA1 *in trans* can mediate the persistent replication of plasmids that contain *cis*-acting elements, other than DS, that function as origins of DNA synthesis.[73–75] Whether the EBNA1-binding sites in DS also contribute *in cis* to the continued replication of plasmids containing *ori*P is unknown. DS alone is an origin of DNA synthesis that does not require EBNA1, and FR and EBNA1 are sufficient for the maintenance of plasmids containing other origins of DNA synthesis. However, DS has evolved to contain multiple binding sites for EBNA1, indicating that EBNA1 likely contributes to its function. The long-term replication of plasmids containing *ori*P requires the action of EBNA1 at FR; however, it is likely that the remarkable efficiency and stability with which *ori*P plasmids replicate in cells are the result of the action of EBNA1 at both FR and DS.

Functional Contributions of EBNA1 to oriP

EBNA1 from the prototypical B95-8 strain of EBV has 641 amino acids[76] and is depicted schematically in Fig. 4. EBNA1 is expressed in all known cells infected latently with EBV, an observation consistent with its requirement for function of the latent origin of replication of the virus, *ori*P.

Dimers of EBNA1 bind DNA at specific, degenerate, 18-bp palindromes.[61,62,77] The domain sufficient for dimerization and specific DNA binding lies within the carboxy-terminal one-third of EBNA1[78,79] (Fig. 4).

[73] P. J. Krysan, S. B. Haase, and M. P. Calos, *Mol. Cell. Biol.* **9,** 1026 (1989).

[74] P. J. Krysan and M. P. Calos, *Mol. Cell. Biol.* **11,** 1464 (1991).

[75] A. L. Kirchmaier and B. Sugden, *J. Virol.* **72,** 4657 (1998).

[76] R. Baer, A. T. Bankier, M. D. Biggin, P. L. Deininger, P. J. Farrell, T. J. Gibson, G. Hatfull, G. S. Hudson, S. C. Satchwell, C. Séguin, P. S. Tuffnell, and B. G. Barrell, *Nature* **310,** 207 (1984).

[77] G. Milman and E. S. Hwang, *J. Virol.* **61,** 465 (1987).

[78] R. F. Ambinder, M. Mullen, Y.-N. Chang, G. S. Hayward, and S. D. Hayward, *J. Virol.* **65,** 1466 (1991).

[79] N. Inoue, S. Harada, T. Honma, T. Kitamura, and K. Yanagi, *Virology* **182,** 84 (1991).

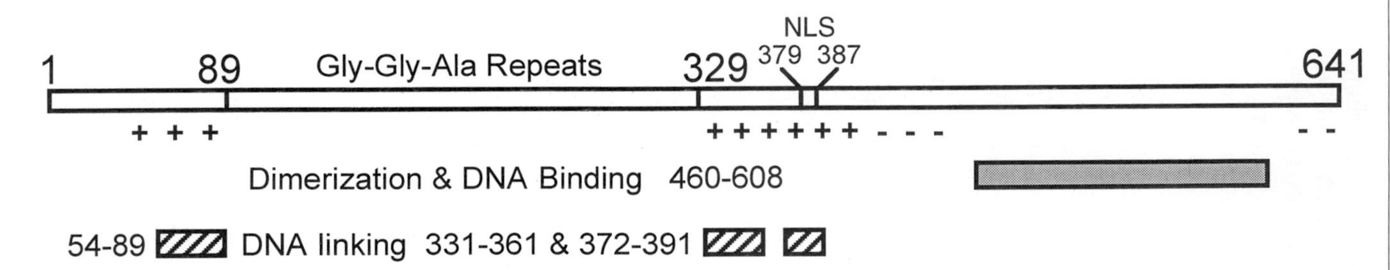

FIG. 4. Schematic representation of the Epstein–Barr viral nuclear antigen 1, EBNA1. EBNA1 from the B95-8 strain of EBV is 641 amino acids in length, including an internal region consisting entirely of glycines and alanines arranged in a repeated array (Gly-Gly-Ala repeats), which appears to prevent its recognition by cytotoxic T lymphocytes,[121] but is dispensable for its functions in cell culture.[2] The carboxy-terminal portion of EBNA1 contains amino acids sufficient for its dimerization and DNA binding (stippled box). Amino acids 379 to 387 have been shown to function as a nuclear localization signal (NLS).[78] Multiple independent domains of EBNA1, found on each side of the Gly-Gly-Ala repeats, can mediate DNA linking (striped boxes). Regions rich in basic residues (+), which correlate with regions with DNA-linking activity, and acidic residues (−) are indicated.

The structure of dimers of this domain by itself and bound to DNA has been determined by crystallography.[80,81] Interestingly, the "core" of this structure is related closely to that of the dimerization and DNA-binding domain of the E2 protein of bovine papillomavirus (BPV),[82] although the proteins do not share sequence identity. E2 contributes to the latent replication of BPV and regulates viral transcription.[83–85] The conservation of structure between the dimerization and DNA-binding domains of E2 and EBNA1 indicates that this structure likely contributes specifically to the common activities of the proteins. Twenty-six binding sites for EBNA1 are found in EBV, 24 of which comprise *ori*P.[61] Through binding to these sites, EBNA1 mediates the latent replication of EBV[2] and regulates transcription from several viral promoters.[30,32] The ability of EBNA1 to activate transcription and support replication is dependent on its ability to bind to DNA.[86–88]

The ability of EBNA1 to bind to DNA is not sufficient for the activation of transcription through FR or support of replication of *ori*P; the dimerization and DNA-binding domain alone mediates neither of these activities.[72,89] Therefore, the amino-terminal two-thirds of EBNA1 contribute to these functions. What biochemical activities, in addition to binding to DNA, does EBNA1 provide *ori*P? Proteins that bind the origins of DNA replication of several DNA viruses have intrinsic DNA-dependent ATPase and helicase activities, including T antigen of SV40,[90] E1 of BPV,[91] and UL9 of herpes simplex virus (HSV).[92] These activities have not been detected in preparations of EBNA1 purified from insect and monkey cells.[63,64] Interaction assays conducted in yeast have identified multiple candidate proteins that interact with EBNA1; however, the significance of association of these proteins with EBNA1 in mammalian cells has yet to be established. One

[80] A. Bochkarev, J. A. Barwell, R. A. Pfuetzner, W. J. Furey, A. M. Edwards, and L. Frappier, *Cell* **83,** 39 (1995).

[81] A. Bochkarev, J. A. Barwell, R. A. Pfuetzner, E. Bochkareva, L. Frappier, and A. M. Edwards, *Cell* **84,** 791 (1996).

[82] R. S. Hedge, S. R. Grossman, L. A. Laimins, and P. B. Sigler, *Nature* **359,** 505 (1992).

[83] M. Ustav and A. Stenlund, *EMBO J.* **10,** 449 (1991).

[84] L. Yang, R. Li, I. J. Mohr, R. Clark, and M. R. Botchan, *Nature* **353,** 628 (1991).

[85] B. A. Spalholz, Y. C. Yang, and P. M. Howley, *Cell* **42,** 183 (1985).

[86] M. Polvino-Bodnar, J. Kiso, and P. A. Schaffer, *Nucleic Acid. Res.* **16,** 3415 (1988).

[87] J. L. Yates and S. M. Camiolo, *Cancer Cells* **6,** 197 (1988).

[88] M. Polvino-Bodnar and P. A. Schaffner, *Virology* **187,** 591 (1992).

[89] T. Middleton and B. Sugden, *J. Virol.* **66,** 1795 (1992).

[90] H. Stahl, P. Droge, and R. Knippers, *EMBO J.* **5,** 1939 (1986).

[91] L. Yang, I. Mohr, E. Fouts, D. A. Lim, M. Nohaile, and M. Botchan, *Proc. Natl. Acad. Sci. U.S.A.* **90,** 5086 (1993).

[92] D. S. Fierer and M. D. Challberg, *J. Virol.* **66,** 3986 (1992).

of these proteins, Rch1, is a component of the nuclear import apparatus.[93] This interaction could contribute to the maintenance of DNAs containing EBNA1-binding sites in cells[55] by localizing EBNA1, and DNAs to which it is bound, to the nucleus. The nuclear pore complex mediates the localization of adenoviral DNA to the nucleus during infection.[94] EBNA1 has also been shown to associate with gC1qR,[95] a protein identified originally as a cell surface receptor for the complement component, C1q.[96] Additional activities have been proposed for gC1qR, some of which place it in the nucleus,[97,98] possibly associated with the nuclear matrix.[99] The fragment of EBV containing *ori*P has MAR (nuclear matrix attachment region) activity[100]; its matrix association may be mediated by the interaction of EBNA1 with gC1qR. Further experimentation is needed to demonstrate whether the interactions identified in yeast contribute to the function of EBNA1.

EBNA1 physically associates DNAs bound site specifically by it.[64] Multiple, independent regions of EBNA1 have been shown to contribute to this activity, termed DNA linking[101,102] (Fig. 4). These regions lie outside the dimerization and DNA-binding domain and can function when fused to the unrelated dimerization and DNA-binding domain of GAL4. Linking of DNAs bound by ENBA1 is mediated by protein : protein interactions between the linking domains of those EBNA1 molecules.[103] When two regions of the same DNA molecule are bound and linked by EBNA1, as is the case for *ori*P, the intervening DNA is looped.[104,105] For several reasons, DNA linking is likely to contribute to the function of EBNA1. First, several cellular and viral DNA-binding proteins that activate transcription and/or

[93] N. Fischer, E. Kremmer, G. Lautscham, L. N. Mueller, and F. A. Grasser, *J. Biol. Chem.* **272,** 3999 (1997).

[94] U. F. Greber, M. Suomalainen, R. P. Stidwill, K. Boucke, M. W. Ebersold, and A. Helenius, *EMBO J.* **16,** 5998 (1997).

[95] Y. Wang, J. E. Finan, J. M. Middeldorp, and S. D. Hayward, *Virology* **236,** 18 (1997).

[96] E. I. Peerschke, K. B. Reid, and B. Ghebrehiwet, *J. Immunol.* **152,** 5896 (1994).

[97] Y. Luo, H. Yu, and B. M. Peterlin, *J. Virol.* **68,** 3850 (1994).

[98] L. Yu, Z. Zhang, P. M. Loewenstein, K. Desai, Q. Tang, D. Mao, J. S. Symington, and M. Green, *J. Virol.* **69,** 3007 (1995).

[99] T. B. Deb and K. Datta, *J. Biol. Chem.* **271,** 2206 (1996).

[100] S. Jankelevich, J. L. Kollman, J. W. Bodnar, and G. Miller, *EMBO J.* **11,** 1165 (1992).

[101] A. Laine and L. Frappier, *J. Biol. Chem.* **270,** 30914 (1995).

[102] D. Mackey, T. Middleton, and B. Sugden, *J. Virol.* **69,** 6199 (1995).

[103] D. Mackey and B. Sugden, *J. Biol. Chem.* **272,** 29873 (1997).

[104] W. Su, T. Middleton, B. Sugden, and H. Echols, *Proc. Natl. Acad. Sci. U.S.A.* **88,** 10870 (1991).

[105] L. Frappier and M. O'Donnell, *Proc. Natl. Acad. Sci. U.S.A.* **88,** 10875 (1991).

replication, including Sp1,[106] E2 of BPV,[107,108] and UL9 of HSV,[109] link DNA. Second, a significant portion of EBNA1, other than residues involved in its dimerization and DNA binding and the Gly-Gly-Ala repeats, is composed of regions with linking activity[102] (Fig. 4). Third, no small deletion within EBNA1, other than those that affect DNA binding, abrogates the ability of EBNA1 to activate replication or transcription.[87] This finding has been interpreted to indicate that EBNA1 contains redundant activating domains. The linking domains are redundant and therefore are fitting candidates for these activating domains. Finally, structure–function analysis of EBNA1 demonstrates a role for linking in the functions of EBNA1.[110] Derivatives of EBNA1 were constructed with various deletions and duplications of the linking domains. The ability of these derivatives to link DNA and to support replication and transcription were strongly correlated. A derivative of EBNA1 lacking all three linking domains fails to activate transcription or support replication; in fact, this derivative dominantly negatively inhibits these activities of wild-type EBNA1.[72] Therefore, DNA linking by EBNA1 is likely to contribute to its activation of transcription and replication.

Several models can explain a role for the ability of EBNA1 to bind to and link DNA in its activation of transcription and replication. The role of DNA linking may be structural, i.e., linking may create a structure that permits additional events to occur. When bound to FR, EBNA1 can activate transcription from a promoter located 10 kbp away.[32] Interaction between linking domains of EBNA1 at FR and protein(s) associated with DNA near the promoter may loop FR to the region of the promoter it is activating. Precedence for this notion comes from the observation that Jun homodimers loop DNA between their binding sites and transcriptional preinitiation complexes (PICs); Jun is necessary for the formation of those PICs and the *in vitro* activity of the promoter.[111] If FR loops to the promoters it activates, EBNA1 could modulate the factor with which it interacts or the loop could alter the structure of chromatin at the promoter. Alternatively, looping may bring multiple copies of EBNA1 into the proximity of the promoter, allowing EBNA1 itself, or factors associated with it, to modulate transcription. Models that posit a structural role of FR linked to a promoter

[106] W. Su, S. Jackson, R. Tjian, and H. Echols, *Gene Dev.* **5,** 820 (1991).

[107] J. D. Knight, R. Li, and M. Botchan, *Proc. Natl. Acad. Sci. U.S.A.* **88,** 3204 (1991).

[108] R. Li, J. D. Knight, S. P. Jackson, R. Tjian, and M. R. Botchan, *Cell* **65,** 493 (1991).

[109] A. Koff, J. F. Schwedes, and P. Tegtmeyer, *J. Virol.* **65,** 3284 (1991).

[110] D. Mackey and B. Sugden, *Mol. Cell. Biol.,* in press.

[111] J. C. Becker, A. Nikroo, T. Brabletz, and R. A. Reisfeld, *Proc. Natl. Acad. Sci. U.S.A.* **92,** 9727 (1995).

in the activation of that promoter are paralleled by models in which FR linked to DS contributes to the function of DS.[104,112]

Linking between FR and DS may modulate the initiation of DNA synthesis dependent on DS. Plasmids containing *ori*P introduced into cells in the absence of EBNA1 replicate for a few cell cycles, but their presence is transient.[69] The loss of these plasmids may be due, in part, to a lack of initiation of DNA synthesis in subsequent cell cycles. Packaging of the plasmids into chromatin could preclude recognition of DS by cellular factors and the resulting initiation of DNA synthesis. Linking between EBNA1 bound to FR and DS increases the apparent affinity of EBNA1 for DS.[104,112] By stabilizing EBNA1 on DS or through forming a looped structure between FR and DS, the chromatin at DS may be maintained in a state that permits its continued, efficient recognition by cellular replication factors.

A plausible model for the contribution of EBNA1 to the segregation and/or maintenance of plasmids containing *ori*P or FR[55,73] is for it to link them to cellular chromosomes. Algorithms designed to predict binding sites for EBNA1 indicate that thousands to hundreds of thousands of sites exist within the human genome.[113] EBNA1 bound to FR and to cellular DNA could link them; both EBV–DNA, *ori*P plasmids, and EBNA1 associate with cellular chromosomes.[53,114–117] It has been shown that the linking domains of EBNA1 are sufficient to localize GFP to metaphase chromosomes.[118] Thus, the association of EBV–DNA and *ori*P plasmids with chromosomes may be mediated by interaction of the linking domains with cellular factors. Whatever the mechanism, association of DNAs containing *ori*P with cellular chromosomes may mediate their efficient partitioning into daughter cells or their maintenance in the nucleus following mitosis (Fig. 5). If replicated plasmids containing *ori*P associate with identical sites on replicated sister chromatids, then, by piggybacking on the chromosomes,

[112] L. Frappier, K. Goldsmith, and L. Bendell, *J. Biol. Chem.* **269,** 1057 (1994).

[113] A. Aiyar, personal communication, 1998.

[114] E. A. Grogan, W. P. Summers, S. Dowling, D. Shedd, L. Gradoville, and G. Miller, *Proc. Natl. Acad. Sci. U.S.A.* **80,** 7650 (1983).

[115] A. Harris, B. D. Young, and B. E. Griffin, *J. Virol.* **56,** 328 (1985).

[116] L. Petti, C. Sample, and E. Kieff, *Virology* **176,** 563 (1990).

[117] H.-J. Delecluse, S. Bartnizke, W. Hammerschmidt, J. Bullerdiek, and G. W. Bornkamm, *J. Virol.* **67,** 1292 (1993).

[118] V. Marechal, A. Dehee, R. Chikhi-Brachet, T. Piolot, M. Coppey-Moisan, and J. C. Nicolas, personal communication, 1998.

[119] J. J. Harrington, B. G. Van, R. W. Mays, K. Gustashaw, and H. F. Willard, *Nature Genet.* **15,** 345 (1997).

[120] L. Gritz and J. Davies, *Gene* **25,** 179 (1983).

[121] J. Levitskaya, A. Sharipo, A. Leonchiks, A. Chiechanover, and M. G. Masucci, *Proc. Natl. Acad. Sci. U.S.A.* **94,** 12616 (1997).

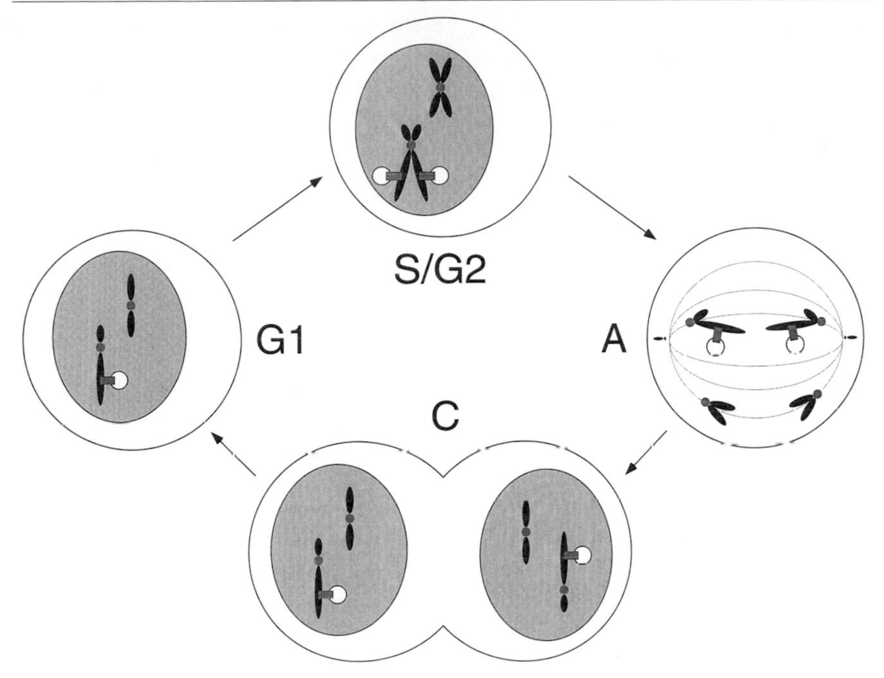

FIG. 5. Model of maintenance and/or segregation of plasmids containing *ori*P by EBNA1. (G1) During interphase, *ori*P plasmids (circles) are linked by EBNA1 (rectangles) to cellular chromosomes. (S/G2) After synthesis and prior to mitosis, the newly replicated *ori*P plasmids are partitioned and linked by EBNA1 to sites on the sister chromatids. (A) During anaphase of mitosis, *ori*P plasmids piggyback on sister chromosomes, resulting in their even segregation into the daughter cells. (C) As the nuclear envelope reforms, prior to cytokinesis, association of *ori*P plasmids with chromosomes maintains them in the nucleus.

the plasmids could be partitioned evenly to daughter cells. Mammalian centromeres, encoded by approximately a megabase of DNA,[119] are orders of magnitude larger than *ori*P. Also, plasmids containing multiple copies of *ori*P are stable,[3] unlike CEN–ARS–CEN plasmids in yeast.[70,71] Therefore it is unlikely that *ori*P encodes a centromere. In this proposed model, segregation of an *ori*P plasmid relies on the cellular mechanisms that segregate the chromosome on which *ori*P piggybacks.

We interpret the results and models described to indicate that *ori*P and EBNA1 have three activities that together mediate the stable replication of plasmids containing *ori*P in proliferating cells expressing EBNA1. First, DS serves as a site recognized by cellular factors that initiate DNA synthesis. Second, EBNA1, by binding and linking FR and DS, organizes chromatin structure to ensure the continued function of DS. Third, by associating *ori*P with cellular chromosomes, EBNA1 mediates the maintenance of *ori*P plasmids in the nucleus and their segregation during mitosis. In the presence of EBNA1, *ori*P plasmids are maintained stably in cells, making them useful vectors for molecular biology.

Acknowledgments

We thank Ashok Aiyar and Paul Lambert for critical review of this paper. This work was supported by Public Health Service Grants CA-2243, CA-09075, CA70723, and CA07175.

[19] Expression of E1 Protein of Human Papillomaviruses in Eukaryotic Cells

By SALEEM A. KHAN, SAIFUDDIN SHEIKH, LAURA SHEAHAN, GERALD VAN HORN, VIDYA GOPALAKRISHNAN, and FRANCIS SVERDRUP

Introduction

Human papillomaviruses (HPVs) contain a circular, double-stranded DNA genome of approximately 8 kb that is usually maintained in an extrachromosomal form at a relatively fixed copy number between 10 and 50. At present, more than 80 different types of HPVs have been isolated. HPVs are responsible for benign epithelial proliferations (warts), and additionally many HPVs are associated with human anogenital cancers, particularly cancer of the cervix.[1] Replication of HPVs requires the viral E1 and E2 proteins and an origin of replication (ori). The ori contains an E1-binding site and multiple E2-binding sites.[2,3] The E1 and E2 proteins encoded by various HPVs are quite homologous. E1 is the replication initiator protein, which is also involved in the regulation of viral gene expression through its interaction with the E2 protein.[4–6] The papillomavirus (PV) E2 protein is an activator as well as a repressor of viral gene transcription.[7,8] Binding

[1] M. Ustav and A. Stenlund, *EMBO J.* **10,** 449 (1991).
[2] P. F. Lambert, *J. Virol.* **65,** 3417 (1991).
[3] L. T. Chow and T. R. Broker, *Intervirology* **37,** 150 (1994).
[4] L. Yang, I. Mohr, E. Fouts, D. A. Lim, M. Nohaile, and M. Botchan, *Proc. Natl. Acad. Sci. U.S.A.* **90,** 5086 (1993).
[5] Y.-S. Seo, F. Muller, M. Lusky, and J. Hurwitz, *Proc. Natl. Acad. Sci. U.S.A.* **90,** 702 (1993).
[6] A. B. Sandler, S. B. Vande-Pol, and B. A. Spalholz, *J. Virol.* **67,** 5079 (1993).
[7] P. G. Fuchs and H. Pfister, *Intervirology* **37,** 159 (1994).
[8] A. A. McBride, H. Romanczuk, and P. M. Howley, *J. Biol. Chem.* **266,** 18411 (1991).

0076-6879/99 $30.00

of E2 protein to the E2-binding sites located immediately upstream of the TATA box of viral E6 and E7 oncogenes promoter results in negative regulation of E6/E7 gene expression.[7] E2 protein also stimulates E1-dependent PV replication by targeting E1 to the ori through the formation of an E1–E2 complex.[9,10] E1 is a 68- to 72-kDa (600 to 650 amino acids) nuclear phosphoprotein, which has ATPase, $3 \rightarrow 5'$-helicase, ori-specific binding and unwinding activities.[3–5] It contains a nuclear localization signal and a domain that interacts with the E2 protein. E1 also interacts with cellular DNA polymerase α and presumably contains domains that interact with other replication proteins.[11] The E1-binding site consists of an AT-rich, 18-bp palindromic sequence that is partially conserved among various PVs.[12] E1 binds to the origin DNA as a hexamer.[13] Attempts have been made to define the structure–function relationship and domain structure of the E1 protein (Fig. 1).[14–23] Generally, these studies have shown that the amino-terminal half of E1 is involved in its DNA-binding activity. The amino-terminal region of E1 also contains a nuclear localization signal. The ATP-binding domain of BPV-1 E1 corresponds to amino acids 433 to 440 and is conserved in all HPV E1 proteins.[4] Both the amino- and the carboxyl-terminal regions of the E1 proteins are involved independently in complex formation with E2, although the C-terminal half of E1 appears to be critical for the stimulation of its DNA-binding and replication activities by the E2 protein.[17,18,22–25] The minimal domain for the helicase activity is not well defined, but probably encompasses the C-terminal two-thirds of the E1 protein.[11,14,19] Although the E1 proteins of PVs are quite homologous and have conserved functions, they also show differences in their replication

[9] I. J. Mohr, R. Clark, S. Sun, E. J. Androphy, P. MacPherson, and M. R. Botchan, *Science* **250,** 1694 (1990).

[10] Y.-S. Seo, F. Muller, M. Lusky, E. Gibbs, H.-Y. Kim, B. Phillips, and J. Hurwitz, *Proc. Natl. Acad. Sci. U.S.A.* **90,** 2865 (1993).

[11] P. J. Masterson, M. A. Stanley, A. P. Lewis, and M. A. Romanos, *J. Virol.* **72,** 7407 (1998).

[12] S. E. Holt, G. Schuller, and V. G. Wilson, *J. Virol.* **68,** 1094 (1994).

[13] J. Sedman and A. Stenlund, *J. Virol.* **72,** 6893 (1998).

[14] F. J. Hughes and M. A. Romanos, *Nucleic Acids Res.* **21,** 5817 (1993).

[15] L. K. Thorner, D. A. Lim, and M. R. Botchan, *J. Virol.* **67,** 6000 (1993).

[16] M. G. Frattini and L. A. Laimins, *Virology* **204,** 799 (1994).

[17] T. R. Sarafi and A. A. McBride, *Virology* **211,** 385 (1995).

[18] F. Muller and M. Sapp, *Virology* **219,** 247 (1996).

[19] K. C. Mansky, A. Batiza, and P. F. Lambert, *J. Virol.* **71,** 7600 (1997).

[20] G. Chen and A. Stenlund, *J. Virol.* **72,** 2567 (1998).

[21] M. C. Ferran and A. A. McBride, *J. Virol.* **72,** 796 (1998).

[22] N. Zou, J.-S. Liu, S.-R. Kuo, T. R. Broker, and L. T. Chow, *J. Virol.* **72,** 3436 (1998).

[23] V. Gopalakrishnan, L. Sheahan, and S. A. Khan, *Virology* **257,** in press (1999).

[24] M. Berg and A. Stenlund, *J. Virol.* **71,** 3853 (1997).

[25] T. Yasugi, J. D. Benson, H. Sakai, M. Vidal, and P. M. Howley, *J. Virol.* **71,** 891 (1997).

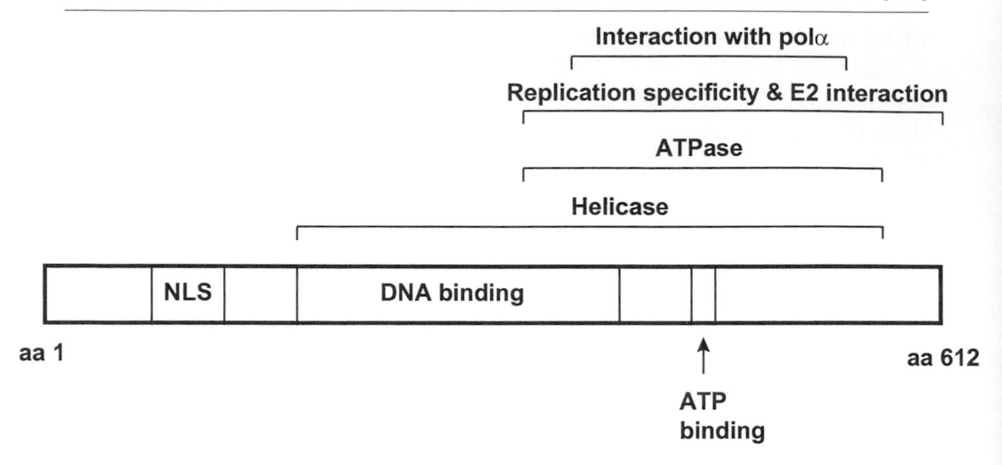

FIG. 1. Functional domains of the PV E1 proteins based on studies in many laboratories.[4,11,14–25] Numbers correspond to those of the HPV-1a E1 protein. The approximate locations of the various domains are as follows: NLS, aa 74–125; DNA binding, 170–390; ATP binding, 440–447; ATPase, 330–570; helicase, 170–570; "replication specificity" and functional E2 interaction, 307–612; and interaction with DNA polymerase α, 360–560. aa, Amino acid(s); NLS, nuclear localization signal.

activities. For example, the E1 protein of HPV type 1a, and that of BPV-1 to some extent, can support the *in vivo* replication of ori plasmids in the absence of the E2 protein.[26] However, the E1 proteins of other HPVs are inactive in replication in the absence of E2.[3] Apart from its importance in studies of viral DNA replication, the E1 protein also represents an important target for antiviral therapy. Several eukaryotic systems have been used for the expression of the PV E1 proteins to study their biological properties. This article describes the expression of the E1 proteins of HPV types 1a and 18 in mammalian and insect cells and the purification of the HPV-1a E1 protein from insect cells using a baculovirus vector.

Materials

Plasmid and Cell Lines

C-33A is a HPV-negative human cervical carcinoma cell line and is maintained in monolayer cultures in Dulbecco's modified Eagle's medium (DMEM) containing 10% (v/v) fetal calf serum supplemented with penicillin/streptomycin/glutamine.[26] Sf21 is an insect cell line derived from

[26] V. Gopalakrishnan and S. A. Khan, *Proc. Natl. Acad. Sci. U.S.A.* **91,** 9597 (1994).

Spodoptera frugiperda. This cell line is maintained in grace complete suspension culture media containing 4% fetal bovine serum and 1% Pluronic F-68. Plasmid pSG5 is a mammalian vector in which the expression of foreign genes is driven from the SV40 early promoter.[27] Plasmid pSGE1-a and pSGE1-18 are pSG5 derivatives expressing the HPV 1a or 18 E1 proteins, respectively.[26,28] Plasmids containing the various HPV-1a and HPV-18 ori have been described previously.[26,28] The anti-FLAG M2 monoclonal antibodies were obtained from Kodak (Rochester, NY).

Methods

Construction of Plasmids Expressing E1 Proteins Fused to the FLAG Epitope

To facilitate the detection and purification of the E1 proteins, two tandem copies of the 8 amino acid long FLAG peptide (DYKDDDDK) are fused at their N-terminal ends. A derivative of pSG5, pSGF2, is generated by inserting a synthetic oligonucleotide containing the initiation ATG codon and encoding two tandem copies of the FLAG epitope into the unique *Bam*HI site of pSG5. The E1 open reading frames of HPVs 1a and 18 are obtained by polymerase chain reaction (PCR) amplification of HPV 1a and 18 DNA using primers containing a unique restriction site to facilitate their cloning into the pSGF2 vector. The PCR reactions amplify the DNA such that it lacks the first, initiation codon of the E1 proteins.

Transient Replication Assays

The human cervical carcinoma cell line C 33A is transfected with plasmid DNA by calcium phosphate coprecipitation.[29] Transfection efficiencies generally range from 25 to 35%. Transient replication assays are done using 7.5×10^5 to 1×10^6 cells in 60-mm plates and 3 or 5 μg of the E1-expressing plasmid (pSGE1 or pSGF2-E1), 1 μg of the E2-expressing plasmid (pSGE2), and 0.5 μg of the ori plasmids. Three to 4 days posttransfection, low molecular weight DNA is isolated by the Hirt extraction procedure.[30] The samples are first treated with an appropriate restriction enzyme to either linearize or cleave the plasmids into two fragments, one of which is complementary to the probe. To distinguish between replicated and unreplicated DNA, one-half of each sample is then treated with an excess

[27] S. Green, I. Isseman, and E. Sheer, *Nucleic Acids Res.* **16,** 369 (1988).
[28] F. Sverdrup and S. A. Khan, *J. Virol.* **68,** 505 (1994).
[29] C. Chen and H. Okayama, *Mol. Cell. Biol.* **7,** 2745 (1987).
[30] B. Hirt, *J. Mol. Biol.* **26,** 365 (1967).

of *Dpn*I to remove the unreplicated, methylated input DNA.[31] DNA samples are analyzed by electrophoresis on 0.7% agarose gels using Tris–borate–EDTA buffer followed by Southern blot hybridization.[32] The DNA is transferred to GeneScreen and the membranes are hybridized to a ^{32}P-labeled pUC19 probe (2–4 $\times$ 10^7 cpm) generated by using the random primer labeling kit (Amersham Pharmacia Biotech, Piscataway, NJ). Specific activity of the probes typically range from 1 $\times$ 10^8 to 1 $\times$ 10^9 cpm/μg of the DNA. Blots are subjected to autoradiography at $-70°$ with intensifying screens. Various bands on the blots are quantitated using an Ambis 100 radioanalytic detector.

Western Blot Analysis

The relative levels of E1 proteins in transfected C-33A cells are measured by immunoblotting. Five micrograms of the pSGF2-E1 plasmid along with various amounts of the pSGE2 plasmid are cotransfected into 1 $\times$ 10^6 C-33A cells by calcium phosphate coprecipitation. Two days posttransfection, cells are lysed using 0.25 ml of lysis buffer (50 m*M* Tris–HCl, pH 8.0, 150 m*M* NaCl, 0.02% sodium azide, and 1% Triton X-100). Protein concentrations in the various samples are determined using the Bio-Rad (Richmond, CA) protein assay kit. Equal amounts of protein are subjected to SDS–polyacrylamide gel electrophoresis on a 10% gel. The proteins are transferred to an Immobilon-P membrane (Millipore, Bedford, MA) and the FLAG-E1 proteins are detected using the anti-FLAG M2 monoclonal antibody (Kodak) and goat antimouse IgG peroxidase conjugate (Sigma, St. Louis, MO). The secondary antibody in the conjugate is detected using an enhanced chemiluminescence reagent from Amersham.

Expression of E1 Proteins of HPV-1a and HPV-18 in Mammalian Cell Lines

To study the requirement of the E1 protein in HPV replication *in vivo,* we have expressed the HPV 1a and 18 E1 proteins using the pSGF2-E1 vectors in a variety of mammalian cell lines. These studies have also been useful in the identification of the domains of the E1 proteins that are involved in their replication specificity and interaction with the E2 protein during replication. The expression of the E1 proteins in transfected cells is monitored by Western blot analysis using the anti-FLAG M2 monoclonal antibody. Transient replication assay involves cotransfection of plasmids expressing the E1 and E2 proteins along with a plasmid containing the ori

[31] K. W. C. Peden, J. M. Pipas, S. Pearson-White, and D. Nathans, *Science* **209,** 1392 (1980).
[32] J. Sambrook, E. F. Fritsch, and T. Maniatis, *in* "Molecular Cloning: A Laboratory Manual," 2nd ed. Cold Spring Harbor Laboratory Press, Cold Spring Harbor, NY, 1989.

of HPVs. Replication of transfected plasmid DNA is usually studied by treating the Hirt fractions with the enzyme *Dpn*I that digests only fully methylated DNA. One or more rounds of replication in mammalian cells generates hemimethylated or unmethylated plasmid DNA that is resistant to digestion by this enzyme.[31] Southern blot analysis of Hirt fractions using an appropriate probe reveals *Dpn*I-resistant ori bands that are a measure of the ability of the E1 (and E2) proteins to support replication of the ori plasmids. Using the technique described earlier, many laboratories, including ours, have shown that for most HPVs both the E1 and the E2 proteins are required for replication of ori plasmids *in vivo*. Similarly, both the E1- and the E2-binding sites located within the ori are required for replication of ori plasmids. This approach has also been used to show that the E1 and E2 proteins of various HPVs are promiscuous in their replication activity and can cross-interact and support replication of plasmids containing heterologous PV origins.[23,24,33] However, in contrast to most other HPVs, we found that the HPV-1a E1 protein by itself is sufficient for the replication of HPV-1a ori plasmids in transfected cells.[26] We have used plasmids expressing the wild-type HPV 1a and 18 E1 proteins as well as N-terminal FLAG and maltose-binding protein (MBP) epitope tags fused to E1 in these experiments. Other investigators have also used either native or N-terminal fusions of different PV E1 proteins with epitopes such as FLAG, EE, and GST and studied the role of this protein in PV replication.[34–38] Interestingly, epitope tagging at the C-terminal end of the E1 protein results in loss of its replication activity, suggesting the presence of a critical domain in this region.[22] We have also used plasmids expressing hybrids between the HPV 1a and 18 E1 proteins in transient replication assays. These results have shown that the replication specificity and E2 interaction domains of these E1 proteins are located in their carboxyl-terminal halves.[23]

The ability of HPV plasmids to replicate episomally in transfected cells has also generated interest in the development of HPV-based plasmids that can support expression of cloned genes. We have developed HPV plasmids that contain the HPV-18 E1 and E2 genes and ori on a single vector molecule that can replicate in transiently transfected cells. Some of these

[33] C.-M. Chiang, M. Ustav, A. Stenlund, T. F. Ho, T. R. Broker, and L. T. Chow, *Proc. Natl. Acad. Sci. U.S.A.* **89,** 5799 (1992).

[34] L. Yang, R. Li, I. J. Mohr, R. Clark, and M. R. Botchan, *Nature* **353,** 628 (1991).

[35] G. L. Bream, C.-A. Ohmstede, and W. C. Phelps, *J. Virol.* **67,** 2655 (1993).

[36] S.-R. Kuo, J.-S. Liu, T. R. Broker, and L. T. Chow, *J. Biol. Chem.* **269,** 24058 (1994).

[37] C. Bonne-Andrea, F. Tillier, G. D. McShan, V. G. Wilson, and P. Clertant, *J. Virol.* **71,** 6805 (1997).

[38] T. A. Zanardi, C. M. Stanley, B. M. Saville, S. M. Spacek, and M. R. Lentz, *Virology* **228,** 1 (1997).

plasmids also contain the hygromycin or neomycin resistance genes to facilitate studies on their stable replication in transfected cells. HPV plasmids expressing reporter genes such as luciferase and β-galactosidase are under development to test the feasibility of using such plasmids for the long-term expression of cloned genes. Similarly, we have developed plasmids containing only the HPV-1a E1 gene and ori and shown them to replicate in transfected cells. The HPV plasmids are able to replicate in several types of epithelial cells, making them potentially useful in gene therapy applications.

Expression of HPV-1a E1 Protein in Insect Cells

To study the biochemical properties of the E1 proteins of PVs, several laboratories have made use of the baculovirus expression system to obtain these proteins in sufficient quantities. The E1 protein is known to be phosphorylated, which is likely to be important for its biological activities. Therefore, expression of these proteins in insect cells is likely to yield proteins that are appropriately modified. The E1 protein of BPV-1 was the first to be isolated and studied for its biochemical activities.[9,34] E1 proteins encoded by HPV types 11 and 33 have also been overexpressed in insect cells.[18,35,36] We have overexpressed the HPV-1a E1 protein with the FLAG epitope present at its amino-terminal end. The replication activity of FLAG-E1 in human cell lines is identical to that of the native protein as demonstrated by transient replication assays. The E1 gene of HPV-1a containing two copies of the FLAG epitope at its N-terminal end is amplified by PCR using the pSGF2-E1 DNA as template, the DNA fragment isolated as a *Not*I fragment, and ligated into the *Not*I site of the pFastBAC1 vector (Life Technologies, Gaithersburg, MD). The cloned DNA is introduced into *Escherichia coli* DH10BAC cells containing the bacmid, which is a baculovirus shuttle vector containing a mini-F replicon (inserted into the polyhedrin locus), a kanamycin resistance gene, and a mini *att*Tn7 site, which is a target for insertion of transposon Tn7 present in the lacZα region. DH10BAC cells also contain a tetracycline resistance helper plasmid that encodes the Tn7 transposase. The transformed cells are plated on L-agar plates containing kanamycin, gentamicin, tetracycline, and X-Gal. White colonies are picked and screened for ampicillin sensitivity to confirm the loss of the pFASTBAC1-E1 plasmid. These colonies represent clones in which the FLAG-E1 gene cassette has presumably transposed into the bacmid. Plasmid preparations containing the composite bacmid DNA are obtained from a few such colonies and used to transfect Sf21 cells by using cellfectin (Life Technologies). Insect cells containing the E1 baculovirus are identified by plaque formation, and a stock of the recombinant FLAG-

E1 baculovirus is prepared by transfecting fresh Sf21 cells. The titer of the viral stock is determined by a plaque assay. Various conditions are tested for the optimal expression of the FLAG-E1 protein by isolating protein lysates from infected cells, followed by SDS–PAGE and Western blotting using the anti-FLAG M2 monoclonal antibodies. The HPV-1a FLAG-E1 protein is optimally expressed at 24 hr after infection of insect cells at a multiplicity of infection of 1.

Purification of HPV-1a E1 Protein from Sf21 Cells

Sf21 cells infected with a baculovirus containing the FLAG-E1 gene are grown in spinner cultures (0.5 to 1 liter) at 37° for 24 hr, and the cells are harvested and washed once with phosphate buffer containing saline. The cells are then resuspended in buffer A [50 m*M* Tris–HCl, pH 8.0, 1 m*M* EDTA, 1 m*M* dithiothreitol (DTT), 0.15 *M* NaCl, 10% glycerol] and complete protease inhibitor (1 tablet/50 ml extraction buffer) containing EDTA, aprotinin, leupeptin and Pefabloc SC (Boehringer Mannheim). The cells are lysed by sonication at a continuous cycle for about 8 min with 30-sec intervals after each 2 min. The lysate is centrifuged, and the supernatant containing the cellular proteins is collected. The supernatant is subjected to ammonium sulfate precipitation, and the precipitate at 40% saturation containing the majority of the FLAG-E1 protein is collected. The precipitate is dialyzed against buffer A and loaded onto an affinity column (2 ml) containing the anti-FLAG monoclonal antibody M2 coupled to Sepharose. The column is washed with buffer A, and the E1 protein is eluted with 0.1 *M* glycine-hydrochloride (pH 3.5). Fractions are collected in tubes containing 1 *M* Tris–HCl (pH 8.0) for neutralization. The FLAG-E1 preparations are analyzed by SDS–PAGE followed by staining of the gels with Coomassie blue and by Western blot analysis. FLAG-E1 protein fractions, which are approximately 90% pure, are stored at −80°.

The FLAG-E1 protein is tested for its associated ATPase and DNA helicase activities. The hydrolysis of ATP is measured in a reaction mixture containing 50 m*M* Tris–HCl, pH 8.0, 10 m*M* $MgCl_2$, 1 m*M* DTT, 0.05 mg/ml bovine serum albumin (BSA) and 1 μl of [α-^{32}P]ATP (3,000 Ci/mmol). After incubation at 37° for 30 min, an aliquot (1 μl) is spotted onto a polyethyleneimine-cellulose thin-layer plate that is developed with 0.5 *M* potassium phosphate buffer (pH 3.5). The plates are dried and subjected to autoradiography. The presence of a faster migrating band corresponding to the released inorganic phosphate is a measure of the ATPase activity of the HPV-1a E1 protein.

The DNA helicase activity of the HPV-1a E1 protein is measured by an oligonucleotide displacement assay. A 5′-^{32}P-labeled 47-mer oligonucleotide containing a central 17 nucleotide long sequence complementary to a region of M13 along with 5′ and 3′ overhangs of 15 nucleotides each is hybridized to single-stranded M13 DNA, and free nucleotides are removed by gel filtration using a Sepharose G-25 column. The FLAG-E1 protein is incubated with the substrate described previously in a reaction mixture containing 20 m*M* Tris–HCl, pH 7.4, 10 m*M* $MgCl_2$, 2 m*M* ATP, and 0.1 mg/ml BSA for 2 hr at 37°, and after the addition of SDS to a final concentration of 0.1%, the products are analyzed by gel electrophoresis on 1.5% agarose or 6% polyacrylamide gels. The helicase activity of E1 is determined from the extent of displaced labeled oligonucleotide on the gel in the presence of the protein. Although the E1 protein binds to DNA containing the E1-binding sites, electrophoretic mobility-shift assays have proven to be difficult in the analysis of this activity. This is presumably due to the weak DNA-binding activity of the E1 protein that may result in the disruption of the DNA–E1 complexes during gel electrophoresis. To circumvent this problem, the DNA-binding activity of E1 is generally determined using the McKay assay.[39] For this, the E1 protein is mixed with a ^{32}P-labeled DNA fragment(s) containing the E1-binding site, and the E1–DNA complexes are recovered using either anti-E1 antibodies or an affinity matrix for epitope-tagged E1 proteins (such as glutathione-Sepharose, EE-Sepharose, or the anti-FLAG M2 antibodies) depending on the type of tag used. After washing the free DNA probe, the DNA protein complex is disrupted by adding SDS and heating, and the samples are subjected to electrophoresis on native polyacrylamide gels followed by autoradiography. The presence of a DNA band of the appropriate size containing the E1-binding site indicates the specific DNA-binding activity of the E1 protein.

The native, EE-, GST-, or FLAG-tagged E1 proteins of BPV-1, HPV-11, and HPV-33 have also been overexpressed and purified by several laboratories using the baculovirus system.[5,9,18,34–38] These E1 proteins were shown to have DNA-binding, ATPase, and helicase activities. The BPV-1 and HPV-11 E1 proteins were also found to be active in the *in vitro* replication of viral DNA.[34,36,37]

Acknowledgment

This work was supported by National Institutes of Health Grant GM-51861.

[39] R. D. McKay, *J. Mol. Biol.* **145,** 471 (1981).

[20] Genetic Manipulation of Herpes Simplex Virus Using Bacterial Artificial Chromosomes

By BRIAN C. HORSBURGH, MARIA M. HUBINETTE, and FRANK TUFARO

Introduction

Herpes simplex virus (HSV) is the prototypic human herpesvirus. Generation of HSV mutants has often relied on drug selection or cotransfection of cells with intact viral and plasmid DNA, usually modified by insertion of a marker gene. Viral mutants containing point mutations, insertions, or deletions are identified by selecting in mammalian cells for resistance to drug, screening for expression of the plasmid marker, screening for viability in complementing cell lines, or direct structural characterization.[1-4] Advances in the development of bacterial artificial chromosomes (BACs) containing a complete copy of the infectious HSV genome have made it possible to exploit bacterial genetics for constructing recombinant herpesviruses.[5] The procedure is rapid and allows for the direct cloning of viral recombinants without the need for multiple rounds of plaque purification in mammalian cells.

The procedure, a modification of Link *et al.*,[6] requires two reagents: (1) an HSV-BAC carrying a marker gene, such as chloramphenicol (cam), and (2) a gene replacement plasmid, e.g., pKO5, containing the desired altered allele flanked by HSV sequences that can target it to a defined locus within the HSV-BAC. This vector also contains a temperature-sensitive origin of DNA replication and marker genes for positive and negative selection, zeocin resistance gene and SacB, respectively. The mutagenesis procedure is illustrated in Fig. 1.

The gene replacement vehicle is electroporated into RecA$^+$ *Escherichia coli* that harbor an HSV-BAC. Bacteria are plated at the nonpermissive temperature for plasmid replication, 43°, on cam/zeocin LB agar plates to select for BAC and plasmid sequences. To obtain colonies that are resistant to the positive marker gene (zeocin), the vector must integrate into either

[1] P. J. Gage, B. Sauer, M. Levine, and J. C. Glorioso, *J. Virol.* **66,** 5509 (1992).
[2] S. H. Qiu, S. L. Deshmane, and N. W. Frazer, *Gene Ther.* **1,** 300 (1994).
[3] F. J. Rixon and J. McLaughlin, *J. Gen. Virol.* **71,** 2931 (1990).
[4] B. Roizman and F. J. Jenkins, *Science* **229,** 1208 (1985).
[5] B. C. Horsburgh, M. M. Hubinette, D. Qiang, M. M. E. MacDonald, and F. Tufaro, *Gene Ther.,* in press.
[6] A. J. Link, D. Phillips, and G. M. Church, *J. Bact.* **179,** 6228 (1997).

0076-6879/99 $30.00

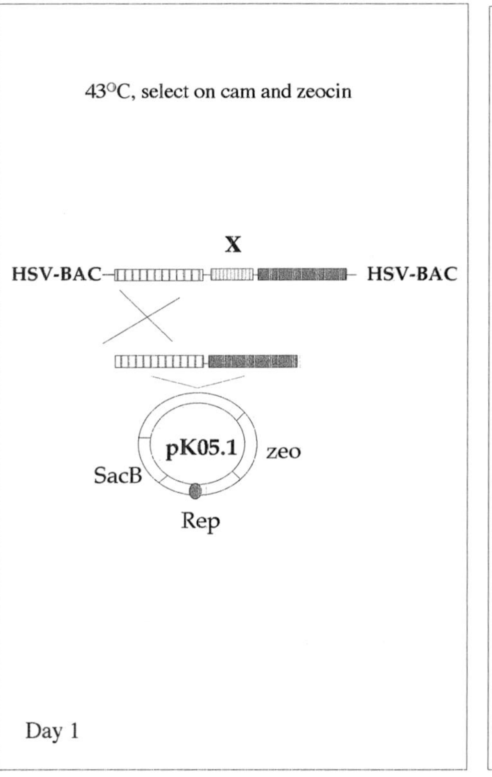
43°C, select on cam and zeocin
X
HSV-BAC
HSV-BAC
pK05.1
zeo
SacB
Rep
Day 1

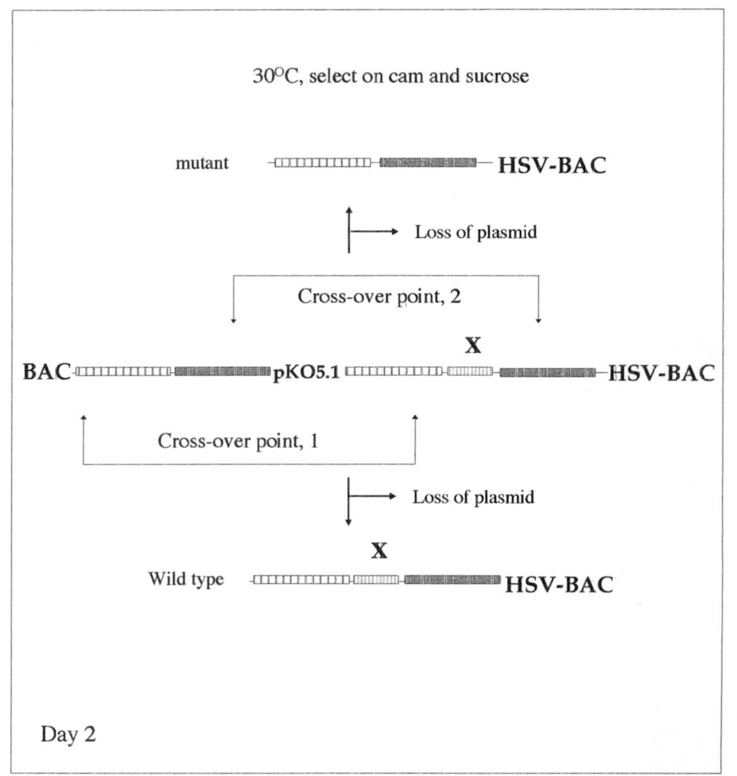
30°C, select on cam and sucrose
mutant
HSV-BAC
Loss of plasmid
Cross-over point, 2
X
BAC
pKO5.1
HSV-BAC
Cross-over point, 1
Loss of plasmid
X
Wild type
HSV-BAC
Day 2

the host chromosome or the HSV-BAC. Because the plasmid contains homologous HSV sequences that flank the mutant allele, it is expected to integrate into the HSV-BAC with higher frequency. The plasmid is forced to integrate via homologous recombination and consequently generates an imperfect tandem duplication. The following day, cam/zeocin-resistant colonies are diluted serially and plated on cam/sucrose LB plates at the permissive temperature, 30°. This results in excision of plasmid sequences by homologous recombination because, at this temperature, the plasmid origin in the HSV-BAC inhibits BAC replication. Excision can occur at homologous sequences that flank the left or the right side of the mutant allele. If integration occurred at sequences to the right of the mutant allele and excision occurred at sequences to the left of the allele, then allelic exchange is observed. Growth on sucrose results in loss of the gene replacement plasmid because SacB is lethal under these conditions. Cam/sucrose-resistant colonies are replica plated onto zeocin LB plates to confirm loss of the pKO5 replacement plasmid. Colony polymerase chain reaction (PCR) is used to identify HSV-BAC mutants from cam^r/$zeocin^s$ bacteria. Recombinant HSV-BAC DNA is isolated from positive clones and transfected into mammalian cells. Cytopathic effects resulting from the replicating HSV should be apparent 48–72 hr posttransfection. Mutant viruses are harvested and analyzed. The whole procedure can be completed within 7 days and is summarized in Table I.

Materials and Methods

Genetic manipulation of HSV in bacteria requires two reagents: an HSV-BAC and a gene replacement plasmid, pKO5. We will describe the details necessary to construct an HSV-BAC and the procedure that permits its manipulation.

FIG. 1. A schematic for deleting gene X from HSV-BAC. (Day 1) The pKO5 vector carrying the altered allele is transformed into *E. coli* haboring HSV-BAC and is plated at the nonpermissive temperature on cam/zeocin LB plates. In this example, the object is to delete gene X from HSV-BAC. SacB and zeocin genes and the temperature-sensitive origin of replication (Rep) are indicated. Cointegrates that arise via homologous recombination are able to survive these conditions; an integration event allows replication of plasmid sequences by the BAC origin. (Day 2) When shifted to the permissive temperature, the plasmid origin is unstable in the BAC and is excised at either crossover point 1 or 2 by homologous recombination, resulting in exchange of allele at frequencies of up to 50%. The counterselectable marker, SacB, is used to select for loss of plasmid sequences. Sucrose-resistant colonies are replica plated onto zeocin plates to confirm the loss of plasmid sequences and the allele replacement event screened by PCR.

TABLE I
SEVEN-DAY GUIDE TO MAKING MUTANT VIRUSES OR VECTORS

Day	Procedure
1	Electroporate pKO5 containing the mutant allele into recA$^+$ *E. coli* harboring pHSV-BAC and incubate at 43°
2	Serial dilute 43° colonies and plate on cm-suc LB plates at 30°
3	Replica plate cm-suc-resistant colonies onto zeocin LB plates at 30° to confirm loss of replacement vector
4	From cm −ve/suc +ve/zeocin −ve colonies, use PCR to identify allele exchange on pHSV-BAC. Set up overnight 2-liter cultures
5	Isolate mutant HSV-BAC DNA and transfect mammalian cells
7–8	Obtain mutant virus

Bacterial Cell Types

All plasmid construction and propagation can be performed using *E. coli* strain DH5α [*endA1 hsdR17*(r_k-m_k^+) *supE44 thi-1 recA1 gyrA*(*Nalr*) *relA1* Δ(*lacZYA-argF*) *U169 deoR*]. The optimal strain for propagation of BAC plasmids is DH10B [F$^-$ *mcrA* Δ(*mrr-hsdRMS-mcrBC*) Δ*lacX74 deoR recA1 endA1 araD139* Δ(*ara, leu*)*7697 galU galK* λ-*rpsL nupG*]. DH10B cells are RecA deficient. Allele replacement experiments use a recombination proficient strain, and any RecA$^+$ strain is suitable. It is essential that the host strain be RecA$^+$ because allele replacement depends on recombination occurring between homologous sequences that are present on the BAC and the pKO5 allele replacement plasmid. Further, it is advisable, although not essential, to use strains that are deficient for EndA. Activity of nonspecific endonuclease I is abolished in these strains and the quality of DNA obtained is superior. Suitable *E. coli* strains for allele replacement include JM105, JM107, RR1, and EMG2.

Media and Growth Conditions

All bacterial strains can be grown in LB medium (10 g Bacto-tryptone, 5 g yeast extract, 5 g NaCl/liter), pH 7.5. Antibiotics are added at concentrations of 20 μg cam/ml or 25 μg zeocin/ml (Invitrogen, CA). Sucrose is added to LB agar medium to a final concentration of 5% (w/v). It is important to note that zeocin is inactivated by salt concentrations greater than 100 m*M* and by acidic or basic conditions. It is therefore necessary to reduce the salt in bacterial medium and to adjust the pH to 7.5 to ensure that the drug remains active.

DNA Purification

KO5 plasmid and BAC DNA can be purified by an alkaline lysis procedure. Two-liter cultures are grown at 30° (pKO5) or 37° (BAC) for 18 hr shaking at 250 rpm. Bacteria are then pelleted by centrifugation, and DNA is isolated using Qiagen plasmid purification kits (Ontario, Canada) or their equivalent, according to the manufacturer's instructions. For added purity, DNA can be banded twice on CsCl gradients. The purified DNA is suspended in 500 μl Tris–HCl (pH 8.0). *Note:* HSV-BAC DNA is infectious, i.e., it can direct the formation of infectious virus following transfection into mammalian cells. Appropriate institutional guidelines for the handling and disposal of biohazardous materials should be strictly followed.

Preparation of Electrocompetent Cells

One colony of *E. coli* is grown to stationary phase in 10 ml of LB. For cells containing HSV-BAC, cam (20 μg/ml) is added to the growth medium to maintain selection. Dilute the culture 1:200 in LB (2.5 ml into 500 ml LB) and grow to an OD_{600} of 0.5–0.6 at 37° by shaking at 300 rpm. Chill the cells on ice for 10 min and transfer to cold conical centrifuge tubes. Pellet the bacteria by spinning at 4° for 20 min at 3000*g* in a suitable centrifuge. Pour off the medium and suspend the pellet in 5 ml distilled H_2O. When bacteria are fully dispersed, add an additional 500 ml ice-cold distilled H_2O. Repeat this procedure once more. Following the second spin, pour off the supernatant and swirl the bacterial pellet to resuspend in the remaining liquid. Add an additional 40 ml of ice-cold distilled H_2O. Transfer this solution to a cold 50-ml conical tube and centrifuge at 2° for 10 min at 3000*g*. Following centrifugation, pour off the supernatant and suspend the bacterial pellet in 40 ml of ice-cold 10% (v/v) glycerol. Centrifuge once more and pour off the supernatant. Estimate the cell pellet volume and add an equal volume of cold 10% glycerol. Aliquot 50–300 μl in cold microcentrifuge tubes, freeze on dry ice, and store at −70° until required.

Electroporation

Approximately 40 μl of electrocompetent cells is mixed with 2 μl of plasmid or BAC DNA in an ice-cold microfuge tube and transferred to a 0.1-cm electroporation cuvette. Electroporation is carried out at 1.8 kV with 25 μF capacitance and 200 Ω resistance. Bacteria are allowed to recover for 1 hr at 30° in 1 ml of LB, plated on LB agar plates containing the appropriate antibiotics or sucrose, and incubated overnight at 30° for pKO5, 37° for BAC, or 43° for the allele replacement procedure.

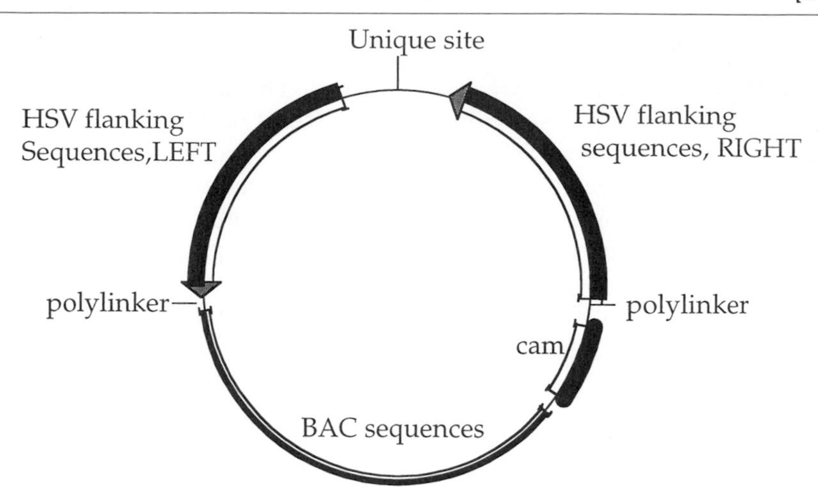

FIG. 2. A plasmid that permits insertion of BAC sequences and marker gene into an HSV locus. cam, chloramphenicol.

Construction of HSV-BAC Plasmid, pHSV-BAC

Construction of an HSV-BAC is straightforward and relies on standard procedures for making recombinant viruses. The first step is to choose the locus in the viral genome for inserting the BAC sequences required for autonomous replication in *E. coli.* We use the thymidine kinase locus because it is possible to select for insertion of BAC sequences by selecting against viral TK activity. pBeloBAC[7] contains the sequences necessary for chromosomal maintenance and a cam marker gene. A plasmid can be constructed containing HSV-1 thymidine kinase (tk) sequences in an inverted orientation (i.e., the 5′ end of the gene and the 3′ end of the gene are both inverted, resulting in the two ends of the tk gene being next to each other). The general design should be

unique site − flanking HSV sequence − pBeloBac
− flanking HSV sequence right

An illustration of this plasmid is shown in Fig. 2.

It is important to design the plasmid to contain a unique restriction site between the left and the right flanking sequences to allow for linearization of the molecule. A TK-BAC plasmid can be constructed as follows.

The *Bgl*II/*Eco*RI fragment from pXholf, HSV strain 17 (Chris Preston,

[7] H. Shizuya, B. Birren, U. J. Kim, V. Mancino, T. Slepak, Y. Tachiiri *et al., Proc. Natl. Acad. Sci. U.S.A.* **89,** 8794 (1992).

University of Glasgow) is subcloned into the *Bam*HI/*Eco*RI sites of pHSVlac,[8] creating plasmid pTK-AB. The tk containing plasmid, BH13, strain 615.9[9], is digested with *Pst*I and *Xho*I to release the 3′ end of the tk gene, and the fragment is subcloned into the *Pst*I/*Xho*I site of pSC1180, creating plasmid TK-CD. Plasmids TK-AB and TK-CD are digested with *Hin*dIII/*Msc*I and *Hin*dIII/*Sal*I, respectively, and the fragments are subcloned into the *Hpa*I/*Sal*I sites of pBelobac to create plasmid pBAC-TK, which contains BAC elements and a cam resistance marker. pBAC-TK is linearized with *Hin*dIII (cuts between TK-AB and TK-CD) and cotransfected into Vero cells with infectious HSV-1 strain F DNA (see later). When cells show 80–100% cytopathic effects, cells are scraped into a sterile 1.5-ml tube. The mixture is frozen and thawed three times, and the virus titer is determined on fresh cell monolayers. TK-deficient viruses are identified by infecting cells at a multiplicity of infection (MOI) of 0.01, adding 20 μM acyclovir (ACV), and covering the monolayer with agar. ACV-resistant plaques arising within several days are tk deficient. These are picked and screened by PCR using primers such as 5′AGGCCGGA TAGCTTGTGC and 5′CGGAACAGAGAGCGTCACA, which will amplify cam sequences. Virus isolates that screen positively by PCR are plaque-purified twice more, and the resulting virus stock is expanded. DNA can be isolated from infected cells, and Southern blot hybridization is performed to confirm the presence of BAC and cam sequences in the correct locus. At least two clones from independent transfections should be analyzed.

The next step is to isolate the HSV recombinant as a BAC plasmid in *E. coli.* To do this, Vero cells are infected with the recombinant, HSV-BAC, at an MOI of 3. Two hours later, the medium is removed and 1 ml of DNAzol or similar reagent (Life Technologies, Inc.) is added per 10-cm plate. The infection should not be allowed to proceed longer than 6 hr because the viral DNA will replicate, resulting in replication intermediates that will yield defective HSV-BAC recombinants. The cells are lysed by gentle pipetting using a wide-bore pipette. Do not vortex. The resulting DNA is precipitated from the lysate by adding 0.5 ml ethanol per milliliter of DNAzol. Samples are mixed by inversion and subjected to centrifugation for 2 min at 12,000*g* or greater at room temperature. The supernatant is removed and the DNA pellet is washed with 70% ethanol. Following centrifugation, the DNA pellet is allowed to air dry. The DNA can be resuspended in 100 μl of TE (10 mM Tris–Cl, 1 mM EDTA, pH 8.0) and stored at −80°.

One microliter of purified DNA is then added to an equal volume of water and electroporated into 40 μl of electrocompetent DH10B *E. coli,*

[8] A. I. Geller, *J. Virol. Methods* **31,** 229 (1991).
[9] B. C. Horsburgh, H. Kollmus, H. Hauser, and D. M. Coen, *Cell* **86,** 949 (1996).

using a suitable electroporation device. Immediately after electroporation, 1 ml SOC is added and the mixture is transferred to a 17 × 100-mm polypropylene tube. The cells are allowed to recover for 1 hr by shaking at 250 rpm at 30° before plating onto LB agar plates containing cam (20 μg/ml). Bacterial colonies arising from overnight incubation at 37° are screened by PCR for BACs that contain HSV sequences. We use primers to four HSV genes: US 6, 5′CCGAATGCTCCTACAACAAG and 5′GTCTT CCGGGGCGAGTTCTG; UL10, 5′GGTGTAGCCGTGCCCCTCAG and 5′GCAGATACGTCCCGCTCAGG; UL 30, 5′ATCAACTTC GACTGGCCCTTC and 5′CCGTACATGTCGATGTTCACC; and UL 40, 5′ACCGCTTCCTCTTCGCTTTC and 5′CCCGCAGAAGGTTG TTGGTG. *Taq* DNA polymerase can be used to amplify the region of interest with 30 cycles (95°, 1 min; 55°, 1 min; 72°, 1 min) using a suitable thermocycler. Bacteria that screen positive with all four primer sets are grown in liquid culture, and DNA is isolated by using any suitable rapid isolation method. Bacterial DNA from these potential HSV-BAC plasmid clones and control viral DNA isolated from the parental HSV recombinant are digested with restriction enzymes and the DNA profiles compared. Their patterns should be identical.

To rescue virus from HSV-BAC plasmid, 2 or 3 μg of DNA from clones that screened positive with all four primer sets is transfected into a 10-cm dish of Vero cells using 18 μl of lipofectamine (Life Technologies, Inc.) or another suitable tranfection reagent. Following incubation at 34° for 2–3 days, virus plaques should be visible. We routinely generate a one-step growth curve from the resulting virus stocks to confirm that the viruses exhibit the expected growth rate (Fig. 3A). Restriction enzyme digests of parental viral DNA and pHSV-BAC DNA can be compared on 0.8% agarose gels to ensure that the entire HSV genome has been cloned (Fig. 3B).

Construction of Gene Replacement Vector

The gene replacement vector, pKO5, a modification of pKO3,[10] contains a temperature-sensitive origin of DNA replication and marker genes for positive and negative selection in bacteria (Fig. 4). The origin of DNA replication is nonpermissive at 43°, and to be maintained, the plasmid must integrate into either the *E. coli* chromosome or the HSV-BAC plasmid. In contrast, at 30°, the plasmid origin of DNA replication is deleterious to chromosomal replication and is frequently excised by homologous recombination. pKO5 also contains a zeocin resistance gene used for positive selec-

[10] A. J. Link, D. Phillips, and G. M. Church, *J. Bact.* **179,** 6228 (1997).

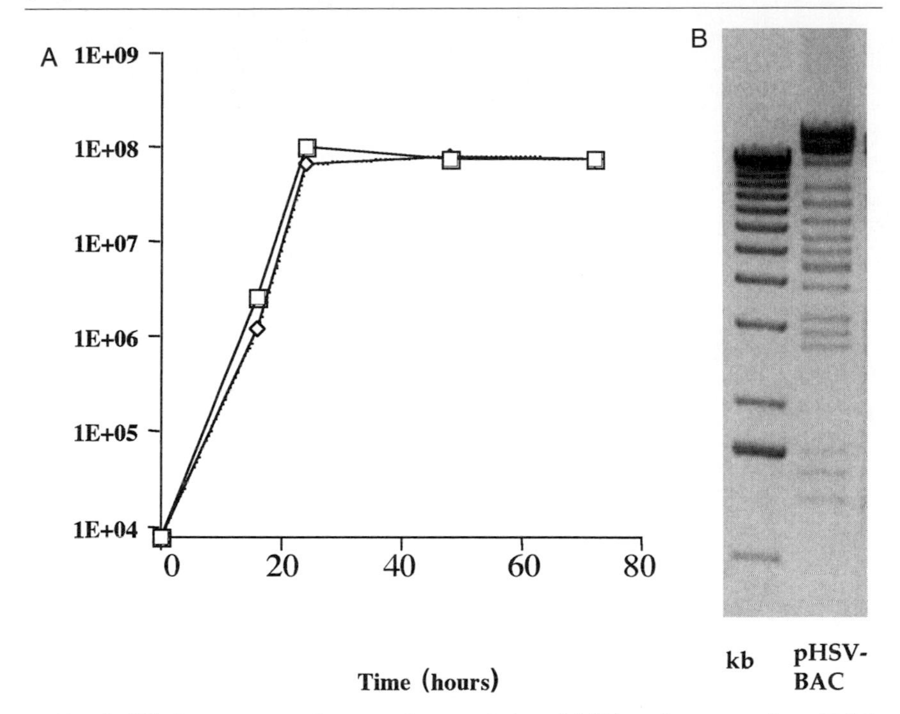

FIG. 3. (A) One-step growth curve of parental virus, 25 (◇), and progeny virus, 19 (□). No significant difference in growth on Vero cells is observed, implying that the passage of HSV-BAC DNA through bacteria is stable. (B) An *Eco*RV restriction enzyme digest of pHSV-BAC DNA.

tion in bacteria and SacB, which encodes levansucrase, a gene whose expression is lethal to *E. coli* growing on media supplemented with 5% sucrose. SacB therefore permits selection for the loss of the plasmid sequences. A multiple cloning region containing *Not*I, *Bam*HI, *Xho*I, *Xba*I, *Sac*I, *Sca*I, *Pac*I, *Pst*I, *Bgl*II, and *Sal*I is present, which facilitates the insertion of a cassette containing a replacement allele such as an insertion, deletion, or point mutation with the appropriate HSV flanking sequences.

Ideally the flanking sequences should be at least 1.5 kb in length. Although flanking sequences as short as 500 bp have been used successfully,[10] the longer the flanking sequence, the better the odds of integration at the desired locus. Furthermore, allele exchange will take place only if integration occurs at sequences to the left of the mutant allele and excision at sequences to the right of the mutant allele or vice versa (Fig. 1). Integration and excision at the same site will not result in allele exchange. Therefore, it is advisable that the flanking sequences be of comparable length. Knock-

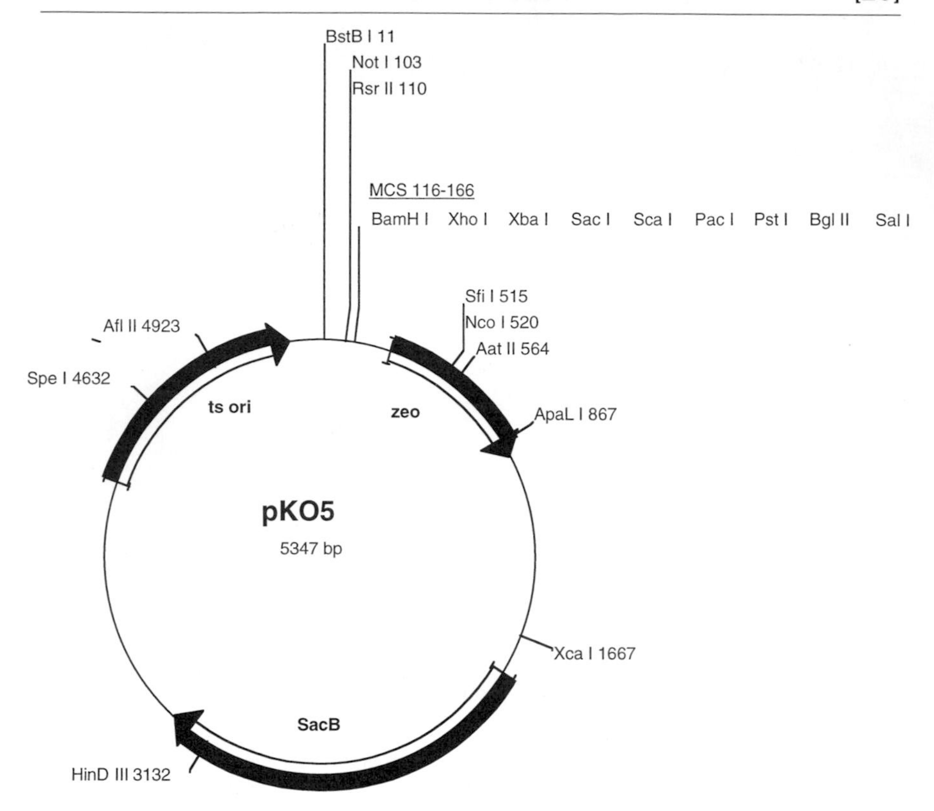

FIG. 4. A schematic of pKO5. zeo, zeocin; ts ori, temperature-sensitive pSC101 origin of replication; MCS, multiple cloning site.

out of viral genes can be achieved by inserting an antibiotic resistance marker gene, e.g., tetracycline, in place of the viral gene. This approach facilitates the selection process.

Although we constructed pKO5 by modifying pKO3 to contain the zeocin resistance gene, other marker genes can be used as long as they are not also present in pHSV-BAC. To generate pKO5, the *Ssp*I–*Dra*I fragment containing the zeocin gene was isolated from the plasmid pZero (Invitrogen) and cloned into the *Nae*I site of pKO3. The cam gene was removed as a *Bst*BI fragment because it can recombine with the cam gene in pHSV-BAC. *E. coli* strains harboring pKO5 can be grown to stationary phase in LB (pH 7.5) at 30° under zeocin selection. The plasmid should be present at ~5–10 copies per cell; thus, DNA isolation protocols should be modified accordingly.

Selective markers in pKO5 must be tested and the plasmid retransformed into bacteria on a regular basis because mutations arise frequently. To test for temperature sensitivity, a single colony containing pKO5 can be grown in a small volume of medium, diluted serially, and plated for single colonies at 30° or 43°. If the test colony is temperature sensitive, colonies arising at 43° should be smaller than those at 30°. The same serial dilutions are performed with sucrose LB versus LB to test for SacB function.

Gene Replacement

Day 1. Electroporate pKO5 (or another suitable gene replacement vector) containing the replacement allele into electrocompetent ($RecA^+$) cells containing pHSV-BAC and allow the cells to recover in 1 ml SOC for 1 hr at 30°. Approximately 1, 10, and 100 μl of cells are plated on prewarmed cam/zeocin plates and incubated at 43° (restrictive temperature). Approximately 0.01 and 0.1 μl of cells are also plated on cam/zeocin plates and incubated at 30° (permissive temperature). Taking into account the dilution factor, the integration frequency is the number of colonies at 43°/number of colonies at 30°. We find that this frequency ranges from 10^{-2} to 10^{-5}. Occasionally, colonies will not arise at 43° due to a low integration frequency. In this instance, colonies that arise at 30° can be diluted serially and plated at 43°.

Day 2–3. From the 43° plate, six colonies are each picked into 1 ml of LB, diluted serially, and plated immediately onto LB agar plates containing 5% (w/v) sucrose and the appropriate antibiotics and incubated at 30° overnight to select for retention of BAC sequences and loss of plasmid sequences. We routinely pick a colony into 1 ml LB, mix, remove 10 μl into another 1 ml LB, mix, and plate 10 or 100 μl at either 30° or 43°. The excision frequency is the ratio of the number of colonies on the 30° sucrose plate to the number of colonies on the 43° cam/zeocin plate.

Day 3–4. Replica plate the 30° colonies from the 5% sucrose plates to (1) 5% sucrose-zeocin plates at 30° to test for loss of the replacement vector and (2) cam plates at 30° to test for maintenance of the HSV-BAC sequences.

To mutate two alleles simultaneously, a replacement vector can be constructed that contains a different positive marker, e.g., tetracycline (tet). Both replacement plasmids are then transformed into *E. coli* containing HSV-BAC and plated onto LB agar plates containing tet, cam, and zeocin at the nonpermissive temperature as described earlier. It is important that the bacterial strain used for the gene replacement process is not already resistant to the markers that you propose to use, e.g., Tn*10* confers tet^r and Tn*5* confers kan^r.

Screening and Characterization

Gene replacement can be confirmed by PCR using primers that flank the targeted allele. Bacterial colonies resistant to both cam and sucrose are picked into PCR reaction buffer. Screening of about 20 colonies should be sufficient. Positive colonies are grown to stationary phase at 37° in 2 liters LB-cam by shaking at 250 rpm. Bacteria are then pelleted by centrifugation in a suitable rotor. DNA can be purified by alkaline lysis using any suitable DNA isolation kit according to the manufacturer's instructions. DNA can also be banded on a cesium chloride gradient according to standard procedures. The purified DNA is then suspended in 200 μl TE (10 m*M* Tris–Cl, 1 m*M* EDTA, pH 8). To confirm that there have been no deletions, restriction enzyme digests can be performed. Southern hybridization and DNA sequencing should be performed to confirm the presence of the desired mutations.

Transfection of pHSV-BAC into Mammalian Cells

Vero cells or other appropriate cell lines are plated the day before transfection at such a density that they will be 70–90% confluent the next day. For Vero cells, this is 1×10^6 cells per 60-mm dish. For generating constructs in which essential viral genes are deleted, the requisite complementing cell line will be required. Vero cells are grown in Dulbecco's modified Eagle's medium (DMEM) supplemented with 10% (v/v) fetal bovine serum (FBS) and 4 m*M* L-glutamine. For each transfection, mix 2–3 μg HSV-BAC DNA with 250 μl OptiMEM (Life Technologies, Inc.). Add 8 μl Lipofectamine Plus reagent (Life Technologies, Inc.) or another suitable lipid-based transfection reagent, adding dropwise while vortexing. Let sit at room temperature for 15 min. In a separate tube, combine 12 μl lipofectamine (Life Technologies, Inc.) and 250 μl OptiMEM. Add the DNA mix to the lipofectamine reagent dropwise while vortexing. Let the mix sit for 15 min at room temperature. During this time, rinse the cells twice with OptiMEM. Leave 2 ml of OptiMEM on the cells after the second rinse. Add the transfection mix to the cells, swirling to distribute evenly. Incubate at 37° for 4 hr to allow for uptake of the DNA into the cells. Following this incubation, remove the transfection mix by aspiration and add 2 ml DMEM/10% FBS. Virus plaques should appear 2–3 days later.

Rescue and Growth of Virus

Recombinant HSV isolated from the resulting plaques does not need to be purified further because clonal selection has occurred in *E. coli.* Cells are harvested when they show 80–100% cytopathic effects. Virus is titered on a fresh monolayer of the appropriate mammalian cell line. Viral stocks can be amplified by infecting cells at an MOI of approximately 0.01. One-step growth curves can be performed by infecting cells at an MOI of 3–10 and harvesting cells at 12, 18, 24, 48, and 72 hr postinfection. The cells are frozen and thawed three times to release virus from the cells, and the virus titers are determined as described earlier.

Preparation of Infectious Viral DNA

Ten confluent T150 flasks are infected at an MOI of 0.01. When cells show 80–100% cytopathic effects (3–4 days), cells are harvested by tapping the flasks vigorously to dislodge the cells and the contents are transferred to 250-ml centrifuge bottles on ice. The mixture is centrifuged at approximately 800*g* for 10 min in an appropriate centrifuge at 4°. The supernatant containing extracellular virus is decanted into clean 250-ml bottles and held on ice until the cell pellets have been processed. Cell pellets are then suspended in 6 ml of 10 m*M* Tris, pH 7.4, 10 m*M* NaCl, 3 m*M* $MgCl_2$, and transferred to a Dounce homogenizer on ice. The cells are allowed to swell on ice for 10 min and are then processed by four strokes with a tight-fitting glass pestle. Pour the homogenate into 50-ml centrifuge tubes, rinse the homogenizer with 10 m*M* Tris, pH 7.4, 10 m*M* NaCl, 3 m*M* $MgCl_2$ and add this rinse to the original homogenate. Centrifuge the mixture at 800*g* for 10 min at 4° to clarify the homogenate. Decant the supernatant, which contains virus released from the cells, into the 250-ml centrifuge bottles containing the extracellular virus, making sure that the pellet is left behind. This combined supernatant is then centrifuged at 16,000*g* and 4° for 1 hr to pellet the virus. Discard the supernatant. The virus pellet is suspended in approximately 4.5 ml of TNE (10 m*M* Tris, pH 8.0, 100 m*M* NaCl, 1 m*M* EDTA) and transferred to a 50-ml conical tube. Add 0.5 ml 10% SDS, and rock the tube gently to lyse the virus without shearing the DNA. Add an additional 11 ml of TNE, followed by 4 ml of 5 *M* $NaClO_4$. Rock gently to mix. To the total volume of 20 ml, add an equal volume of chloroform : isoamyl alcohol ($CHCl_3$: IAA, 24 : 1). Invert the tube gently for 10 min to emulsify. Do not vortex. When this is finished, the tube can be centrifuged at approximately 800*g* for 5 min at 4°. The $CHCl_3$ extraction is repeated three more times.

Transfer the aqueous phase to $\frac{3}{4}$-inch-wide dialysis tubing (8000 M_r cutoff) and dialyze against 0.1× SSC. Change buffer at least twice and dialyze overnight at 4°. When this is complete, reduce the volume of liquid in the dialysis tubing by laying the tubing containing the DNA on a bed of polyethylene glycol (PEG) 8000 and cover with additional PEG-8000. After 2 hr, most of the liquid will have been absorbed by the PEG. Rinse the dialysis tubing with sterile water to remove the PEG and place bag back to dialyze for an additional 2 hr at 4°, changing buffer twice. Following dialysis, the DNA solution is removed using a sterile, wide-bore pipette and stored in sterile tubes. Determine DNA concentration by measuring absorbance at 260 and 280 nm.

Example of HSV-BAC Technology

To illustrate the usefulness of this methodology, we deleted a nonessential gene, vhs, from HSV-BAC (Fig. 5A). First, we cloned a 6.6-kb *Hin*dIII/*Eco*RI fragment (90,146–96,752; relative to the strain 17 sequence[11]) that contains vhs (91,168–92,637) into pBlueScript (Stratagene, La Jolla, CA). This plasmid was digested with *Bam*HI and *Sca*I, made blunt ended, and ligated, thereby destroying both restriction sites. This plasmid, pΔvhs, has a ~300 nucleotide deletion within the vhs coding region that ablates vhs function. pΔvhs was digested with *Eco*RV and *Sst*I to release a fragment, containing the modified vhs sequences flanked by at least 1 kb of HSV sequences, which was subsequently cloned into pKO5. pKO5Δvhs was introduced into $RecA^+$ RR1 cells containing pHSV-BAC by electroporation, and the resulting bacteria were plated at 43°. The following day, 6 colonies were suspended in 1 ml of LB, and 10 μl from each tube was removed and added to a new tube containing 1 ml of LB. Fifty microliters of this was plated on cam/sucrose LB agar plates and incubated at 30° overnight. The following day, $cam/sucrose^r$ colonies were replica plated onto zeocin plates and incubated at 30° overnight. No colonies grew on these plates, indicating that the replacement vector had been lost from the cells. PCR was performed on $cam/sucrose^r$, $zeocin^s$ colonies using primers that lie outside the deletion. Several PCR-positive colonies were picked for further analysis. Southern blots confirmed the presence of the deletion (Fig. 5B). To confirm that the vhs function was ablated, one clone, p25-34Δvhs, was transfected into Vero cells and virus isolated. Vero cells were then infected with either HSV-BAC-Δvhs or wild-type virus at an MOI of 20 or mock infected. The infections, performed in the presence of actinomycin D, were allowed to proceed for 3 hr. The cells were then radiolabeled with 50 μCi of $[^{35}S]$methionine for 1 hr in methionine-free medium. The monolayer was then lysed in 100 μl of Nonidet P-40 lysis buffer, added to 10 μl of loading buffer, boiled for 10 min, and subjected to electrophoresis on a 10% polyacrylamide gel. Figure 5C shows that the protein synthesis of mock and mutant viruses were comparable. In contrast, the host cell protein synthesis with the wild-type virus was compromised severely. These

[11] D. J. McGeoch, M. A. Dalrymple, A. J. Davison, A. Dolan, M. C. Frame, and D. McNab *et al., J. Gen. Virol.* **69,** 1531 (1988).

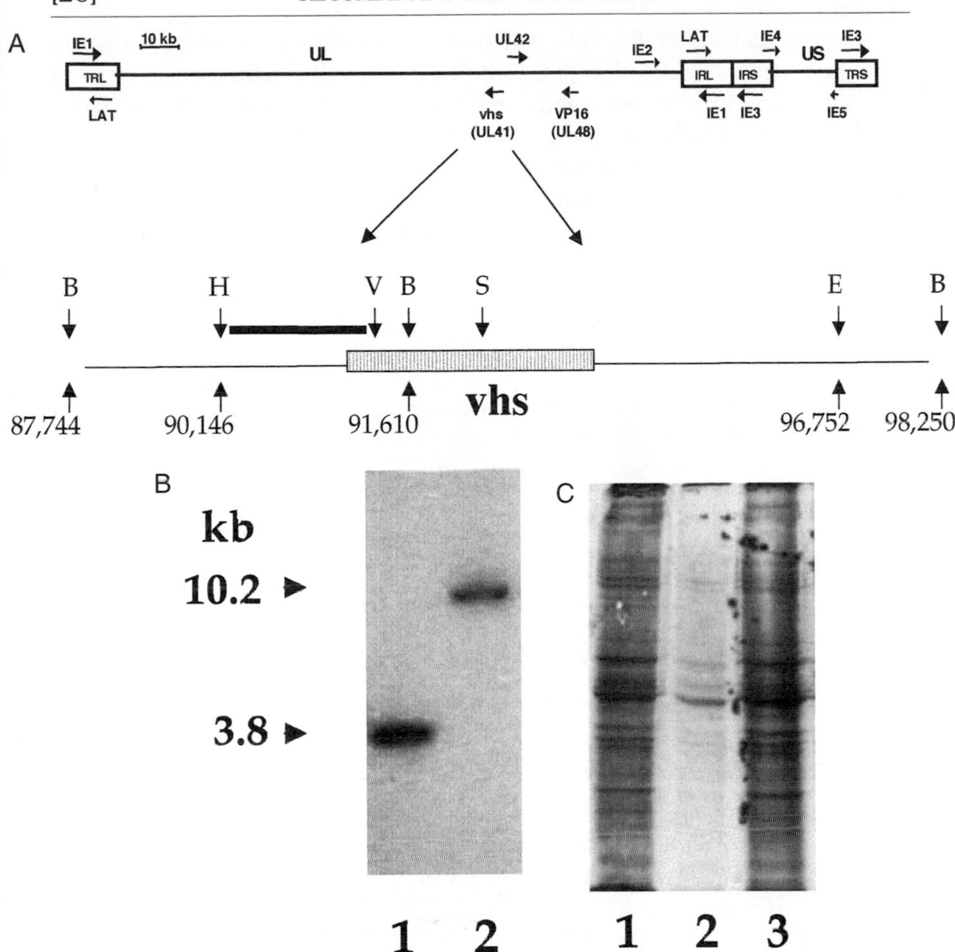

FIG. 5. Deletion of a *Bam*HI–*Sca*I fragment from vhs (UL41). (A) The HSV genome and an expansion of the vhs locus. Restriction enzyme sites are B, *Bam*HI; E, *Eco*RI; H, *Hin*dIII; S, *Sca*I; and V, *Eco*RV. Nucleotide positions relative to strain 17 are indicated. The *Hin*dIII–*Eco*RV fragment that was used as a probe is illustrated as a black bar. (B) A Southern blot confirming loss of the *Bam*HI–*Sca*I fragment from vhs. *Bam*HI-digested DNA was probed with the radiolabeled *Hin*dIII–*Eco*RV fragment. Lane 1, wt virus; and lane 2, HSV-BAC-Δvhs. The sizes of the fragments are indicated. (C) Analysis of host cell protein synthesis after infection of wild-type HSV-1 (lane 1), HSV-BAC-Δvhs (lane 2), or mock treated (lane 3). Infected cell lysates were separated on a 10% SDS–polyacrylamide gel.

data show that a specific 300 nucleotide deletion was introduced into the intended locus. The deletion was confirmed by Southern blotting. Extensive RE analysis (not shown) confirmed that the only change in restriction profile was the expected change. The HSV-BAC-Δvhs virus grew as well as its parent in a one-step growth curve (data not shown). As expected, the only difference between these two viruses was in their ability to inhibit host cell protein synthesis. In conclusion, using this procedure, it took 7 days to produce the desired recombinant HSV.

Acknowledgments

We thank Dong Qiang for excellent technical assistance, Andrew Wieczorek for helpful discussions, and George Church for pKO3.

[21] Protein and RNA Affinity Selection Techniques Using HIV-1 Tat

By MITCHELL E. GARBER, PING WEI, GEORG CADERAS, and KATHERINE A. JONES

Introduction

The human immunodeficiency virus type 1 (HIV-1) Tat protein is a small nuclear protein expressed early in the virus life cycle that functions as a potent and highly selective activator of HIV-1 transcription.[1–4] Genetic and biochemical studies have shown that Tat regulates step in early elongation of RNA polymerase (RNAPII) transcription and functions through the TAR element, which folds into an RNA hairpin structure at the 5′ end of the emerging viral transcript. Studies indicate that Tat interacts directly with human cyclin T1 (CycT1),[5] a subunit of the Tat-associated kinase (TAK) complex.[6–8] The TAK complex also contains the nuclear kinase,

[1] K. A. Jones and B. M. Peterlin, *Annu. Rev. Biochem.* **63,** 717 (1994).
[2] K. A. Jones, *Genes Dev.* **11,** 2593 (1997).
[3] M. Emerman and M. Malim, *Science* **280,** 1880 (1998).
[4] B. R. Cullen, *Cell* **93,** 685 (1998).
[5] P. Wei, M. E. Garber, S.-M. Fang, W. H. Fischer, and K. A. Jones, *Cell* **92,** 451 (1998).
[6] J. Peng, Y. Zhu, J. Milton, and D. H. Price, *Genes Dev.* **12,** 755 (1998).
[7] C. H. Herrmann and A. P. Rice, *Virology* **197,** 601 (1993).
[8] C. H. Herrmann and A. P. Rice, *J. Virol.* **69,** 1612 (1995).

0076-6879/99 $30.00

CDK,[9–11] as well as other subunits of the positive transcription elongation factor (P-TEFb) complex.[12–13] CDK9 has been shown to phosphorylate the carboxyl-terminal domain (CTD) of RNAPII,[11–15] and previous work has established that both CDK9 and RNAPII CTD are essential for Tat-mediated transactivation (for review, see Jones[2]). Interestingly, Tat binds cooperatively with hCycT1 to TAR RNA, and the hCycT1 : Tat complex, unlike free Tat alone, recognizes sequences in the loop of TAR RNA that are critical for Tat transactivation.[5] Through its ability to interact with CycT1, Tat can recruit the multisubunit TAK/P-TEFb complex to nascent viral RNA transcripts. This step may be critical to activate (or derepress) CDK9 activity,[16–17] resulting in phosphorylation of the RNAPII CTD and efficient elongation of viral transcription.

This article describes the *in vitro* transcription and TAR RNA competition protocols used to study the mechanism of TAR-dependent activation of the HIV-1 promoter by Tat. In addition, we present an RNA affinity selection procedure that was used to establish that Tat induces loop-specific binding of TAK/P-TEFb complexes to TAR RNA in nuclear extracts. Finally, we outline the glutathione *S*-transferase (GST)–Tat protein affinity selection protocol, modified from the procedure of Herrmann and Rice,[7,8] used to purify human cyclin T1 (hCycT1) from HeLa nuclear extracts in quantities sufficient for microsequence analysis and subsequent molecular cloning of the hCycT1 gene.[5] These approaches are currently used by us and others[18] to assess the composition of the TAK/P-TEFb complex and to study the regulation of TAK activity in various cells. More generally, the techniques described here can be adapted readily for studies of other nuclear proteins that interact with enzymes (e.g., kinases, phosphatases, acetyltransferases, deacetylases) to regulate transcription or other important processes in the cell.

[9] Y. Zhu, T. Pe'ery, J. Peng, Y. Ramanathan, N. Marshall, T. Marshall, B. Amendt, M. Mathews, and D. H. Price, *Genes Dev.* **11,** 2622 (1997).

[10] H. Mancebo, G. Lee, J. Flygare, J. Tomassini, P. Luu, Y. Zhu, C. Blau, D. Hazuda, D. Price, and O. Flores, *Genes Dev.* **11,** 2633 (1997).

[11] X. Yang, C. H. Herrmann, and A. P. Rice, *J. Virol.* **70,** 4576 (1996).

[12] N. Marshall and D. H. Price, *Mol. Cell. Biol.* **12,** 2078 (1992).

[13] N. Marshall and D. H. Price, *J. Biol. Chem.* **270,** 12335 (1995).

[14] N. Marshall, J. Peng, Z. Xie, and D. H. Price, *J. Biol. Chem.* **271,** 27176 (1996).

[15] X. Yang, M. Gold, D. Tang, D. Lewis, E. Aguilar-Cordova, A. P. Rice, and C. H. Herrmann, *Proc. Natl. Acad. Sci. U.S.A.* **94,** 12331 (1997).

[16] C. Parada and R. G. Roeder, *Nature* **384,** 375 (1996).

[17] L. Garcia-Martinez, G. Mavenkal, J. Neveu, W. Lane, D. Ivanov, and R. B. Gaynor, *EMBO J.* **16,** 2836 (1997).

[18] C. H. Herrmann, R. G. Carroll, P. Wei, K. A. Jones, and A. P. Rice, *J. Virol.*, in press (1998).

In Vitro Transcription Assay to Study TAR-Dependent Transactivation by Tat

This section outlines three procedures that are essential to study transactivation by HIV-1 Tat *in vitro:* (1) preparation of a HeLa nuclear extract, (2) preparation of recombinant GST–Tat proteins in bacteria, and (3) conditions to study Tat transactivation and squelching by exogenous TAR RNA using a "runoff" transcription assay. The procedure described here uses relatively large quantities of HeLa cells, which are required for biochemical purification of nuclear proteins, but can be scaled down for general analytical purposes.

Preparation of HeLa Cell Nuclear Extracts

Materials

1-liter Sorvall 53 bottles, 250-ml GSA bottles
45-ml Bellco (Dounce) homogenizer with size B ball type pestle
10-ml wide-bore glass tissue culture pipette
Opaque and clear (polycarbonate) centrifuge tubes

Solutions

Phosphate-buffered saline (PBS)/$MgCl_2$: 1× PBS containing 5 m*M* $MgCl_2$

10× PBS: 80 g NaCl (1*M*)/2g KCl (26 m*M*)/9.2 g Na_2HPO_4 (anhydrous; 65 m*M*)/2 g KH_2PO_4 (15 m*M*). Add doubly distilled water to 1 liter.

Buffer H: 10 m*M* Tris–HCl, pH 7.9 (or HEPES, pH 7.9), 10 m*M* KCl, 0.75 m*M* spermidine, 0.15 m*M* spermine, 0.1 m*M* EDTA, and 0.1 m*M* EGTA. Added fresh: 2 m*M* dithiothreitol (DTT), 0.1 m*M* phenylmethylsulfonyl fluoride (PMSF), 1× protease inhibitors (leupeptin/pepstatin/aprotonin), and 2 mg/ml benzamidine.

Buffer D: 50 m*M* Tris–HCl, pH 7.5, 10% sucrose, 0.42 *M* KCl, 5 m*M* $MgCl_2$, 0.1 m*M* EDTA, and 20% (v/v) glycerol. Added fresh: 2 m*M* DTT, 0.1 m*M* PMSF, and 2 mg/ml benzamidine.

TM buffer: 50 m*M* Tris–HCl, pH 7.9, 12.5 m*M* $MgCl_2$, 1 m*M* EDTA, and 20% glycerol

100× protease inhibitors: 100 μg/ml pepstatin A, 400 μg/ml leupeptin, 1 mg/ml aprotinin, and 2 mg/ml soybean trypsin inhibitor. Store in 1-ml aliquots at $-20°$.

Procedure

1. Harvest 8 liters of HeLa S3 cells (it is important to use exponentially growing cells at a density of 3–6 × 10^5 cells/ml of suspension culture).

Cells are spun in 1 liter Sorvall 53 bottles at 4200 for 15 min at 4° in a Beckman J-6B centrifuge. Use a reduced brake, if available, to avoid disrupting the cell pellets. Decant the media carefully and wash once in approximately 200 ml 1× PBS/$MgCl_2$ using a wide-bore 10-ml glass tissue culture pipette to resuspend the pellets gently. Transfer the resuspended cells to 250-ml GSA bottles and spin again at 4200 for 12 min.

2. Decant the supernatant and resuspend the cell pellet in 4 packed cell volumes (PCV) of buffer H (approximately 40 ml for 8 liters of cells). Swell on ice for 15 min.
3. Homogenize the cells by douncing 8 times with a B-type pestle on ice (cell lines other than HeLa may rupture more readily, consequently it is advised to use a microscope to check for cell lysis at this point). Spin out nuclei in the opaque polycarbonate tubes at 5000 for 8 min in the SS34 rotor. Decant the supernatant carefully to avoid losing the nuclear pellet, which may be loose.
4. Resuspend the nuclei in 15 ml buffer H using a wide-bore glass pipette, pool into a single centrifuge tube, and spin again at 5000 for 8 min. Decant off the wash buffer and freeze the nuclear pellet at −80° until 36–40 liters have accumulated as required for a full-scale preparation. Alternatively, if more cells are harvested, or if a smaller preparation is needed, proceed to step 5.

   Chill the SW28 buckets and rotor in the cold room.
   Precool the centrifuge(s) to 4°
   Pull out stored nuclei and thaw. Add DTT, PMSF, and benzamidine to 4 PCV of buffer D.
5. Resuspend the frozen nuclei in 4 PCV buffer D (approximately 200 ml for 36 liters of HeLa nuclear pellets). If nuclei have been frozen, dounce the suspension with five strokes. Stir on ice for 30 min, transfer to SW28 tubes, and spin at 25,000 for 60 min.
6. Collect the high-speed supernatant (HSSN) and note the volume. Weigh out 0.33 g solid ammonium sulfate per milliliter of HSSN, grind up the ammonium sulfate in a mortar and pestle and slowly add to the HSSN with gentle stirring over a period of 15–20 min on ice in the cold room. Allow the ammonium sulfate to dissolve completely and continue stirring for an additional 20–30 min on ice.
7. Transfer the HSSN to opaque polycarbonate tubes and spin at 17,000 for 20 min in an SS34 rotor. Decant off the ammonium sulfate supernatant and remove any residual material with a Pasteur pipette. The nuclear protein pellet can be stored on ice at 4° overnight, if desired.
8. Resuspend the pellets in 15–20 ml (for 36 liter HeLa cells) of TM 0M containing 2 m*M* DTT and 0.1 m*M* PMSF. Spin the protein at

15,000 for 5–10 min to pellet any insoluble material before sequential dialysis of the extract to remove excess ammonium sulfate.

Preparation of Recombinant GST–Tat Protein

Dilute a 50-ml overnight culture (BL21 DE3) with L broth to 500 ml, grow bacteria to a density of 0.9 A_{600} and induce with 0.1 m*M* IPTG for 7 hr at 30°. Resuspend cells in lysis buffer (1× PBS containing 10 m*M* DTT, 5 m*M* EDTA, 1 m*M* PMSF, 2 mg/ml benzamidine, 2 μg/ml pepstatin A, 8 μg/ml leupeptin, 20 μg/ml aprotinin, 40 μg/ml soybean trypsin inhibitor, and 200 μg/ml lysozyme), sonicate for 30 sec, and freeze and thaw once. Clarify the cell lysate by centrifugation at 4° in a SW41 rotor for 30 min at 30,000 rpm. Bind the supernatant to 500-μl glutathione beads for 1 hr at 4°, wash three times in 1× PBS containing 10 m*M* DTT and 0.1 m*M* PMSF, rinse once in Tat elution buffer (50 m*M* Tris–HCl, pH 7.5, 100 m*M* NaCl, 10 m*M* DTT, and 0.1 m*M* PMSF), and elute in Tat elution buffer containing 10 m*M* glutathione. Add glycerol to 10% (final concentration) and store the Tat protein at −80° prior to use. Cleave the GST peptide with thrombin at room temperature for 2 hr.

Assay for TAR-Dependent Transcriptional Activation by Tat in Vitro

Several different procedures have been described to study transactivation by HIV-1 Tat *in vitro*.[19–21] Our protocol differs from some of these in that it does not require an ATP preincubation step or incubation of extracts with citrate to repress basal elongation activity in the absence of Tat. When coupled with the just-described method for preparing nuclear extracts, Tat activates transcription very strongly in this system (Fig. 1).

Solutions. Each transcription reaction (30 μl, final volume) contains the following components: HeLa nuclear extract (100 μg) in 14 μl TM buffer containing 200 m*M* KCl; 2 μl HIV Tat in Tat elution buffer; 1.2 μl poly(dI-dC), 0.5 mg/ml; 1.2 μl poly(rI-rC), 0.5 mg/ml; 1 μl linearized pHIV2-CAT (150 ng), cut with *Nco*I; 1 μl linearized pHIV2Δ34-CAT (100 ng), cut with *Nco*I; 1.5 μl 20× rNTP mix (50 m*M* DTT, 2 m*M* EDTA, 0.2 *M* phosphocreatine, 12 m*M* rATP, 12 m*M* rGTP, 12 m*M* rCTP, 0.8 m*M* rUTP); and 0.2–0.4 μl [α-^{32}P]rUTP (800 Ci/mmol, 20 μCi/μl).

[19] H. Kato, H. Sumimoto, P. Pognonec, C. H. Chen, C. A. Rosen, and R. G. Roeder, *Genes Dev.* **6,** 655 (1992).

[20] R. A. Marciniak, B. J. Calnan, A. D. Frankel, and P. A. Sharp, *Cell* **63,** 791 (1990).

[21] C. T. Sheline, L. H. Milocco, and K. A. Jones, *Genes Dev.* **5,** 2508 (1991).

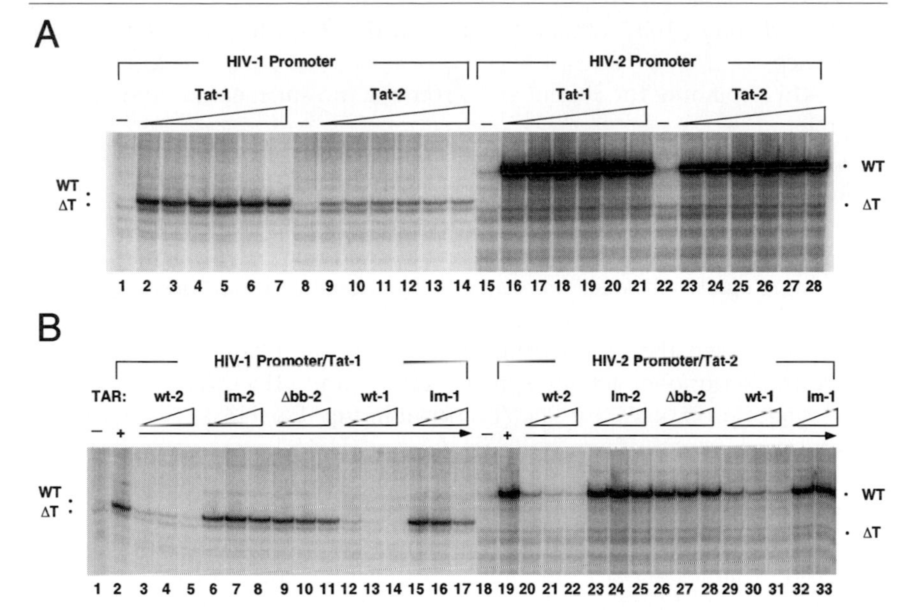

FIG. 1. Exogenous TAR RNA specifically inhibits Tat transactivation of the HIV-1 promoter *in vitro.* (A) Tat transactivation of the HIV-1 and HIV-2 promoters *in vitro.* The runoff transcription was done in the absence (−) or presence (+) of 20, 40, 80, 160, or 320 ng GST-cleaved Tat-1 or Tat-2 protein as indicated. Runoff transcripts from either HIV-1 or HIV-2 templates are labeled WT. As an internal control, the HIV-2 Δ34 template, which lacks a functional TAR element, results in a runoff transcript labeled ΔT. (B) Specific inhibition of Tat activation with the addition of TAR RNA *in trans.* SP6- or T7-derived TAR RNA transcripts were added exogenously into the *in vitro* Tat activation reaction immediately preceding the addition of DNA template and ribonucleotides. Runoff RNA transcripts were generated in the absence (−) or presence (+) of 80 ng of Tat-1 or Tat-2 as indicated. RNA competitors were at 8, 16, and 32 pmol for TAR-2 wild-type (wt-2), double loop mutant (lm-2), or double bulge deletion (Δbb-2) and 16, 32, and 64 pmol for TAR-1 wild-type (wt-1) or loop-mutant (lm-1).

Protocol

1. Incubate the mixture for 30 min at 30°
2. Terminate the transcription reactions by adding 100 μl of stop solution per reaction (1% SDS, 20 m*M* EDTA, 0.1 *M* NaCl, 100 μg/μl yeast tRNA).
3. Extract the mixture once with phenol:chloroform [1:1 (v/v)] and once with chloroform.
4. Add 0.5 volume of ammonium acetate and 2.5 volumes of ethanol to the supernatant (for 120 μl supernatant, use 60 μl 7.5 *M* ammonium

acetate and 300 μl 100% ethanol) and allow the RNA to precipitate for 30 min on ice.

5. After spinning for 30 min at 4°, remove the supernatant and air dry the pellet.

6. Dissolve the pellet in 8–10 μl loading dye (98% deionized formamide, 10 m*M* EDTA, 0.04% bromphenol blue, 0.04% xylene cyanol).

7. Perform electrophoresis on a 6% denaturing polyacrylamide gel (41 cm × 20 cm × 0.4 mm) at 40 W constant power for 2–3 hr.

8. After electrophoresis, dry the gel and visualize by autoradiography.

We have used the *in vitro* runoff transcription assay to evaluate the relative transcriptional activities of the HIV-1 and HIV-2 Tat proteins at their cognate promoters (Fig. 1). As shown in Fig. 1A, HIV-1 Tat strongly activates transcription from both HIV-1 and HIV-2 promoters *in vitro*. Phosphoimager scanning of this gel indicates that HIV-1 and HIV-2 Tat activate the HIV-2 promoter approximately 45-fold under the conditions described here. The HIV-1 promoter was also induced strongly by HIV-1 Tat (35-fold; lanes 23–28) but was activated only weakly by HIV-2 Tat (5.4-fold; lanes 9–14). Tat did not stimulate transcription from a mutant HIV-2 promoter that lacks the *cis*-acting TAR element (ΔTAR), indicating that Tat activity is TAR dependent in this assay. The relative inability of HIV-2 Tat to activate transcription from the HIV-1 promoter is consistent with previous *in vivo* findings and has been attributed to differences in the ARM region of the two Tat proteins that weaken the binding of HIV-2 Tat to the HIV-1 TAR RNA structure.

To assess the ability of exogenous HIV-1 and HIV-2 TAR RNAs to inhibit (or squelch) Tat transactivation *in vitro*, we expressed wild-type and loop or bulge mutant versions of these two TAR RNAs synthetically, as described previously.[5] The wild-type and mutant TAR RNAs used in the competition experiment are outlined in Fig. 2. Analysis of the ability of these RNAs to compete for Tat transactivation *in vitro* (Fig. 1B) revealed that the wild-type HIV-2 TAR RNA was approximately twice as effective a competitor as HIV-1 TAR RNA, consistent with the fact that the HIV-2 TAR RNA contains two functional RNA stem–loop structures. The HIV-1 TAR RNA was a poor inhibitor of HIV-2 transcription (compare lanes 29–31 and 20–22), consistent with the fact that HIV-2 Tat binds poorly to HIV-1 TAR RNA. Bulge deletions or point mutations affecting four loop residues in either TAR structure eliminated the ability of either RNA to inhibit Tat activation *in trans*. Thus the inhibition of Tat activity by exogenous HIV-1 and HIV-2 TAR RNAs is specific and requires sequences in both the loop and the bulge of the RNA hairpin structure.

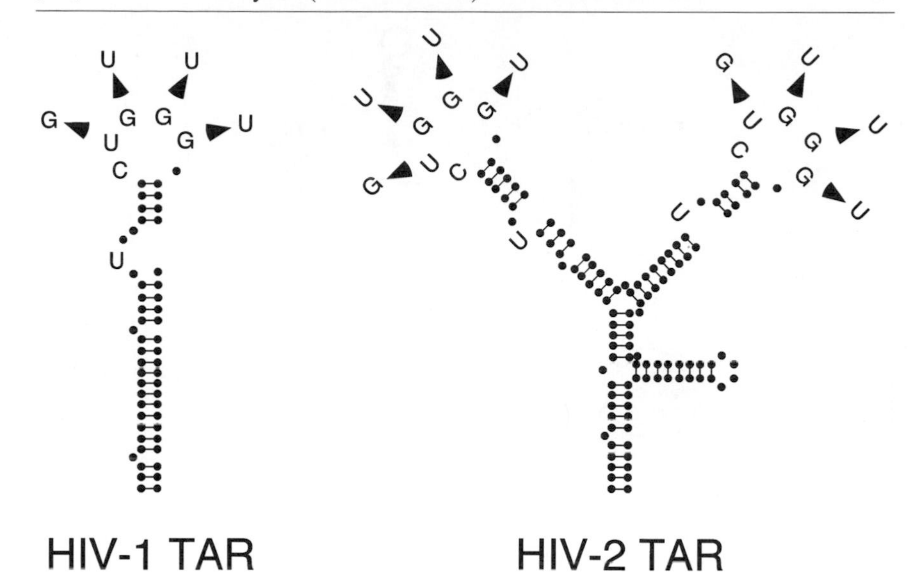

FIG. 2. Structure of the wild-type and mutant HIV-1 and HIV-2 TAR RNAs used in the competition experiment in Fig. 1B.

RNA and Protein Affinity Selection Procedures to Isolate hCycT1-Containing TAK/P-TEFb Complexes from Nuclear Extracts

The Tat-associated kinase (TAK) complex can be isolated readily from HeLa nuclear extracts by affinity selection with GST–Tat (amino acids 1–48) following the procedure of Herrmann and Rice.[7,8] Our modifications to this approach, and the scaled-up procedure necessary to purify hCycT1 from HeLa nuclear extracts, are outlined in this section. We have also used an affinity selection procedure in which the Tat:TAK complex is purified from crude HeLa nuclear extracts by TAR RNA affinity chromatography, as outlined schematically in Fig. 3. This general approach has been used successfully to isolate nuclear proteins that bind to retroviral transport RNA elements.[22] An example of the application of this approach to detect hCycT1 and CDK9 is presented in Fig. 4. In this experiment, HIV-2 Tat was incubated with HeLa nuclear extract, and the Tat:TAK complex was isolated by RNA affinity selection. CDK9, hCycT1, and other subunits

[22] H. Tang, G. M. Gaietta, W. H. Fischer, and F. Wong-Staal, *Science* **276,** 1412 (1997).

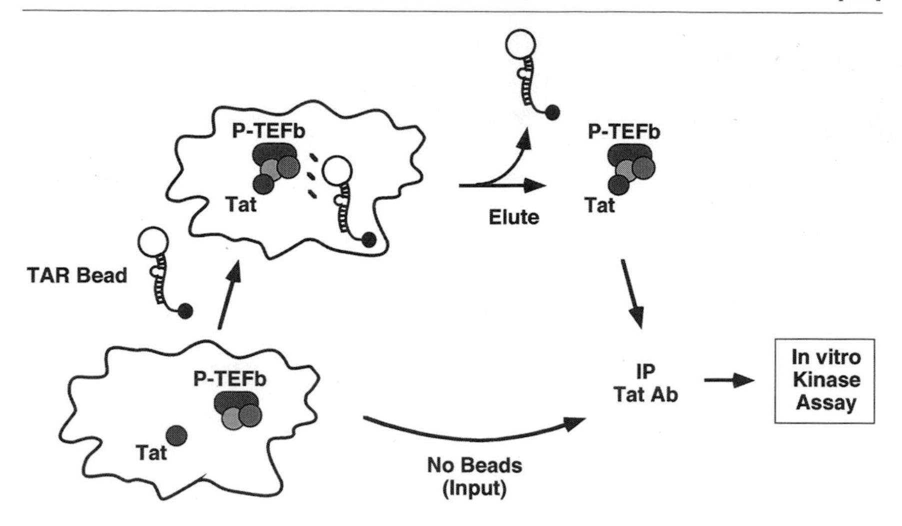

FIG. 3. Schematic diagram of the TAR RNA affinity selection strategy used to examine Tat-dependent binding of nuclear TAK/P-TEFb to TAR RNA.

of TAK/P-TEFb could be visualized by incubation of the beads with [γ-^{32}P]ATP, which permits autophosphorylation of the kinase and adventitious labeling of the cyclin and Tat-2 proteins. The Tat : TAK complex can also be purified to near homogeneity by incubation with wild-type TAR RNA-coupled beads, and the eluted TAK complex recovered efficiently following GST-Tat affinity selection (data not shown). Previous studies have shown that TAK/P-TEFb does not associate with activation domain mutant Tat proteins. In addition, the Tat : TAK complex in nuclear extracts does not associate with loop mutant TAR RNA beads (Fig. 4, lane 2). Through these studies, as well as Western blot analysis of CDK9 and hCycT1 in these different fractions, we have been able to ascertain that TAK/P-TEFb does not bind to TAR RNA in the absence of Tat and that the nuclear Tat : TAK complex, like the complex formed between recombinant hCycT1 and Tat, binds to TAR RNA in a loop-specific manner. Details of these procedures follow.

HIV-1 TAR RNA Affinity Chromatography

Solutions

Tat kinase buffer (50 m*M* Tris–HCl, pH 7.5, 10 m*M* $MgCl_2$, and 5 m*M* DTT)

Binding buffer (50 m*M* Tris–HCl, pH 8.0, 0.5% NP-40, 5 m*M* DTT)

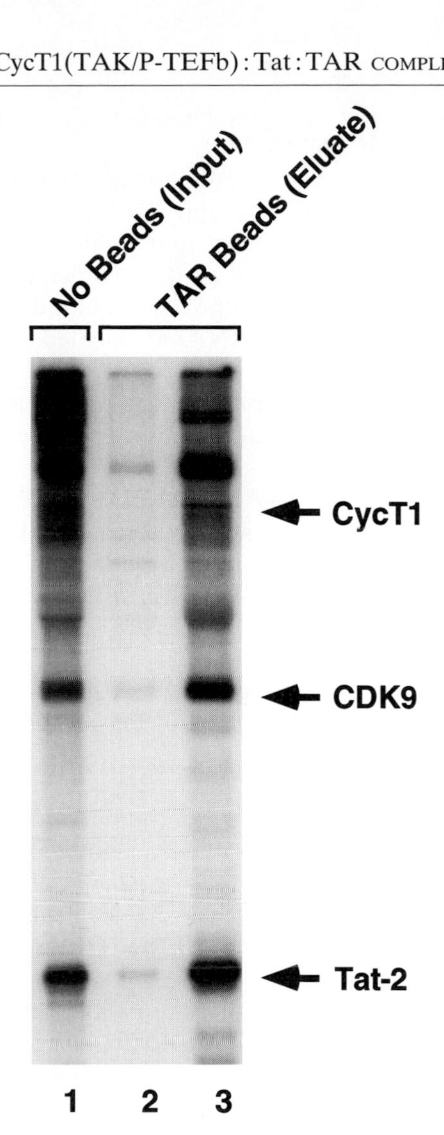

FIG. 4. Purification of TAK/P-TEFb from crude nuclear extracts using TAR RNA affinity chromatography. The HeLa nuclear extract and HIV-2 Tat were incubated in the absence (lane 1) or presence of either loop mutant (lane 2) or wild-type (lane 3) HIV-2 TAR RNA coupled to streptavidin agarose beads. TAR RNA-binding proteins were eluted with high salt and immunoprecipitated with anti-Tat antibody, and the Tat-bound proteins were incubated with [γ-^{32}P]ATP as described in the legend to Fig. 3. Labeled proteins were separated on SDS–PAGE and visualized by autoradiography.

Tat elution buffer (50 mM Tris–HCl, pH 7.5, 100 mM NaCl, 10 mM DTT, and 0.1 mM PMSF)

Procedure. Large-scale synthesis (400 μl reaction volume) of biotinylated TAR RNA is performed with Sp6 RNA polymerase as described,[5] except that reactions contain 100 nmol unmodified rUTP (1.0 μl of 0.1 M) and 120 nmol biotin-rUTP (12 μl of a 10 mM stock from Clontech). This results is an approximately 85% incorporation of biotin-rUTP and no more than one biotin per TAR molecule. RNAs are labeled to low specific activity by the addition of 10 μCi of [α-^{32}P]rUTP to the synthesis reaction in order to quantitate the efficiency of coupling of TAR RNA to the beads and the extent of RNA release from the beads on incubation with crude nuclear extract.

For the batch purification of TAK/P-TEFb by TAR RNA affinity chromatography, the following components were assembled, per reaction: 40 μl NE (500 μg protein), 12 μl TM buffer containing 100 mM KCl and 0.2% NP-40, 11 μl Tat elution buffer, 1 μl GST-cleaved Tat (200 ng), and 6.8 μl doubly distilled water.

1. Incubate for 30 min at 30°. Add 20 μl of 0.1 M NaCl and 28 μl doubly distilled water. Allow to stand at 0° for an additional 40 min. Spin to pellet.
2. Bind 10 μg biotin-TAR RNA to 25-μl streptavidin beads in 1× SSC. The TAR RNA-bound beads are equilibrated and resuspended in 80 μl 0.5× TM containing 75 mM KCl and 0.05% NP-40.
3. Add 120 μl nuclear extract mix to a slurry (80 μl volume) of TAR RNA-coupled beads. Rotate in a silanized tube for approximately 14 hr at 4°. Roughly 90% of the TAR RNA remains bound to the beads during the 14-hr incubation as measured by radioactivity.
4. Wash in 120 μl of 0.5× TM buffer containing 75 mM KCl and 0.05% NP-40. To elute, add 120 μl TM buffer containing 1 M KCl and 0.05% NP-40, incubate on ice with gentle agitation for 30 min, and spin to pellet the beads.
5. For the flow through and wash fractions, dilute the 120-μl aliquot of the supernatant fraction to 600 μl with binding buffer containing 0.25% gelatin. For the 1 M elution, dilute to 600 μl with binding buffer containing 120 mM NaCl and 0.25% gelatin. Add roughly 2 μl crude Tat polyclonal serum and rotate at 4° for 1 hr.
6. Add and incubate the protein A-Sepharose beads (15 μl), equilibrated in binding buffer, with the antibody/antigen complex at 4° for 1 hr. Wash beads three times in binding buffer containing 0.03% SDS and once in Tat kinase buffer. Initiate the kinase reaction with the addition of 25 μl of Tat kinase buffer containing 2.5 mM $MnCl_2$,

1 μM cold ATP, and 10 μCi [γ-^{32}P]ATP, as described previously.[7,8] Following a 1-hr incubation at room temperature, boil beads in 50 μl 1× SDS–PAGE sample buffer, separate the proteins by 10% SDS–PAGE, and visualize the gel by autoradiography.

Purification of TAK/P-TEFb from Nuclear Extracts by GST–Tat Affinity Selection

In this procedure, a crude HeLa nuclear extract is incubated in batch with GST–Tat bound to glutathione beads and washed extensively to purify the TAK/P-TEFb complex.

1. Dilute 10 μl NE (10 mg/ml) 1:1 with TM0.1 *M* KCl and preclear by successive incubation with (1) 15 μl glutathione-S-Sepharose beads and (2) 15 μl GST (or mutant GST–Tat)-coupled beads. Carry out incubations at 4° for 30–60 min per step.
2. Incubate 2 μg of purified GST or GST–Tat proteins (proteins had been eluted from beads with glutathione, but not dialyzed) with 15 μl glutathione beads (as a 50% slurry) in EBC buffer (50 m*M* Tris–HCl, pH 8, 120 m*M* NaCl, 0.5% NP-40) at 4° for 15–30 min. Wash the beads twice with 150 μl EBC buffer containing 5 m*M* DTT and 0.075% SDS and then wash once with EBC buffer containing 5 m*M* DTT.
3. Incubate GST–Tat beads with precleared NE at 4° for 60 min. Wash the complexes three times with 10 20× (bead volume) EBC buffer containing 5 m*M* DTT and 0.03% SDS and then wash once with Tat kinase buffer (TKB; 50 m*M* Tris–HCl, pH 7.5, 5 m*M* $MnCl_2$, and 5 m*M* DTT).
4. Analyze the washed beads by SDS–PAGE (after boiling beads in sample buffer) and also by *in vitro* kinase assays: Incubate a 50-μl reaction containing TKB, 5 μCi of [γ-^{32}P]ATP, 2 μM ATP, and the bead complexes for 30 min at room temperature. Wash the complexes twice with TKB buffer and boil in Laemmli buffer before analysis on SDS–PAGE.

It is important to optimize the efficiency of coupling of GST–Tat proteins to the beads and to minimize contamination of the preparation by bacterial proteins, as these could obscure the detection of nuclear proteins that bind Tat.

Scale-up of this procedure 80-fold yielded around 1 μg of hCycT1 (p87), which was run out on seven to eight lanes of a preparative SDS–PAGE for microsequencing analysis as described.[5]

Conclusions

This article reviewed methods that have been used to isolate the Tat-associated kinase (TAK/P-TEFb) complex on a large scale from crude nuclear extracts and to assess the composition of the Tat:TAK complex that forms a stable complex with TAR RNA. Using these procedures, we obtained sufficient quantities of purified hCycT1 for microsequence analysis and molecular cloning of the hCycT1 gene. These assays can be adapted readily to further study the TAK/P-TEFb proteins in extracts from lymphoid cells and, in general, may be useful for studying protein associations and nucleic acid interactions for any soluble nuclear factor. We used the reiterative RNA and protein affinity selection procedure to assess the composition and activity of the TAK:TAR complex. Because Tat targets only a subset of nuclear P-TEFb complexes, this approach is required to discriminate among the different CDK9- and hCycT1-containing complexes in the cell. When coupled to an enzymatic assay, such as the *in vitro* kinase assay used to study TAK/P-TEFb, this approach is highly sensitive and can yield considerable useful information using only small quantities of nuclear extracts. We are using these procedures to examine the ability of DNA-binding enhancer factors to associate through their transcriptional activation domains with nuclear enzymes, including kinases and protein acetyltransferases.

Author Index

Numbers in parentheses are footnote reference numbers and indicate that an author's work is referred to although the name is not cited in the text.

A

B

D

G

H

I

J

K

L

M

N

O

Q

R

S

T

U

V

W

X

Y

Z

Subject Index

A

B

C

D

H

I

L

M

N

O

P

R

S

T

W

Y

ISBN 0-12-182207-9